Modern Methods of Plant Analysis

New Series Volume 12

Editors
H.F. Linskens, Erlangen/Nijmegen/Amherst
J.F. Jackson, Adelaide

Volumes Already Published in this Series:

Volume 1: Cell Components
1985, ISBN 3-540-15822-7

Volume 2: Nuclear Magnetic Resonance
1986, ISBN 3-540-15910-X

Volume 3: Gas Chromatography/
Mass Spectrometry
1986, ISBN 3-540-15911-8

Volume 4: Immunology in Plant Sciences
1986, ISBN 3-540-16842-7

Volume 5: High Performance Liquid Chromatography
in Plant Sciences
1987, ISBN 3-540-17243-2

Volume 6: Wine Analysis
1988, ISBN 3-540-18819-3

Volume 7: Beer Analysis
1988, ISBN 3-540-18308-6

Volume 8: Analysis of Nonalcoholic Beverages
1988, ISBN 3-540-18820-7

Volume 9: Gases in Plant and Microbial Cells
1989, ISBN 3-540-18821-5

Volume 10: Plant Fibers
1989, ISBN 3-540-18822-3

Volume 11: Physical Methods in Plant Sciences
1990, ISBN 3-540-50332-3

Volume 12: Essential Oils and Waxes
1991, ISBN 3-540-51915-7

Forthcoming:

Volume 13: Plant Toxin Analysis
ISBN 3-540-52328-6

Volume 14: Seed Analysis
ISBN 3-540-52737-0

Essential Oils and Waxes

Edited by
H.F. Linskens and J.F. Jackson

Contributors

R.P. Adams M.E. Crespo S.G. Deans H.E.M. Dobson P. Dunlop
C.A.J. Erdelmeier A. Ghosh E.G. Hammond R.B. Inman
J.F. Jackson J. Jiménez M. Kawakami A.D. Kinghorn A. Kiritsakis
A. Kobayashi S. Kokkini H.F. Linskens P. Markakis J. Metzger
S. Misra C. Navarro L.A.C. Pieters C. Tringali T.A. van Beek
G. Vernin A.J. Vlietinck Y. Yokouchi

With 102 Figures

Springer-Verlag
Berlin Heidelberg New York
London Paris Tokyo
Hong Kong Barcelona
Budapest

Professor Dr. HANS FERDINAND LINSKENS
Goldberglein 7
W-8520 Erlangen, FRG

Professor Dr. JOHN F. JACKSON
Department of Viticulture, Enology and Horticulture
Waite Agricultural Research Institute
University of Adelaide
Glen Osmond, S.A. 5064
Australia

ISBN 3-540-51915-7 Springer-Verlag Berlin Heidelberg New York
ISBN 0-387-51915-7 Springer-Verlag New York Berlin Heidelberg

The Library of Congress Card Number 87-659239 (ISSN 0077-0183)

Typesetting: International Typesetters Inc., Makati, Philippines

31/3145-543210 – Printed on acid-free paper

Introduction

Modern Methods of Plant Analysis

When the handbook *Modern Methods of Plant Analysis* was first introduced in 1954 the considerations were:

1. the dependence of scientific progress in biology on the improvement of existing and the introduction of new methods;
2. the difficulty in finding many new analytical methods in specialized journals which are normally not accessible to experimental plant biologists;
3. the fact that in the methods sections of papers the description of methods is frequently so compact, or even sometimes so incomplete that it is difficult to reproduce experiments.

These considerations still stand today.

The series was highly successful, seven volumes appearing between 1956 and 1964. Since there is still today a demand for the old series, the publisher has decided to resume publication of *Modern Methods of Plant Analysis*. It is hoped that the New Series will be just as acceptable to those working in plant sciences and related fields as the early volumes undoubtedly were. It is difficult to single out the major reasons for success of any publication, but we believe that the methods published in the first series were up-to-date at the time and presented in a way that made description, as applied to plant material, complete in itself with little need to consult other publications.

Contributing authors have attempted to follow these guidelines in this New Series of volumes.

Editorial

The earlier series *Modern Methods of Plant Analysis* was initiated by Michel V. Tracey, at that time in Rothamsted, later in Sydney, and by the late Karl Paech (1910–1955), at that time at Tübingen. The New Series will be edited by Paech's successor H.F. Linskens (Nijmegen, The Netherlands) and John F. Jackson (Adelaide, South Australia). As were the earlier editors, we are convinced "that there is a real need for a collection of reliable up-to-date methods for plant analysis in large areas of applied biology ranging from agriculture and horticultural experiment stations to pharmaceutical and technical institutes concerned with raw material of plant origin". The recent developments in the fields of plant biotechnology and genetic engineering make it even more important for workers in the plant sciences to become acquainted with the more sophisticated methods, which sometimes come from biochemistry and biophysics, but which also have been developed in commercial firms, space science laboratories, non-university research institutes, and medical establishments.

Concept of the New Series

Many methods described in the biochemical, biophysical, and medical literature cannot be applied directly to plant material because of the special cell structure, surrounded by a tough cell wall, and the general lack of knowledge of the specific behavior of plant raw material during extraction procedures. Therefore all authors of this New Series have been chosen because of their special experience with handling plant material, resulting in the adaptation of methods to problems of plant metabolism. Nevertheless, each particular material from a plant species may require some modification of described methods and usual techniques. The methods are described critically, with hints as to their limitations. In general it will be possible to adapt the methods described to the specific needs of the users of this series, but nevertheless references have been made to the original papers and authors. While the editors have worked to plan in this New Series and made efforts to ensure that the aims and general layout of the contributions are within the general guidelines indicated above, we have tried not to interfere too much with the personal style of each author.

There are several ways of classifying the methods used in modern plant analysis. The first is according to the technological and instrumental progress made over recent years. These aspects were used for the first five volumes in this series describing methods in a systematic way according to the basic principles of the methods.

A second classification is according to the plant material that has to undergo analysis. The specific application of the analytical method is determined by the special anatomical, physiological, and biochemical properties of the raw material and the technology used in processing. This classification will be used in Volumes 6 to 8, and for some later volumes in the series.

A third way of arranging a description of methods is according to the classes of substances present in the plant material and the subject of analytic methods. The latter will be used in this volume and for later volumes of the series, which will describe modern analytical methods for alkaloids, drugs, hormones, etc.

Naturally, these three approaches to developments in analytical techniques for plant materials cannot exclude some small overlap and repetition; but careful selection of the authors of individual chapters, according to their expertise and experience with the specific methodological technique, the group of substances to be analyzed, or the plant material which is the subject of chemical and physical analysis, guarantees that recent developments in analytical methodology are described in an optimal way.

Volume Twelve – Essential Oils and Waxes

Essential oils and waxes are often classified as secondary plant products, and as a result many may be led to believe that they are of only secondary importance. However, nothing could be further from the truth. Perusal of a well-known monthly publication listing research papers of leading plant science journals shows that, currently, papers on essential oils comprise at least 10% of titles published. A total approximately 2000 papers are scanned per month. Furthermore, as the world population moves into an era with increasing demand for "organically" grown products with minimal use of manufactured chemicals both in agriculture and for other purposes, these secondary plant products are assuming a growing importance. Many of the essential oils have extremely useful properties and can be put to use in many ways. For example, terpinen-4-ol, an antiseptic agent, is produced synthetically in Europe at great expense ($25000 per tonne). It also occurs naturally in Australia as an essential oil of tea tree (*Melaleuca alternifolia*) and is sold as tea tree oil obtained by distillation of the plant material. At present only about 50 tonnes of tea tree oil are produced annually, but it is possible that this will increase dramatically with the setting-up of tea tree plantations and as the demand increases for a natural antiseptic perceived as an "organic" cure or additive. Eucalyptus oil is another useful, so-called secondary plant product which is finding commercial markets around the world. Approximately 650 tonnes of eucalyptus oil are produced each year, a large amount considering that this involved the steam distillation of more than 15000 tonnes dry weight of eucalyptus leaves per annum.

By definition essential oils are highly volatile substances isolated by a physical method or process from an odiferous plant of a single botanical species (*Encyclopaedia Britannica,* 15th Edition, 1986). The oil bears the name of the plant species from which it is derived, e.g., rose oil, peppermint oil, etc. These oils were termed essential because they were thought to represent the very essence of odor and flavor. Isolation methods in the past have included steam distillation, enfleurage (i.e., extraction using fat), maceration, solvent extraction, or mechanical pressing. The oils are generally stored in the plant in glands whose function is not well understood. Chemically the essential oils are for the most part tropanes, multiple isoprenoid units, aromatics, heterocyclics, and terpenes. The genetics and biochemistry of the biosynthesis of the essential oils is reasonably well understood in some cases. For example, it is known that limonene is synthesized from geranyl pyrophosphate in the oil glands of *Mentha* spp., and something of the genetics of this conversion is understood. Further, we know that the various oxygenated derivatives of limonene that make up the remainder of *Mentha* oil are synthesized from limonene by microsomal preparations from oil glands, which show cytochrome P-450-type mixed function oxygenase activity. Investigations of this type are underway for many of the essential oils and waxes, and so there is a great demand for reliable methods of analysis for these compounds.

Given the growing importance of essential oils and waxes, this Volume deals with the analysis of a broad spectrum of these compounds from many plant origins. Classical methods in lipid analysis, such as acid value, hydrogen value, iodine value, Reichert-Meissl value, saponification value, thyocyanogen test, optical rotation, viscosity, refraction index, etc., are not included; they can be found in the standard methods books. We restrict ourselves to modern, recently developed analytical methods, and to oily and waxy substances of plant origin.

Commercial oils of long-standing importance such as olive oil are, of course, dealt with here, but we include also chapters on lesser-known oils such as that from thyme, tea, ginger, desert trees, eucalypt, garlic, mint, cedar, and juniper. In addition analysis of spices, seasoning, seaweeds, perfumes, liquors, and of atmospheric monoterpene hydrocarbons are to be found here, together with a treatment of pharmacological and allergenic activity analysis of plant essential oils. The volatiles of flower and pollen may be of importance in attracting bees and other insects to certain plants for pollination purposes; this topic is accordingly dealt with herein. Waxes perhaps are harder to find amongst topics in the current scientific literature, but nevertheless epicuticular waxes are included here as an ever-important topic, as well as presentation of aspects of analysis of waxes which render certain sands water-repellent in Australia! In the driest of earth's continents, this is certainly a topic of interest.

Acknowledgments. The editors express their thanks to all contributors for their cooperation, including keeping to production schedules, as well as to Dr. Dieter Czeschlik and Ms. C. Wißmeier, Ms. H. Böning, editorial office and production department of Springer, Heidelberg, for their commitment and work with this volume and others to follow in the series *Modern Methods of Plant Analysis*. The constant help and steady interest of Ms. José Broekmans, Nijmegen, is greatly acknowledged.

Nijmegen/Siena and Adelaide, Summer 1991

H. F. Linskens
J. F. Jackson

Contents

Cedar Wood Oil – Analyses and Properties
R. P. ADAMS (With 7 Figures)

Analysis of Croton Oil by Reversed-Phase Overpressure-Layer Chromatography
A. D. KINGHORN and C. A. J. ERDELMEIER (With 5 Figures)

Rotation Locular Countercurrent Chromatography Analysis of Croton Oil
L. A. C. PIETERS and A. J. VLIETINCK (With 4 Figures)

Oils and Waxes of Eucalypts
Vacuum Distillation Method for Essential Oils
R. B. INMAN, P. DUNLOP, and J. F. JACKSON (With 1 Figure)

Analysis of Epicuticular Waxes
S. MISRA and A. GHOSH (With 9 Figures)

Analysis of Flower and Pollen Volatiles
H. E. M. DOBSON (With 3 Figures)

List of Contributors

ADAMS, ROBERT P., Plant Biotechnology Center, Baylor University, P.O. Box 97372, Waco, Texas, 76798-7372, USA

CRESPO, MARIA ESPERANZA, Dept. of Pharmacology, School of Pharmacy, University of Granada, E-18071 Granada, Spain

DEANS, STANLEY G., Dept. of Biochemical Sciences, Scottish Agricultural College, Auchincruive, Ayr KA6 5HW, Scotland, UK

DOBSON, HEIDI E. M., Dept. of Chemical Ecology, University of Göteborg, Reutersgatan 2c, S-41320 Göteborg, Sweden

DUNLOP, P., Dept. of Physical and Inorganic Chemistry, University of Adelaide, Adelaide, South Australia 5064, Australia

ERDELMEIER, CLEMENS A. J., Research and Development, Dr. Willmar Schwabe Arzneimittel, Postfach 41 09 25, W-7500 Karlsruhe 41

GHOSH, AMITABHA, Dept. of Chemistry, Bose Institute, 93/1 A.P.C. Road, Calcutta 700 009, India

HAMMOND, E. G., Food Science and Human Nutrition, Dairy Industry Building, Iowa State University, Ames, Iowa 50011, USA

INMAN, ROSS B., Institute for Molecular Virology, University of Wisconsin, 1525 Linden Drive, Madison, Wisconsin 53706, USA

JACKSON, JOHN F., Dept. of Viticulture, Enology and Horticulture, Waite Agricultural Research Institute, University of Adelaide, Glen Osmond, S.A. 5064, Australia

JIMÉNEZ, JOSE, Dept. of Pharmacology, School of Pharmacy, University of Granada, E-18071 Granada, Spain

KAWAKAMI, MICHIKO, Shion Junior College, 6-11-1 Ohmika-cho, Hitachi-shi Ibaraki-ken 319-12, Japan

KINGHORN, A. DOUGLAS, Dept. of Medicinal Chemistry and Pharmacognosy, College of Pharmacy, m/c 781, University of Illinois at Chicago, Box 6998, Chicago, Illinois 60680, USA

KIRITSAKIS, APOSTOLOS, School of Food Technology and Nutrition, Technological Educational Institution, (TEI) of Thessaloniki, Thessaloniki, Greece

KOBAYASHI, AKIO, Laboratory of Food Chemistry, Dept. of Food and Nutrition Ochanomizu University, 2-1-1, Ohtsuka, Bunkyo-ku, Tokyo 112, Japan

KOKKINI, STELLA, Laboratory of Systematic Botany and Phytogeography, School of Biology, Faculty of Sciences, University of Thessaloniki, GR-540 06 Thessaloniki, Greece

LINSKENS, HANS FERDINAND, Goldberglein 7, W-8520 Erlangen, FRG

MARKAKIS, P., Dept. of Food Technology, School of Food Technology and Human Nutrition, Technological Educational Inst. (TEI), Thessaloniki, Greece

METZGER, JACQUES, Organic Chemistry Laboratory A, Faculty of Sciences and Techniques of Saint-Jérôme, Avenue Escadrille, Normandie-Niémen, P.O. Box 561, F-13397 Marseille CEDEX 13, France

MISRA, SUNITI, Dept. of Marine Science, University of Calcutta, 35 B.C. Road, Calcutta 700 019, India

NAVARRO, CONCEPCION, Dept. of Pharmacology, School of Pharmacy, University of Granada, E-18071 Granada, Spain

PIETERS, LUC A.C., Dept. of Pharmaceutical Sciences, University of Antwerp (UIA), Universiteitsplein 1, B-2610 Antwerp, Belgium

TRINGALI, CORRADO, Dipartimento di Scienze Chimiche, Università di Catania, Viale A. Doria 8, I-95125 Catania, Italia

VAN BEEK, TERIS A., Dept. of Organic Chemistry, Phytochemical Section, Agricultural University, Dreijenplein 8, N-6703 HB Wageningen, The Netherlands

VERNIN, GASTON, Organic Chemistry Laboratory A, Faculty of Sciences and Techniques of Saint-Jérôme, Avenue Escadrille Normandie-Niémen, P.O. Box 561, F-13397 Marseille CEDEX 13, France

VLIETINCK, ARNOLD J., Dept. of Pharmaceutical Sciences, University of Antwerp (UIA), Universiteitsplein 1, B-2610 Antwerp, Belgium

YOKOUCHI, YOKO, National Institute for Environmental Studies, 16-2, Onogawa, Tsukuba, Ibaraki 305, Japan

Olive Oil Analysis

A. KIRITSAKIS and P. MARKAKIS

1 Introduction

Olive oil is extracted from the fruit of the olive tree, *Olea europea* L., one of the oldest-known cultivated trees in the world.

The olive tree is widely cultivated in Southern Europe, hence its name *Olea europea*. Of the existing olive trees today 98% are grown around the Mediterranean Sea.

The olive tree was not native in the New World. Spanish and Portuguese settlers and missionaries brought the olive tree to California and to other subtropical areas of the Western Hemisphere (Kinman 1922). Italian immigrants brought the olive tree to Australia. A real effort for the development of olive cultivation in South Africa started in 1925. Recently, there has been a significant interest in developing olive cultivation in China, India, and other countries.

Olive culture played an enormous role in the early civilizations (Goor 1966). Athens was named in honor of the goddess Athena, who brought the olive tree to the city. The olive tree played an important role in areas such as diet, religion, and medicine.

The production of olives and olive oil has nearly doubled during the last 50 years (Kiritsakis and Min 1989). The aggregate value of olive products is estimated to be 2.8 billion US dollars.

Olive oil is an important commodity in the daily diet of the Mediterranean peoples, the Greeks having the highest per capita consumption, 20.8 kg yearly, although it varies considerably among locations within Greece. The Spaniards consume 10.0 kg/person/year, followed by the Italians at 8.1 kg/person/year (Sellianakis 1984).

Olive oil is almost unique among vegetable oils in that it can be consumed without any refining treatment. Olive oil of good quality is characterized by a fragrant and delicate flavor and high nutritional and health value (Kiritsakis and Markakis 1987). For these reasons, olive oil commands a higher price on the international market than other vegetable oils. Consequently, adulteration of olive oil with other oils or fats is a temptation.

2 Quality Tests of Olive Oil-Determination of Acidity and Oxidation

2.1 Acidity (IOOC 1968; AOCS 1978)

Acidity is the basic criterion of grading the quality of olive oil. Edible olive oil has acidity lower or equal to 3.3% (as oleic acid), while industrial olive oil has acidity higher than 3.3%. Total acidity is essentially a measure of the free fatty acids present in the oil.

A. Equipment. Conical flasks, 250 ml, buret, 50 ml, balance.

B. Reagents

1. Ethanol 95%, neutralized as follows. To 50 ml ethanol, add 2 ml of phenolphthalein indicator and enough 0.1 N NaOH to produce a faint permanent pink color just before using
2. Phenolphthalein indicator, 1% in 95% ethanol
3. NaOH solution, 0.1 N

C. Procedure. Weigh 28.2 g of sample in a 250 ml flask. Add 50 ml of hot neutralized alcohol and 2 ml of indicator. Titrate with NaOH under vigorous shaking until a faint pink color appears which lasts more than 1 min.

Acidity is calculated as % free fatty acids (FFA) expressed as oleic acid:

$$\text{FFA, as oleic (\%)} = \frac{\text{ml NaOH} \times \text{N} \times \text{mEq oleic (0.282)} \times 100}{\text{weight of sample (g)}}.$$

FFA may also be expressed as acid value (AV), which is the mg KOH necessary to neutralize a 1-g sample.

$$\text{AV} = \text{\% FFA (as oleic)} \times 1.99$$

Comment. An apparatus called FIAtron is commercially available for the rapid determination of total acidity. It uses n-butanol/water as a solvent and methyl red as indicator. The agreement between the FIAtron procedure and conventional titration is good.

2.2 Oxidation

Several methods (Lea 1931; Wheeler 1932; IOOC 1966; AOCS 1978), exist for measuring the total hydroperoxides as the primary products of the oxidation of olive oil.

2.2.1 Peroxide Value (AOCS 1978)

A. Equipment. Conical, glass-stoppered flasks, 250 ml, buret 50 ml, balance.

B. Reagents.

1. Glacial acetic acid-chloroform solution, 3:2 by volume.
2. Potassium iodide solution. Prepare saturated solution by dissolving excess KI, in freshly boiled distilled water. Store in dark. Test before use by adding 0.5 to 30 ml acetic acid-chloroform; then add 2 drops 1% starch solution. If a blue color is formed which requires more than 1 drop of 0.1 N sodium thiosulfate solution to decolorize, prepare fresh KI solution

3. Sodium thiosulfate, 0.1 N and 0.01 N
4. Soluble starch indicator 1%, in distilled water

C. Procedure. Weigh 5 g olive oil into a 250 ml glass-stoppered conical flask. Add 25 ml acetic acid-chloroform solution and 1 ml saturated KI preferably using a Mohr-type pipet. Swirl the flask until the sample is dissolved in the solution. Allow the solution to stand in the dark for exactly 1 min. Add 75 ml distilled water. Titrate with 0.1 N sodium thiosulfate, adding it gradually with constant and vigorous shaking. Continue the titration until the yellow color has almost disappeared. Add about 2 ml of starch indicator solution. Continue the titration, shaking the flask vigorously near the end point to liberate all the iodine from the chloroform layer. Add the thiosulfate dropwise until the blue color just disappears. If less than 0.5 ml of 0.1 N sodium thiosulfate is used, repeat the titration using 0.01 thiosulfate.

Run blank (all reagents, no sample) daily. Blank value must not exceed 0.1 ml thiosulfate and must be subtracted from the sample value.

Peroxide value (PV), is expressed as milliequivalents of peroxides per kg sample:

$$PV = \frac{(S-B) \times (N) \times (1000)}{\text{weight of sample (g)}},$$

where S and B are titration values for sample and blank, respectively, and N is the normality of thiosulfate.

2.2.1.1 Peroxide Value (Colorimetric Determination, Asakawa and Matsushita 1978)

A. Reagents

1. KI-silica gel, Merck N 60, suitable for column chromatography. Prepare KI-silica gel by dissolving 100 g silica gel (70–320 mesh) in 100 ml KI solution (10% N). Filter and dry the wet silica gel in a clean air stream. Dry the silica gel further in an oven at 110 °C for 30 min. Keep it in the dark.
2. Hexane
3. Ethyl alcohol 50%
4. HCl 0.01 N
5. Soluble starch, solution 1%, containing 20% NaCl

B. Procedure. Place 200 mg or smaller quantity of sample in a test tube. Add 1 ml of hexane and 1.5 g of reagent (KI-silica gel) and swirl rapidly. Incubate the sample for 5 min at 30 °C in a dry-block heater. Add 2 ml of ethyl alcohol (50%) and shake vigorously. Add 15 ml 0.01 N HCL and 0.5 ml starch solution and keep shaking. Transfer the solution to a centrifuge tube and centrifuge for 3 min at 3000 rpm. Read the absorbance at 560 nm immediately.

Run a blank along with the sample. Prepare a standard curve. A linear relationship between the absorbance and sample weight, from 20 to 200 mg, must be observed.

The PV, expressed as mEq O^2/kg olive oil, is calculated using the formula:

$$PV = \frac{A}{0,425} \times \frac{1}{\text{weight sample(g)}},$$
(mEq/kg)

where A = absorbance at 560 nm.

Terao and Matsushita (1977) proposed another colorimetric method for measuring hydroperoxides in photooxidized oils. Cadmium acetate was used and the absorbance of the color formed was measured at 350 nm.

Maurikos et al (1972) used a polarographic method to determine the PV in virgin olive oil. The supporting electrolyte was LiCL in methanol benzene with a dropping mercury electrode.

2.2.2 Conjugated Diene Measurement

The oxidation of polyunsaturated fatty acids results in the formation of a conjugated double bond system along with the appearance of hydroperoxides. Several workers (Montefredine and Luciano 1968; Bartolomeo and Sergio 1969; Ninnis and Ninni 1968; Jiminez and Gutierrez 1970) found that there is a good correlation between the absorbance of olive oil at 232 nm, which is due to diene conjugation, and the degree of oxidation.

A. *Reagents.* Isooctane (2,2,4-trimethyl pentane), pure or spectral grade.

B. *Procedure.* Weigh into petti-caps 10 mg olive oil. Place the petti-caps into 30 ml test tubes. Add 10 ml pure isooctane and cover the test tubes with stoppers. Mix the solution well and filter it if it is not clear. Fill a 1-cm pathcell and measure the absorbance in a spectrophotometer at 232 nm using isooctane as a blank. The absorbance of the blank compared to distilled water must not exceed 0.07. If necessary, dilute the sample so that the absorbance at 232 nm is between 0.2 and 0.8.

Calculate absorptivity $\alpha = A/bc$,

where, A = absorbance, b = cell length, and c = g sample per liter of dilution used to measure A.

This value, α_{232}, must be corrected for esters. The ester correction is 0.07. The value

$$\% \text{ conjugated diene} = (\alpha_{232} - 0.07) \times 0.91$$

is considered a measure of the oxidation of olive oil.

2.2.3 Specific Absorbances of Olive Oil

It has been mentioned that conjugated hydroperoxides formed as primary oxidation products of olive oil absorb at 232 nm. The secondary oxidation products (aldehydes and ketones) absorb at higher wavelengths (270 nm). Conjugated diene and trienes, formed during the refining or bleaching of olive oil, also absorb at 270 nm.

The values of the specific absorbances ($K^{1\%}_{232\,1cm}$, $K^{1\%}_{270\,1cm}$) of olive oil are used for evaluating its quality (IOOC 1968). Low absorbance values, at these wavelengths, correspond to good olive oil quality.

A. Reagents

1. Pure isooctane or cyclohexane. The minimum transmittance must be 40% at 220 nm and 95% at 250 nm when the solvents are compared to distilled water.
2. Basic alumina of chromatographic quality (size 30 to 130 μ). Activate alumina by heating it for 3 h at 380–400 °C.

B. Procedure. Filter olive oil if not clear at ambient temperature. Weigh accurately 1 g of sample in a 100 ml volumetric flask. Add solvent up to the mark and mix vigorously. Read the absorbance at 232 and 270 nm against pure solvent as blank.

The wavelengths of 262, 268, and 274 nm are used to differentiate extra, fine, and semi-fine olive oils on the basis of the ΔK value:

$$\Delta K = K_{268} - \frac{K_{262} + K_{274}}{2}.$$

Olive oil with a specific absorbance K 270 > 0.20 is considered virgin only if that absorbance, after passing through active alumina, is <0.11.

2.2.4 Thiobarbituric Acid Test (TBA)

The TBA test has been applied by several workers (Gutierrez and Romero 1960; Casillo 1968; Kiritsakis 1982) to determine the secondary oxidation products of olive oil. Casillo (1968) reported that the TBA test detects the rancidity of olive oil at a lower level than other tests (PV, Kreis test).

The TBA test is based in the formation of a chromogen through the condensation of one molecule of malonaldehyde with two molecules of TBA. Malonaldehyde forms readily when trienoic or tetraenoic acids are present (Dugan 1976). It may also be formed from dienoic acid (linoleic).

Tarladgis et al. (1962) reported that the structure of TBA is altered by acid and heat treatment as well as by the presence of peroxides, and recommended that blank determinations be carried out in conjunction with the test.

A. Reagents

1. Acetic acid, glacial
2. Carbon tetrachloride
3. 2-Thiobarbituric acid (TBA) reagent

Dissolve 0.67 g of TBA with distilled water. Keep the mixture in a water bath until the acid is completely dissolved. Transfer the solution into a 100 ml volumetric flask. Allow to cool and bring to volume with distilled water.

The TBA reagent consists of an equal volume of TBA solution and glacial acetic acid.

B. Procedure. Dissolve 3 g of oil with 3 ml carbon tetrachloride and add 10 ml TBA reagent. Mix vigorously. Transfer to a separatory funnel. Take the aqueous phase, which contains the malonaldehyde, and put it in a test tube. Keep the test tube in a water bath (100 °C) for 30 min. Let the mixture cool and measure the absorbance at 532 or at 535 nm. Use as blank TBA reagent.

Results may be expressed either as absorbance values or as malonaldehyde units (based on a reference curve).

Comment. The determination of the oxidative rancidity of olive oil should not be based only on one method (Kiritsakis, in press), since each one of them has a weak point. The use of two or more methods gives a better indication for the oxidative deterioration of the oil.

3 Sensory Evaluation of Olive Oil

Sensory evaluation of olive oil is another criterion for estimating its quality. The International Olive Oil Council (IOOC) proposed the following vocabulary of taste and flavor for the sensory evaluation of olive oil.

Almond. This flavor may appear in two forms; that which is typical of fresh almond, or that which is peculiar to dried, sound almonds.

Apple. Flavor of olive oil which is reminiscent of this fruit.

Atrojado (musty). Characteristic flavor of oil obtained from olives stored in piles which have undergone an advanced stage of fermentation.

Bitter. Characteristic taste of oils obtained from green olive fruit or olives turning color.

Brine. Flavor of oil extracted from olive fruit preserved in saline solutions.

Cucumber. Flavor produced when an oil is hermetically packed for too long, particularly in tin containers. This is attributed to the formation of the compound 2,6-nonadienal.

Earthy. Characteristic flavor of oil obtained from olive fruit which has been collected with earth or mud on it.

Esparto. Characteristic flavor of oil obtained from olives pressed in new esparto mats.

Flat or Smooth. Flavor of olive oil with very weak sensory characteristics due to the loss of their aromatic compounds.

Fruity. Flavor which is reminiscent of both the odor and taste of sound, fresh fruit picked at its optimum stage of ripeness.

Grass. Characteristic flavor of certain olive oils reminiscent of recently mown grass.

Green Leaves (bitter). Flavor of oil obtained from excessively green olives or olives that have been crushed with leaves and twigs.

Grubby. Characteristic flavor of oil obtained from olives which have been heavily attacked by olive fly (*Dacus oleae*).

Harsh. Characteristic sensation of certain oils reminiscent of dried grass.

Heated or Burnt. Characteristic flavor of oils caused by excessive and/or prolonged heating during processing, particularly when the paste is thermally processed.

Metallic. Flavor that is reminiscent of metal. Characteristic of oils which have been in prolonged contact, under suitable conditions, with foodstuffs or metallic surfaces during crushing, mixing, and pressing, or storage.

Muddy Sediment. Characteristic flavor of oil recovered from the decanted sediment in vats and underground tanks.

Mustiness-Humidity. Characteristic flavor of oils obtained from fruit in which large numbers of fungi and yeasts have been developed.

Old. Characteristic flavor of oil that has been kept too long in storage containers. This flavor also appears in oils which have been packed for an excessively long period.

Pomace. Characteristic flavor that is reminiscent of the flavor of olive pomace.

Pressing Mat. Characteristic flavor of oil obtained from olives that have been pressed in dirty mats, where fermented residues have been left.

Rancid. Characteristic flavor common to autoxidized olive oil.

Ripely Fruity. Flavor of olive oil obtained from ripe fruit.

Rough. Characteristic perception in certain oils. They produce a thick, pasty mouth-feel sensation when they are tasted.

Soapy. Flavor that is reminiscent of that produced by green soap.

Sweet. Pleasant taste, not exactly sugary, found in olive oil in which the bitter, astringent, and pungent attributes do not predominate.

Vegetable Water. Characteristic flavor acquired by the oil as a result of poor decantation and prolonged contact with vegetable water (aqueous phase of extract).

Winey-Vinegary. Characteristic flavor of certain oils reminiscent of wine or vinegar. It is mainly due to the formation of acetic acid, ethyl acetate, and ethanol in large amounts.

Table 1 represents a grading sheet for a sensory evaluation of olive oil. The relationships between the different designations of olive oil proposed by the IOOC and certain sensory and laboratory criteria are shown in Table 2.

For a "global" quality evaluation of virgin olive oil the IOOC proposed the following equation:

$$\text{O.Q.I. (Over-all Quality Index)} = 2.55 + 0.91\text{SE} - 0.78\text{AV} - 7.35\text{K}_{270} - 0.66\ \text{PV},$$

where

SE = sensory evaluation (from 3.5–9.0)
AV = acid value (from 0.1–3.3)
K_{270} = absorbance at 270 nm (from 0.08–0.25)
PV = peroxide value (from 1.0–20.0).

Table 1. Grading sheet for sensory evaluation of olive oil (IOOC 1987)

Defects	Sensory characteristics	"Overall" evaluation (points)
None	Olive fruity	9
	Olive fruity and fruitiness of other fresh fruit	8 7
Slight and barely perceptible	Weak fruitiness of any type	6
Perceptible	Rather imperfect fruitiness, anomalous odors and tastes	5
Considerable, on the border of acceptability	Clearly imperfect, unpleasant odors and tastes	4
Grave and/or serious, clearly perceptible	Totally inadmissible odors and tastes for consumption	3 2 1

Table 2. Designations of olive oil as related to sensory and laboratory criteria (IOOC 1984, 1987; Kiritsakis and Markakis 1987)

Sensory	Olive Oil					
Characteristics	Virgin olive oil extra	Virgin olive oil fine	Virgin olive oil semi-fine	Virgin olive oil lampante	Refined olive oil	Olive oil
Odor	Absolutely perfect	Absolutely perfect	Good	Off smelling	Acceptable	Good
Taste	Absolutely perfect	Absolutely perfect	Good	Off smelling	Acceptable	Good
Color	Light yellow to green	Light yellow to green	Light yellow to green	Yellow to brownisn green	Light yellow	Light yellow to green
Appearance at 20 °C	Limpid	Limpid	Limpid	—	Limpid	Limpid
FFA as oleic acid (%) (max)	1.0	2.0	3.3	>0.5	0.5	1.5
PV (max) (mEq 0_2/kg oil)	20	20	20	20	10	20
Moisture and volatile matter (%) (max)	0.2	0.2	0.2	0.3	0.1	0.1
Impurit. insol. in light petrol. (%) (max)	0.1	0.1	0.1	0.2	0.05	0.05
K_{270}(max)[a]	0.20	0.20	0.30	0.30	1.10	0.90
ΔK (max)[b]	0.01	0.01	0.01	0.01	0.15	0.15

[a] K_{270} is absorbance $E^{1\%}_{1cm}$ at 270 nm.
[b] $\Delta K = K_{268} - (K_{262} + K_{274}) / 2$.

4 Determination of Certain Constituents of Olive Oil

4.1 Chlorophyll Determination (AOCS 1978)

The chlorophyll content of olive oil can be determined by measuring its absorbance at 630, 670, and 710 nm by a spectrophotometer.

A. Reagents. Carbon tetrachloride, redistilled.

B. Procedure. Fill up the cell sample with oil heated at 30 °C and read the absorbance using carbon tetrachloride as blank. Take all the readings for the three wave lengths. All absorbance values should preferably fall between 0.3 and 0.8.

Calculations. While the method is proposed for use with the Beckman B spectrophotometer, the Beckman DU or Cary instruments or othespectrophotometers may be used following the manufacturer's instructions for making the absorbance readings.

Beckman Model B

$$\text{Chlorophyll, ppm} = \frac{A_{670} - \dfrac{A_{630} + A_{710}}{2}}{0.0964\ L},$$

where A = absorbance, L = path length in cm.

Rahmani and Csallany (1985) proposed a rapid and precise HPLC method for determining the chlorophylls a and b and pheophytins a and b in olive oil. The oil sample is diluted 20 times (w/v) with the solvent mixture, hexane-isopropanol (98.5:1.5) and then injected into a μ-Porasil column.

4.2 Determination of Phenols

The phenols in olive oil may be determined by the Gutfinger (1981) method.

A. Reagents
1. Aqueous methanol 60%
2. Folin-Ciocalteau (2N) reagent
3. Na_2CO_3 solution, 35% w/v
4. Hexane

B. Procedure. Weigh 10 g olive oil into a 250 ml Erlenmeyer flask and dissolve it with 50 ml hexane. Add 20 ml aqueous methanol (60%) and mix vigorously for 2 min. Remove the methanolic phase and put it in a beaker each time after the two phases are separated. Bring the combined extracts to dryness in a vacuum rotary evaporator at 70 °C. Dissolve the residue in 1 ml methanol and store it at low temperature until use. Bring 0.1 ml from the methanolic extract into a 10 ml volumetric flask. Add 5 ml distilled water, 0.25 ml Folin-Ciocalteau (2N) and mix well for 3 min. Add 1 ml Na_2CO_3

and fill with distilled water up to the mark. Measure the absorbance of the blue color formed, after 1 h, at 725 nm wavelength. If total phenol content is expressed as caffeic acid, or reference curve must be prepared using this acid.

5 Moisture Determination

The moisture content of olive oil can be determined by drying it in an oven, by the distillation procedure (AOCS 1978), or by the infrared (IR) technique.

5.1 Moisture Determination by Infrared Balance

The moisture determination is carried out on a special balance furnished with an infrared radiation lamp. The weight loss during evaporation is determined by a balance system and is directly expressed in % of the sample weight.

6 Determination of Soap Content

The soap content in olive oil is usually determined by Codex Alimentarius Commission (1969) method, proposed by IOOC.

A. Equipment
1. Test tubes approximately 150 x 40 mm of borosilicate glass fitted with ground glass stoppers and flattened at their ends
2. Microburet 5 ml
3. Steam bath

B. Reagents
1. Distilled acetone, containing 2% water
2. Hydrochloric acid, 0.01 N
3. Bromophenol blue indicator, 1% solution in 95% (v/v) ethanol

C. Procedure. Prepare the test solution by adding 0.5 ml of the bromophenol blue indicator to each 100 ml of the aqueous acetone just before use and titrate with 0.01 N acid or alkali until it turns yellow. Wash the tubes with the test solution. Weigh 40 g of the oil into the test tube. Add 1 ml of water, warm on the steam bath, and shake vigorously. Add 50 ml of the neutralized aqueous acetone and, after warming on the steam bath, shake the vessel well and allow the contents to stand until they separate into two layers.

If soap is present in the oil, the upper layer will be colored green to blue. Then add 0.01 N acid preferably from a microburet, until the yellow color is restored. Continue the process of warming and shaking until the yellow color of the upper layer remains permanent.

Results. Dissolved soap, as sodium oleate, % $= \frac{0.0304\ V}{W}$, where V = volume in ml of 0.01 N acid required, and W = weight of sample in grams.

7 Olive Oil Adulteration — Adulteration and Genuineness Tests

Olive oil, a natural fruit product of fine aroma, pleasant taste, and high nutritional value, is often adulterated with other oils or fats. Oils known to be widely used for this purpose include : olive pomace oil, corn oil, peanut oil, cottonseed oil, sunflower oil, soybean oil, and poppy seed oil. In addition, castor oil, pork fat (lard), as well as other animal fats, have occasionally been used in small quantities. Adulteration with reesterified oils, and with industrial grade colza (rapeseed oil) has been reported. Some cases of adulteration are even hazardous to the public health.

Firestone (1987) reported that many brands of olive oil available in the US market contained undeclared esterified olive oils, pomace olive oils, and seed oils as well.

7.1 Methods and Techniques of Detecting Adulteration

Several physical and chemical tests have been proposed for detecting adulteration of olive oil (IOOC 1966, 1984; Ninnis and Ninni 1966; Codex Alimentarius Commission 1969; Fedeli 1977).

Adulteration changes certain constants (refractive index, specific weight, iodine value) of olive oil. Adulteration also changes the squalene content. Olive oil contains squalene which ranges from 136–708 mg/100 g of oil. Any value outside these limits indicates adulteration of olive oil (Cuisa and Morgante 1974).

Ninnis and Ninni (1966) proposed a method for determining olive oil adulteration based on the absorbance at two wavelengths: 210 and 268 nm.

Gas chromatography has also been used for estimating the admixture of olive oil with vegetable oils (Colakoglu 1966; Gegiou and Georgouli 1980). A combination of gas and liquid chromatography permits the detection of as little as 5% of a foreign oil (Kapoulas and Passaloglou - Emmanouilidou 1981).

Tsimidou et al. (1987) used reversed phase HPLC to investigate the liquid fraction of oils obtained after low temperature crystallization, and suggested that this procedure may be applied to assist the analysis for detecting olive oil adulteration. Casadei (1987) used also HPLC to detect adulterated and misbranded olive oil products with hazel-nut oil. According to him, HPLC is appropriate to detect olive oil adulterated with esterified oil.

One of the criteria proposed by the IOOC (1984) to detect olive oil purity refers to saturated fatty acid content at position 2 of the triglyceride. Reesterified oils differ in their fatty acids composition at position 2. Hydrolysis by means of pancreatic lipase, which specifically attacks position 2, allows the differentiation between virgin and reesterified olive oils.

The saturated fatty acid level (sum of palmitic and stearic acids) at position 2 should not exceed 1.5% in virgin olive oil, 1.8% in refined olive oil or in pure olive oil, and 2.2% in refined olive-pomace oil (IOOC) 1984).

Paganuzzi (1980, 1981) reported that quantitation of erythrodiol and uvaol provides a good indication for differentiation between pressed and solvent-extracted olive oils.

Since soybean oil is much richer in γ- and δ-tocopherols and cottonseed oil in γ-tocopherol than olive oil, quantitative determination of these tocopherols has been suggested as a method for determining the adulteration of olive oil with the less expensive soybean or cottonseed oils (Tiscornia and Bertini 1972; Gutfinger and Letan 1974).

The detection of elaidic acid, by IR spectroscopy, in olive oils reveals the adulteration with reesterified (Pallotta 1976) and hydrogenated oils.

Analysis of the entire unsaphonifiable fraction (e.g., sterols, tocopherols, triterpenic alcohols) will probably provide additional tests for sharpening the differentiation among vegetable oils (Cuisa and Morgante 1974; Pallotta 1976; Fedeli 1977; Itoh et al. 1981).

The determination of β-sitosterol content in olive oil, as well as the value of the ratio:

$$\frac{\beta\text{-sitosterol}}{\text{campesterol} + \text{stigmasterol}}$$

is used for detecting olive oil adulteration with seed oils (Pallotta 1976).

According to Gutfinger and Letan (1974), large quantities of stigmasterol indicate the presence of soybean oil in olive oil.

Solvent-extracted oils may be distinguished from oils obtained by the pressure process, by applying simple procedures such as the Bellier, Carocci-Buzi, and Vizern, or determining the erythrodiol content by GLC (Fedeli 1977).

Details for some tests detecting adulteration of olive oil are given below:

7.2 Adulteration Tests Based on Color Formation

7.2.1 Nitric Acid Test (Synodinou-Konsta method)

One of the first tests used to detect the adulteration of olive oil is that of Hauchecorne. The initial Hauchecorne method was later modified by Synodinou and Konsta.

A. *Reagents*
1. Nitric acid 1.40 (specific gravity)
2. Tonsil earth

B. *Procedure.* Place 30 ml of sample in a 50 ml stoppered cylinder. Add 3 g Tonsil, mix vigorously and filter. Remove 10 ml extract in a 50 ml cylinder and add 10 ml nitric acid. Mix well for 30 s and observe the color immediately and 2–5 min later.

In virgin olive oil the color is hay to hay-yellow (negative reaction)

In virgin-refined olive oil mixtures the color is hay-yellow to yellow-brown (positive reaction).

Any other color indicates the presence of seed oil or olive residue oil in the olive oil (positive reaction).

Comment. Substances responsible for the color development have not been completely identified. There is evidence, however, that these substances are formed either as a result of the various industrial treatments, or as a result of autoxidation.

7.2.2 Halphen Test

The Halphen test detects adulteration of olive oil with cottonseed oil. It is characteristic of the cyclopropenoic acids, malvic and stercoulic, found in cottonseed oil.

A. Reagents

1. Halphen reagent. Prepare it by mixing equal volumes of amyl alcohol and sulfur solution, 1% in carbon disulfide
2. NaCl-saturated solution

B. Procedure. Place 10 ml of olive oil and 10 ml Halphen reagent in a 250 ml spherical flask with a vertical air condenser. Mix the contents vigorously and place the flask in a saturated aqueous NaCl boiling solution (110–115 °C). Keep it there for 1–2 h.

The formation of a red color at the end of this period indicates the presence of cottonseed oil.

The depth of color is to a certain extent proportional to the amount of cottonseed oil in the sample.

Comment. The method is sensitive but not always reliable. Even if the reaction is negative, adulteration with cottonseed oil is possible.

7.2.3 Sesame Oil Tests

There are different tests (Baudouin, Villavecchia, Fabris), commonly used for the detection of sesame oil in olive oil. They are based on the color formation due to the presence of the phenolic substances sesamol, sesamolin, and sesamin in the sesame oil (Fig. 1).

7.2.3.1 Fabris Test

This method is based on the reaction between the phenolic substance sesamin and furfural.

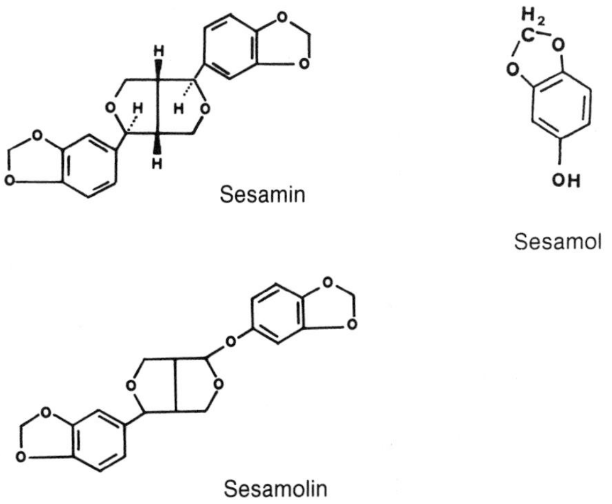

Sesamin

Sesamol

Sesamolin

Fig. 1. Structures of sesamin, sesamol and sesamolin

A. Reagents
1. Furfural solution 0.35% in acetic anhydrite
2. Concentrated sulfuric acid

B. Procedure. Place 10 ml olive oil and 5 ml furfural solution in a volumetric stoppered tube. Close the tube and shake vigorously for about 1 min. Remove the mixture into a separating funnel and let it stand until the contents are separated into two layers. Place part of the lower layer in a porcelain crucible. Add six to seven drops saturated sulfuric acid and mix occasionally by moving the crucible.

The reaction is considered positive when a yellow-green color appears.

Comment. The formation of crimson color (Baudouin) and any color (Villavecchia) in the lower layer, indicates the presence of sesame oil in olive oil.

7.3 Adulteration Tests Based on Residue or Clouding Formation

7.3.1 Carocci-Buzzi Test

This test detects adulteration of olive oil with olive pomace oil.

A. Reagents
1. KOH alcoholic solution (8.5%)
2. Diluted acetic acid (1+2) and concentrated acetic acid
3. Alcohol (70%)

B. Procedure. Place 1 ml of filtered sample in a spherical 10 ml flask with vertical condenser. Add 5 ml alkaline KOH solution. Saponify the mixture until clear. Allow to cool and add 1.5 ml acetic acid (1 + 2), two or three drops concentrated acetic acid, and 50 ml hot (50 °C) alcohol (70%). Shake the flask vigorously, insert a thermometer on it and leave it in a place where the temperature is no lower than 18 °C for 12 h.

Flake formation is an indication of olive pomace oil in the sample. Clouding with no flake formation does not necessarily prove the absence of olive pomace oil in olive oil.

7.3.2 Vizern-Guillot Test (Codex Alimentarius Commission 1969)

The Vizern-Guillot test detects adulteration of virgin and refined olive oil with semi-drying oils (Iodine Value 100–150). All seed oils fall into this category.

A. Equipment
1. Stoppered 50 ml Erlenmeyer flask
2. Bath

B. Reagents
1. Hexane or, if not available, light petroleum ether with distillation point between 40 and 60 °C and bromine value less than 1.
2. Bromine reagent. Prepare it by adding drop by drop, while shaking, 4 ml of chemically pure bromine into 100 ml of hexane or light petroleum. The reagent should be chilled to 0 °C and kept in a melting ice bath until use.

C. Procedure. Filter and dry the oil to be tested. Add 1 ml of the oil in the Erlenmeyer flask and dissolve in 10 ml of hexane. Place the stoppered flask in the ice bath. Add 10 ml of bromine reagent, 4 min later, in small quantities at a time, while shaking and maintaining the temperature at 0 °C. Leave the flasks in the ice bath for 1 h.

A flocculent precipitate will form if semi-siccative oil is present. The solution remains clear and transparent in genuine olive oils.

Comment. The method is not considered reliable for olive oils with high iodine value.

7.3.3 Bellier-Marcille Index

The Bellier-Marcille index detects adulteration of olive oil, mainly with peanut oil. It is the temperature (°C) at which precipitation (appearance of clouding) of the salts of the oil fatty acids commences.

A. Equipment
1. Test tubes 220 × 26–27 mm
2. Condenser consisting of a glass tube with stopper
3. Thermometer graduated in 1/40 from 8 to 25 °C, inserted in a stopper.

B. Reagents
1. Aqueous ethanolic potassium hydroxide solution. Prepare it by dissolving 42.5 g of pure KOH in 72 ml of 95% (v/v) ethanol.
2. Ethanol solution, 70% (v/v)

3. Aqueous acetic acid solution 1:2 (v/v). 1.5 ml of this solution will neutralize exactly 5 ml of the aqueous ethanolic potassium hydroxide solution.

C. Procedure. Place 1 ml of filtered oil into a test tube. Add 5 ml of the aqueous ethanolic KOH solution. Connect the condenser and heat moderately, rotary agitating from time to time until saponification is complete. Disconnect the condenser after cooling, and add 1.5 ml of the aqueous acetic acid solution and 50 ml of the ethanol solution. Insert the thermometer. Place test tube in a water bath at 23–25 °C. If a flocculent precipitate forms, leave standing for 1 h at the same temperature and filter into a test tube. Place the test tube for a moment in a beaker containing water of a temperature at about 10 °C less than the expected in the Bellier index. Withdraw and insert a number of times (cooling should be at the rate of about 1 °C per min). Repeat this procedure until clouding appears. Record the temperature. Allow the temperature to increase to dissolve the precipitate. Cool the test tube. The cooling should be slow and the shaking more frequent as the temperature approaches that recorded the first time.

The Bellier index is the temperature (°C) at which the clouding reappears.

Comment. The Bellier-Marcille index for virgin and refined olive oil should not exceed 17.

7.4 Detection of Oil Adulteration by a UV Lamp

An ultraviolet (UV) lamp is used for detecting adulteration of olive oil with refined olive oil or refined olive pomace oil.

A. Procedure. Place 50 ml olive oil in a 100 ml beaker of 5 cm in diameter. Observe the color of the sample under the UV lamp in the dark.

Flourescent light yellow to orange-yellow color is characteristic for virgin olive oils. Fluorescent light blue-gray to blue green is present in refined oils except for olive pomace oil. Bright blue fluorescent color is characteristic for refined olive pomace oils. Blue fluorescent color appears in olive oils containing olive pomace oil and in refined olive pomace oil.

7.5 Detection of Olive Oil Purity by Infrared Spectrophotometry

Infrared (IR) spectrophotometry can be used to detect the purity of olive oil. The different oils show characteristic absorbance in the IR region (Fig. 2).

A. Procedure. Adjust the spectrophotometer. Dilute the sample with carbon tetrachloride or carbon disulfide. Place the sample (0.1 mm thick) in the cell. Record the absorbance.

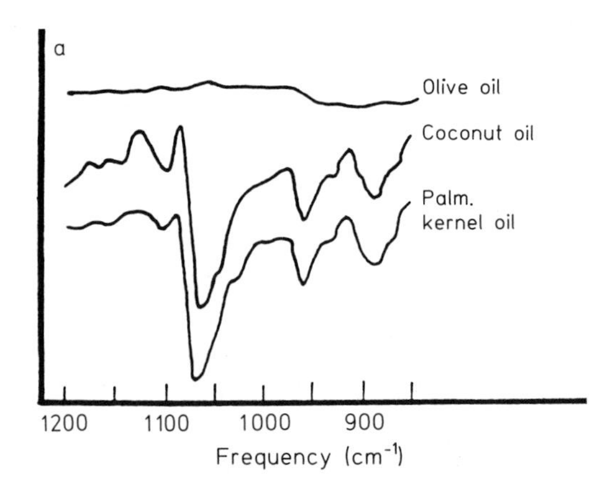

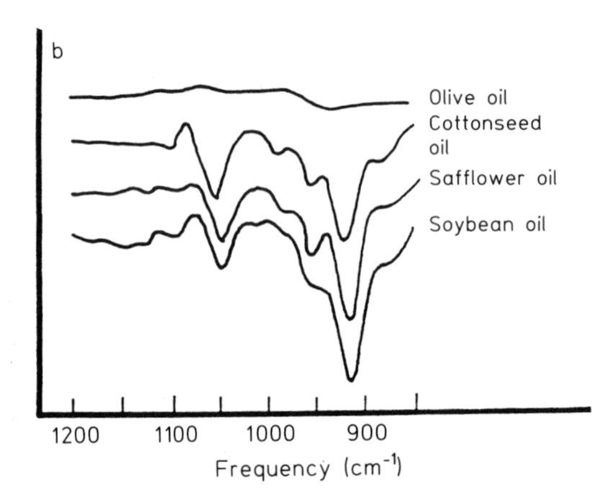

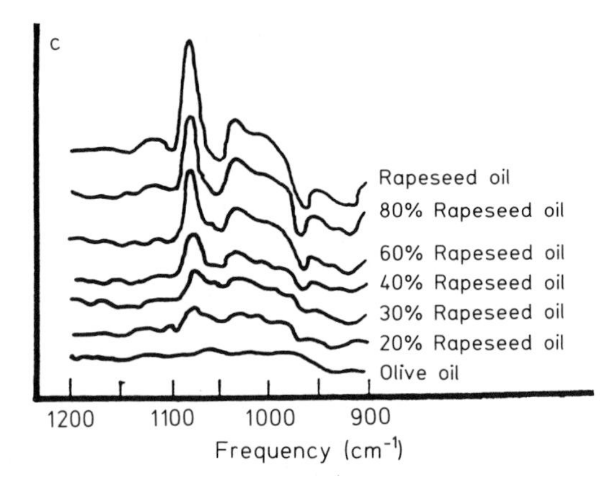

Fig. 2a-c. Infrared (IR) spectra of virgin and adulterated olive oil. (Mehlenbacher 1960)

References

Am Oil Chem Soc (AOCS) (1978) Official and tentative methods. Third edition including additions and revision. Champaign, Illinois
Asakawa A, Matsushita S (1978) Colorimetric determination of peroxide value with potassium iodide-silica gel reagent. J Am Oil Chem Soc 55:619
Bartolomeo D, Sergio R (1969) Physicochemical features and acidic composition of some meridional virgin olive oils. Riv Ital Sostanze Grasse 46:467
Casadei E (1987) First results on detection of adulterated olive oil products with hazel-nut and/or esterified oils by HPLC of triglycerides. Riv Ital Sostanze Grasse 64:373
Casillo R (1968) Quality evaluation of virgin olive oils. Thiobarbituric acid test . Riv Ital Sostanze Grasse 45:753
Codex Alimentarius Commission WHO/FAO (1969) Methods of analysis for edible fats and oils. CAC/RM 9/14–69, Rome
Colakoglu M (1966) Donnes analytiques nouvelles sur les huiles d'olive. Rev Fr Coprs Gras 4:261
Cuisa W, Morgante A (1974) Polycyclic aromatic hydrocarbons in olives. Quad Merced 13:31
Dugan LR (1976) Lipids. In: Fennema O (ed) Principles of food science, Part I Food chemistry. Marcel Dekker, New York
Fedeli E (1977) Lipids of olives. Prog Chem Fats Other Lipids 15:57
FIAtron Process systems (1989) On-line free fatty acid analysis. Bulletin FIAchem APA 1002A, Oconomowoc WI USA
Firestone D (1987) Control of olive oil adulteration and misbranding in the United States. Riv Ital Sostanze Grasse 64:293
Gegiou D, Georgouli M (1980) Detection of reesterified oils. Determination of fatty acids at position –2 in glycerides of oils. J Am Oil Chem Soc 57:313
Goor A (1966) The place of the olive in the Holy Land and its history through the ages. Econ Bot 20:223
Gutfinger J (1981) Polyphenols in olive oils. J Am Oil Chem Soc 58:966
Gutfinger J, Letan A (1974) Studies of unsaponifiables in several vegetable oils. Lipids 9:658
Gutierrez Gonzalez-Quijano R, Romero AV (1960) Estudios sobre el enranciamiento del aceite de oliva. Comparacion entre las diferences pruedas para la determinacion del grado de rancidez. Grasas Aceites 11:67
International Olive Oil Council (IOOC) (1966) Trade standards for virgin olive oils. COI No 2/66/15–II, Madrid, Spain
International Olive Oil Council (IOOC) (1968) Common methods for analysis of olive oils. T 14/DOC No4/Corr 1, Madrid, Spain
International Olive Oil Council (IOOC) (1984) International trade standard applying to olive oils and olive-residue oils. COI/T 15/No1, Madrid, Spain
International Olive Oil Council (IOOC) (1987) Organoleptic assessment of virgin olive oil. COI/T 20/Doc no 3, Madrid, Spain
Itoh T, Yoshita K, Yatsu T, Tamura T, Matsumoto T (1981) Triterpene alcohols and sterols of Spanish olive oil. J Am Oil Chem Soc 58:545
Jimenez OJM, Gutierrez Gonzalez-Quijano R (1970) Packaging of olive oil in commercial type containers. Changes in absorptivity at 232 and 270 nm. Grasas Aceites 21:329
Kapoulas VM, Passaloglou-Emmanoulidou S (1981) Detection of adulteration of olive oil with seed oils by a combination of column and gas liquid chromatography. J Am Oil Chem Soc 58:694
Kinman CF (1922) Olive growing in the southwestern United States. Department of Agricul Farmer's Bulletin No 1249, California
Kiritsakis A (1982) Quality studies on olive oil. Thesis Michigan State University. East Lansing Michigan, USA
Kiritsakis A (1990) The olive oil. Am Oil Chem Soc, Champaign, Illinois (in press)
Kiritsakis A, Markakis P (1987) Olive oil, a review. Adv Food Res 31:453
Kiritsakis A, Min D (1989) Flavor Chemistry of Olive Oil. In: Min D, Smouse T (eds) Flavor chemistry of lipid foods, AOCS Champaign, Illinois, pp 196–221
Lea CH (1931) The effect of light on the oxidation of fats. Proc Soc Lond 108 B:175
Maurikos P, Gegkou D, Iliopoulos G (1972) Polarographic study of peroxides in olive oil. Chem Chron 1:118 (In Greek)

Mehlenbacher VC (1960) The analysis of fats and oils. The Garrard Press, Champaign, Illinois
Montefredine A, Luciano L (1968) Quality characteristic of virgin Italian olive oils. Boll Lab Chim Prov 19:784
Ninnis LN, Ninni ML (1966) L'importance pour l'analyse de l'huile d'olive de l'absorption dans l'ultra-violet (190 a 220 nm). Rev Fr Corps Gras 13:1
Ninnis LN, Ninni ML (1968) Stabilité thermique de l' huile d'olive et sa prévision par son absorption dans l'ultra-violet a 232 et a 268 nm. Rev Fr Corps Gras 15:441
Paganuzzi V (1980) Distribution of alcoholic components in the unsaponifiable fraction of olive drupes. Riv Ital Sostanze Grasse 57:291
Paganuzzi V (1981) Influence of origin and conservation state on the triterpene dialcohols relative content of untreated olive oils. Fette Seifen Anstrichm 84:115
Pallotta U (1976) Analytic problems in the ascertainment of olive oil genuinesses. In:Paoletti R, Jacini G, Porsellati R (eds) Lipids Vol 2 Technology. Raven, New York, pp 38–393
Rahmani M, Csallany AS (1985) Mise au point d'une méthode de chromatographie liquide a haute performance pour la determination des pigments chlorophylliens dans les huiles végétales. Rev Fr Coprs Gras 6–7:1
Sellianakis G (1984) The effect on Greek olive cultivation due to the fact that Spain and Portugal became members of EEC. Eleourgiki, Athens Greece (In Greek)
Tarladgis BG, Pearson AM, Dugan LR (1962) The chemistry of the 2-thiobarbituric acid test for the determination of oxidative rancidity in foods. J Am Oil Chem Soc 39:34
Terao J, Matsuchita S (1977) Products formed by photosensitized oxidation of unsaturated fatty acid esters. J Am Oil Chem Soc 54:234
Tiscornia E, Bertini G (1972) Recent analytical data in chemical composition and structure of olive oil. Riv Ital Sostanze Grasse 49:3
Tsimidou M, Macrae R, Wilson I (1987) Authentication of virgin olive oils using principal components analysis of tryglyceride and fatty acid profiles. Food Chem 25:227
Wheeler DH (1932) Peroxide formation as a measure of autoxidative deterioration. Oil Soap 9:89

Analysis of Essential Oils of Tea

A. KOBAYASHI and M. KAWAKAMI

1 Introduction

Tea, one of the most popular beverages in the world, is made from only one plant species, *Thea sinensis*. However, the spread of the tea-drinking custom throughout the world has resulted in various types of tea being developed by different manufacturing processes. The most suitable cultivars of the tea plant are also selected for a specific tea product. Nowadays, tea is classified from its manufacturing process as nonfermented tea (or green tea, mainly produced in China, Japan, and other southeastern Asian countries), semi-fermented tea (or oolong tea, mainly produced in China and Taiwan) and fermented tea (or black tea, mainly produced in India, Sri Lanka, Indonesia, and some African countries). The "fermentation" during tea manufacturing is not a microbiological process but an enzymatic reaction occurring in the tea leaf, and the grade of fermentation is determined from the production of catechin polymers produced by the action of oxidases in the tea leaf.

Theoretically, three different types of tea can be manufactured from a tea tree; however, their characteristics as a beverage differ greatly according to aroma as well as taste and color (Yamanishi 1978, 1981). As the aroma is a major factor in determining the quality and character of tea and tea products, the analysis of essential oils in tea is usually directed to tea flavor analyses. For this purpose, an exhaustive separation of the essential oils is not required, but the recovery of the original tea aroma becomes an important issue during the isolation of the essential oil. We have to take great care not to contaminate any artifact and not to change any unstable constituent to another product during the separation process.

Complexity is another distinctive feature of the essential oil constituents. As the range of fluctuation in the odor threshold value (or odor strength) of each constituent is as large as 10^{10} or more, even a minor component should not be overlooked during the isolation and identification of the components in the essential oil.

These characteristic requirements for analyzing tea essential oils lead us to (1) pay much attention to the separation method (2) apply high resolution gas chromatography combined with mass spectrometry to isolate and identify each component, and (3) compare the complicated gas chromatographic data by mathematically treating with computer programs.

2 Isolation of Essential Oils

2.1 Steam Distillation Under Reduced Pressure (SDR)

Steam distillation under normal pressure is the basic isolation method to obtain essential oils. However, it has the disadvantage of alteration of the components, as the sample is maintained under severe conditions. In contrast, under reduced pressure, the sample is maintained at a low temperature (40–60 °C), and any change to the constituents can be minimized.

Using rotary evaporation apparatus (Yamanishi et al. 1970) is an easy and efficient method for steam distillation, as shown in Fig. 1. Two hundred grams of powdered tea mixed with 1.5 l of distilled water (60 °C) in a flask are evaporated under at 20 mm Hg for 90 min. Around 1.2 l of the distillate obtained is then saturated with sodium chloride and extracted with ethyl ether. After being dried with sodium sulfate, the ether extract is concentrated at 40 °C. The yield of the aroma concentrate is not very high, as shown in Table 1, but the aroma is almost the same as that of the original tea.

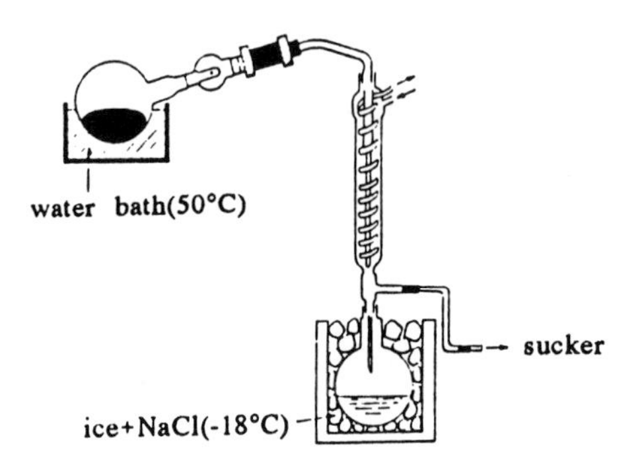

Fig. 1. Apparatus of steam distillation under reduced pressure (SDR)

Table 1. Quantity of aroma extracts from black tea by different methods

	Quality mg/100 g tea			
	Dimbulla-1	Dimbulla-2	Uva-1	Uva-2
SDR	3.47	5.12	4.92	6.44
SDE	15.59	20.65	13.97	17.14

2.2 Simultaneous Distillation and Extraction (SDE)

Simultaneous distillation and extraction by a modified Likens-Nickerson apparatus (Schultz et al. 1977) can yield a high amount of concentrate. A disadvantage of this method is alteration of the components, because the sample is heated at 100 °C and the extract in the solvent is also heated to the boiling point of the solvent. From this point of view, ethyl ether is more suitable than dichloromethane as the extracting solvent. The gas chromatograms obtained by the SDE and SDR methods are shown in Fig. 2, which illustrates that the different isolation processes give different gas chromatograms.

The advantage of the SDE method is its easy operation and short time of isolation. For example, 100 g of powdered tea sample mixed with 300 ml of distilled hot water in a sample flask is heated at 100 °C, and 50 ml of ethyl ether in a solvent flask is held at 50 °C. Distillation and extraction proceed simultaneously and continuously for 60 min, and the extract dried with sodium sulfate is concentrated at 40 °C.

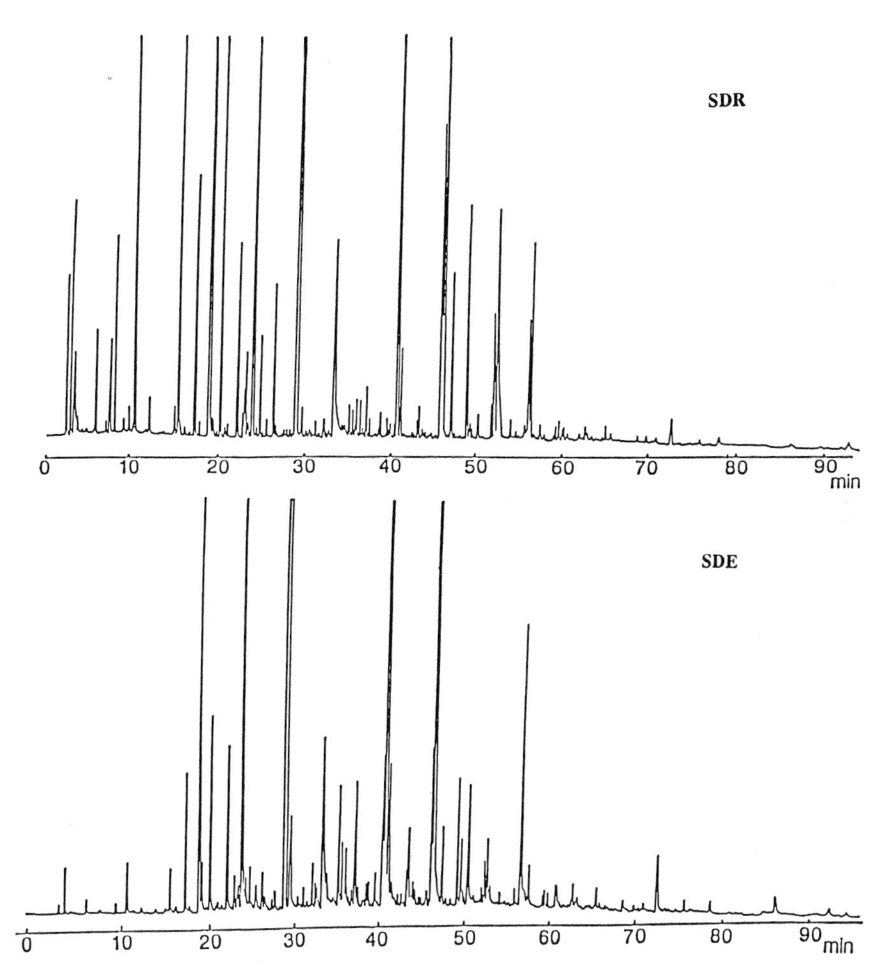

Fig. 2. Gas chromatograms of black tea (Assum) aroma volatiles obtained by SDR and SDE. GC condition:column, PEG 20 M capillary 50 m x 0.25 mm i.d.; oven temperature held at 60°C for 4 min and programmed to 180 °C at 2 °C/min; carrier gas flow rate He 30 cm/s

2.3 Head Space Gas Analysis

The organoleptic evaluation of tea and tea drinks is performed by sniffing or tasting. Our olfactory organ is stimulated by a complicated vapor-phase mixture (head space gas), whose constituents are not always the same as that of the essential oil. The head space gas analysis is simple, but condensation of trace amounts of volatile compounds in the vapor is necessary.

2.3.1 Adsorption Method

Trapping on Tenax GC, a porous polymer for gas absorption (Tsugita et al. 1979) is suitable for condensing the tea headspace gas, because the volatile produced by this method resembles in flavor what we drink as an infusion. The volatile components are adsorbed on 200 mg of Tenax GC (60–80 mesh) by an enforced draught of head space gas. N_2 gas (30 ml/min) is swept into 50 g of powdered green tea sample in a U-shaped glass tube maintained at 60 °C for 60 min. After being trapped, the Tenax tube is set up for GC and is heated at 200 °C with a coiled heater around the tube to inject directly into the GC or GC/MS system.

2.3.2 Purge and Trap Injection (PTI) Method

Cooled trapping of the head space volatiles (Badings et al. 1985) and subsequent heat desorption can also be employed to inject the concentrated mixture directly for GC. An example of an automatic GC-injection system with purging, cooled trapping and heat desorption is shown in Fig. 3. Fifteen milliliters of a hot water extract of green tea (prepared from 2.5 g of tea and 200 ml of hot water) is purged for 15 min, and volatiles are injected for GC by this method.

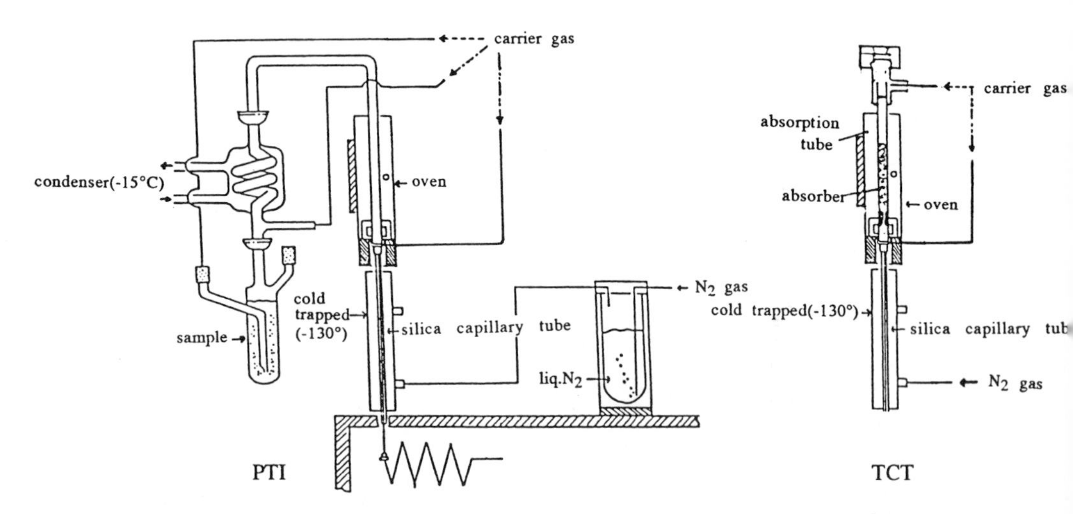

Fig. 3. Apparatus of PTI and TCT. ←- - - Pathway for purge; ←• – • – pathway for sweep; ← •• – •• – heat-deabsorption and carrier gas pathway

As shown in Fig. 4, the gas chromatogram from the adsorption method (lower) shows a wider range of volatile compounds than that from the PTI method. The difference may arise from the relatively short time for trapping and from the cooling system to condense moisture in the latter method.

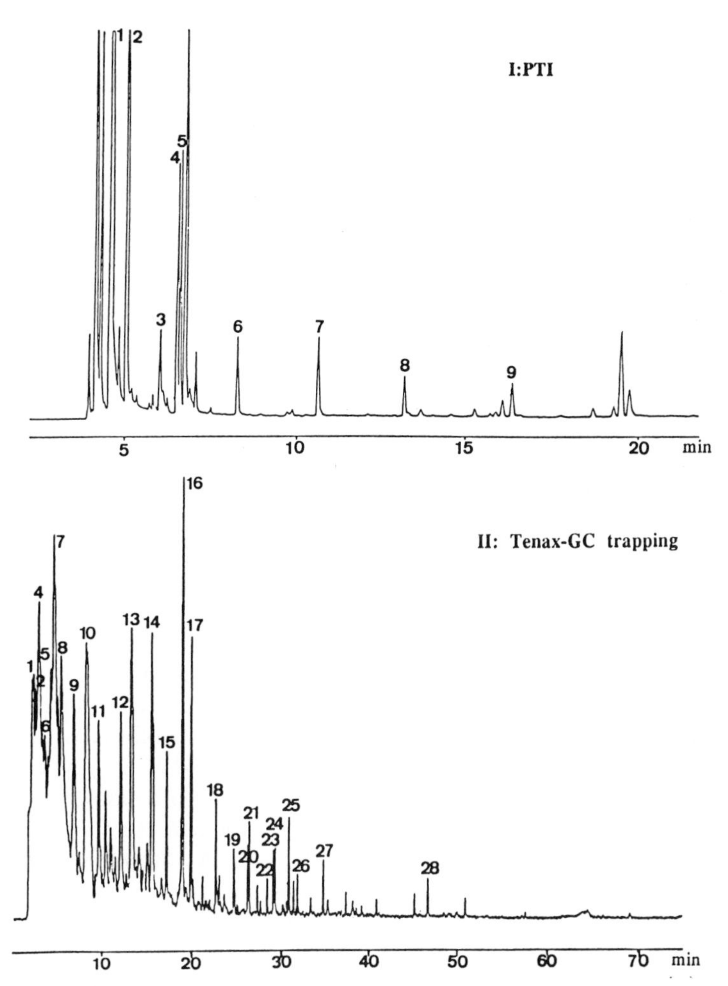

Fig. 4. Gas chromatograms of green tea (Sen-cha) head space vapor obtained by PTI and Tenax-GC trapping. GC condition: **I:PTI,** Column, FS-wcot CP-wax 52CB capillary 50 m × 0.25 mm i.d., Oven temperature held at 40 °C for 10 min and programmed to 70 °C at 2 °C/min, and to 150 °C at 4 °C/min, carrier gas flow rate, He 1 ml/min. **II:Tenax-GC trapping,** Column, FFAP capillary 50 m × 0.25 mm i.d. Oven temperature held at 60 °C for 4 min and programmed to 180 °C at 2 °C/min. **Peak identities: 1** dimethyl sulfide; **2** 2-methylpropanal; **3** ethyl acetate; **4** 2-methylbutanal; **5** 3-methylbutanal; **6** pentanal; **7** toluene; **8** hexanal; **9** 4-methyl-3-penten-2-one; **10** 1-penten-3-ol; **11** 2-amylfuran; **12** pentanol **13** cyclohexanone; **14** (Z)-2-pentenol **15** hexanol; **16** (Z)-3-hexanol; **17** (E)-2-hexenol; **18** 1-octen-3-ol **19** (E)-2,(E)-4 heptadienal; **20** benzaldehyde; **21** (E)-3,(Z)-5-octadienone; **22** linalool; **23** octanol; **24** (E)-3,(E)-5-octadienone; **25** 2,6,6-trimethyl-2-hydroxycyclohexanone; **26** β-cyclocitral; **27** (Z)-3-hexenyl hexanoate; **28** benzyl alcohol

2.3.3 On Column Injection

As described later, the chemically bonded FS-WCOT column offers high stability against chemical and physical changes during the analytical process, and the splitless system has become familiar for GC injection. Therefore, a simple head gas injection method is also possible without any sophisticated injection system as shown in Fig. 3.

Ten grams of powdered tea sample is mixed with 50 ml of hot distilled water in a 250-ml flask with a needle cap, and is incubated in a water bath for 15 min. Forty milliliters of the headspace gas is slowly collected in a syringe through the fused silica needle about for 10 min, and is slowly injected into a GC column chilled with liquid nitrogen, holding for 3 min in the splitless mode. After the coolant has been removed, GC analysis starts.

3 Separation and Identification of Essential Oil Components from Tea

3.1 Gas Chromatography

3.1.1 Column

A capillary column or open tubular (OT) column provides good resolution and short retention times. With the development of OT column preparation, we can generally use a 0.25 mm i.d.x 50 m long flexible fused silica OT column, which is easy for handling and gives reasonable resolution in the GC-MS system as well as in the routine GC analysis. The stationary phase is chemically bonded to the inner silica wall. This technique provides a longer life for the OT column and allows a wider bore (not capillary) for preparative GC or sniffing of the GC effluents. Therefore, the most popular GC column for tea essential oil analysis is the chemically bonded, fused silica and wall coated open tubular (FS-WCOT) type.

The high resolution efficiency of the OT column does not require many types of stationary phase for each organic compound. Presently, we can use only two different types of liquid phase, a polar and a nonpolar one. As the main constituents of the tea essential oil are more or less polar compounds, the polyethylene glycol type of polar liquid phase is most widely applied for GC analysis.

3.1.2 Detectors

The flame ionization detector (FID) gives a peak area proportional to the total mass of the essential oil. Most GC is now combined with a computing integrator, which calculates peak areas and their retention times, and determines the concentration of each component approximately from its peak area percentage.

As a variation of the FID system, a flame photometric detector (FPD) is applicable for detecting sulfur-containing organic compounds, and an alkaline FID or flame thermionic detector (FTD) for detecting nitrogeneous compounds. The former is useful for green tea analysis because of the presence of disulfides in the essential oil. On the other hand, the latter is used for detecting thermally generated compounds formed during tea manufacturing. The mass spectrometer can also be used for detection, but since it describes each peak area as a result of recalculating the total ion current, its proportional figure is different from that of FID.

3.1.3 Retention Index

GC gives little information for the identification of each compound appearing as a single peak on the gas chromatogram. Retention time varies with a subtle change of GC conditions and relative retention time (relative to that of an internal standard or a main peak in its gas chromatogram) is also unreliable. Kovats' index (KI) is one of the most popular retention indices (Kovats 1958) because, when the stationary phase is the same, there is good coincidence among KI values calculated under different GC conditions, even under programmed temperature conditions (Van den Dool and Knatz 1963). Fortunately, the use of an OT column decreases the selection of stationary phases and we can use KI data compilations in which the generally selected stationary phases are PEG 20M (Carbowax 20M) and methyl silicone (OV-101). The calculation of KI value starts with the injection of a standard hydrocarbon mixture (C_6-C_{20} for PEG 20M, and C_6-C_{30} for OV-101) under the same linear programming conditions. Their retention times are recorded directly or manually by the computing integrator. When peak A in the sample appears between hydrocarbon peaks M and M+1, KI of peak A (KI_A) is calculated from the following equation:

$$KI_A = 100M + 100\left(\frac{T_A - T_M}{T_{M+1} - T_M}\right),$$

where T_A, T_M, and T_{M+1} are the retention times of peaks A, M, and M + 1, respectively, and M is the number of carbon atoms of n-paraffin.

The software of the integrator can detect each peak, calculate each KI value automatically, and print out it with other GC data. A data book based on retention indices is also available (Sadtler Research Laboratories 1985).

3.2 Gas Chromatography-Mass Spectrometry (GC-MS)

The application of GC-MS to plant analysis has already been described in this series (Linskens and Jackson 1986). In this section, we will take up some topics constituting the characteristic features of tea analysis, and methods developed since that publication.

As the molecular weights of the compounds are less than 400, the quadrupole mass spectrometer is now widely used as well as the double-focus type, which can be applicable for more sophisticated methods, such as high resolution MS. The coupling between GC and MS is also simple. The flexible FS-WCOT column can be introduced to the MS ion source directly without any condensation process for the GC effluent because of the low flow rate of the carrier gas (1.0–1.5 ml/min).

The acquisition of MS data and deduction of molecular structures with capillary GC-MS became possible when computers with a large memory capacity became available. Because each capillary GC peak appears for only a few seconds and MS scanning should correspond to such a short time, a large amount of MS data is accumulated in the computer for one run of GC-MS measurements.

With low-resolution MS, the m/z value of each ion is digitalized as an integral number. As the components of the essential oil have a relatively simple chemical structure, a library search is efficient enough to identify each compound as well as to interpret each mass spectrum. The software for a library search is usually an option with the GC-MS computer system. The compilations of MS data with numerical (MS Spectrometry Data Centre 1983) or graphical (MacLafferty and Stanffer 1989) expressions are also available to identify the molecular structure.

High resolution MS can measure an m/z value as low as 10^{-4} mass unit, from which the atomic composition of the corresponding ion can be calculated; therefore, the molecular formula of the compound is given from the high resolution MS data of its molecular ion.

Instead of the total ion current for each components, the ion current of the specific m/z value can be selected to make a chromatogram (called an MS chromatogram), which can show the specific compound on the gas chromatogram. For example, an MS chromatogram of m/z = 136 indicates the peaks of monoterpenes and monoterpene alcohols, and their KI values differ one from the other.

3.3 Gas Chromatography-Infrared Spectrometry-Mass Spectrometry (GC-IR-MS)

The development of the Fourier transfer infrared spectrometer (FTIR) has allowed the linear combination of GC, FTIR and MS (Smith 1984). IR spectra give information about the presence of a functional group, and the stereochemistry concerning a double bond or cyclic structure. The mass spectrum is based on a decomposition pattern of a compound, while the IR spectrum is a specific physical characteristic of a compound, and its complexity of absorption bands in the region of 650–1300 cm^{-1} is valuable for identification. The information from MS and IR supplement each other, and their combination should be successful for the identification of more complex structures in future. The FTIR/library search is also available with the GC-IR-MS computer system, and a data book of FTIR spectra is also useful (Pouchert 1989).

4 Components of the Essential Oil

4.1 Comparison of the Main Components in Various Teas

There are no less than 300 components already identified in tea aroma (Yamanishi 1981; Straten and Maarse 1983). The essential oil from fresh tea leaves is relatively different in composition according to the variety, plucking season, distribution and climate. The main components in various teas are shown in Table 2 in addition to KI (CW 20M as the stationary phase) values and the five strongest m/z values from MS.

Terpene alcohols such as linalool, its oxides, and geraniol, (Z)-3-hexenol, methyl salicylate, and benzyl alcohol are the main components in tea leaf oil (Takei et al. 1978; Kawakami and Yamanishi 1981). The aroma patterns of various kinds of tea with their characteristic components are compared in Fig. 5.

Nonfermented tea or green tea is inhibited in its enzymatic action by heating at an early stage of processing, and is classified into two types: Chinese pan-fired, such as Longjing in Chinese, and Japanese steamed, such as Sen-cha and Gyokuro in Japanese. The steamed type of green tea well maintains its original fresh leaf aroma (Kawakami and Yamanishi 1981; Yamaguchi and Shibamoto 1981). On the other hand, the aroma of the pan-fired type of green tea drastically changes during the parching process (Kawakami and Yamanishi 1983). Sesquiterpenes and their alcohols, especially the cadelene type of terpenes, increase (Nose et al. 1971) and low-volatile compounds such as (Z)-3-hexenol having a greenish aroma and methyl salicylate decrease. Oxidation of linalool also proceeds by parching.

Semi-fermented tea is classified into three types, i.e., pouchong tea (10–20%), Ti-Kuan-Yien (20–30%) and oolong tea (40–50%) by the degree of oxidation of catechin (shown in parentheses). Pouchong tea, with the manufacturing process of solar-withering, indoor-withering, parching, and rolling, is particularly produced in Taiwan, for which the aroma pattern is very specific. Nerolidol and indole are present in large amounts. Benzyl cyanide, jasmine lactone, and methyl jasmonate, that increase during the withering process, are characteristic components of pouchong tea (Yamanishi et al. 1980; Tokitomo et al. 1984; Kobayashi et al. 1985; Kawakami et al. 1986). Recently, unsaturated lactones such as (Z)-octen-4-olide have also been identified as characteristic components of pouchong tea (Nobumoto 1990). Oxidation of linalool and esterification of (Z)-3-hexenol also occur in the same process (Takeo 1982).

Black tea manufactured by the withering, rolling, fermentation, and drying process is 100% fermented tea. The essential oil constituents of black tea vary according to the cultivar (Aisaka et al. 1978). Linalool and its derivatives and geraniol, the most significant terpene alcohols, characterize the variety shown in Table 3. The terpene alcohol of Uva Ceylon tea consists of 27% linalool and its oxides. On the other hand, Chinese Keemun contains a high amount of geraniol, approaching 30%. Ceylon Dimbulla shows an intermediate character between Uva and Keemun, suggesting that Uva tea is made from a hybrid of the two cultivars above. (E)-2-Hexenal, (Z)-3-hexenol, and methyl salicylate are also significant for the fresh black tea aroma.

Microbial fermented tea produced in Eastern Asia is a primitive type of tea and is currently becoming rare. This tea is classified into two different types, i.e., pickled tea and piled tea. After being steamed or parched, pickled teas such as Thailand Miang (Kawakami et al. 1987a), and Japanese Goishi-cha (Kawakami et al. 1987b) and Awa-cha (Kawakami et al.1989) undergo a pickling process with lactic acid fermentation mainly by bacteria such as *Leuconostoc mesenteroides*. On the other hand, piled tea such as Chinese Zhuan-cha and Japanese Koku-cha (Kawakami et al. 1987c) undergo piling with fermentation

Table 2. KI and MS/MZ value of compounds identified in tea volatiles

Compound	Identification	KI	MW	MS/MZ
2-Methylpropanal	N F	814	72	43 41 72 27 39
Butanal	F	864	72	44 42 41 72 39 57
2-Ethylfuran	N F	936	96	81 96 53 39 95
Pentanal	N F M	960	86	44 41 58 57 43
Hexanal	N S F	1064	100	44 56 41 43 57
2-Pentenal	N F	1104	84	55 84 83 39 41
4-Methyl-3-penten-2-one	N S F M	1110	98	83 55 98 43 39
1-Penten-3-ol	N S F M	1139	86	57 41 39 58 55 86
Heptanal	N F M	1160	114	70 44 43 41 55 57
(E)-2-Hexenal	N S F M	1192	98	41 42 55 69 83
Pentanol	N S F M	1231	88	42 55 41 70 57
(Z)-2-Pentenol	N S F	1296	86	57 41 39 44 68
6-Methyl-5-hepten-2-one	L S F M	1303	126	43 41 55 69 108
Hexanol	N S F M	1332	102	56 55 42 41 43 69
(Z)-3-Hexenol	L N S F M	1361	100	41 67 82 55 69
Nonanal	N S F	1367	142	57 41 43 56 98
(E)-2,(E)-4-Hexadienal	F M	1368	96	81 39 96 53 67
(E)-2-Hexenol	N S F M	1381	100	57 41 82 67 44
Acetic Acid	N S M	1405	60	43 45 60 42 29
Linalool oxide I	L N S F M	1409	170	59 43 94 68 111
1-Octen-3-ol	N S F M	1427	128	57 72 43 41 85
Furfural	N S F M	1427	96	96 95 39 29 37
Linalool oxide II	L N S F M	1440	170	59 43 68 55 94
(E)-2,(E)-4-Heptadienal	L N S F M	1457	110	81 57 39 41 53
2-Ethylhexan-1-ol	N S M	1462	130	57 43 41 70 83
Benzaldehyde	N S F M	1482	106	77 106 105 51 50
Linalool	L N S F M	1522	154	71 41 93 55 69
Octanol	N S F M	1534	130	56 55 69 70 84
(E)-3,(E)-5-Octadien-2-one	N F M	1541	124	95 43 81 124 109 53
6-Methyl-(E)-3,(E)-5-heptadien-2-one	N S F M	1551	124	109 81 43 124 53
2,6,6-Trimethyl-2-hydroxycyclo-hexan-1-one	N S F M	1555	156	71 43 58 95 110
1-Ethyl-2-formylpyrrole	N F M	1567	123	123 94 122 108 106
3,7,-Dimethyl-1,5,7-octatrien-3-ol	L N S F	1573	152	71 82 43 67 55 95
β-Cyclocitral	N S F M	1578	152	137 152 123 109 67
Acetophenone	L N F M	1587	120	105 77 120 91 51
Phenylacetaldehyde	S F	1592	120	91 92 120 65 51
(Z)-3-Hexenyl hexanoate	N S F	1625	198	82 67 43 99 71
Furfuryl alcohol	N F	1626	98	98 41 97 81 53
Safranal	N F M	1596	150	107 91 121 150 79
2,6,6-Trimethyl-2-cyclohexen-1,4-dione	N M	1647	152	68 96 152 39 109 137
α-Terpineol	L N S F M	1662	154	59 93 121 81 136
Benzyl acetate	F	1680	150	108 91 43 90 150
Linalool oxide III	N S F M	1698	170	68 59 94 43 67
α-Farnesane	N S F	1712	204	93 69 41 55 107
Methyl salicilate	L N S F M	1720	152	120 92 152 121 65
Linalool oxide IV	L N S F M	1727	170	68 59 94 43 67
(E)-2,(E)-4-Decadienal	F	1764	152	81 41 67 55 95
Nerol	L N S F	1766	154	69 41 39 93 68 29
Phenethyl acetate	S F	1772	164	104 43 91 105 65
α-Ionone	N S F M	1801	194	121 93 43 136 192
Hexanoic acid	N S F M	1807	116	60 73 41 43 87

Table 2. (*continued*)

Compound	Identification	KI	MW	MS/MZ
Geraniol	L N S F M	1812	154	69 41 39 68 53 93
Geranyl acetone	N S F M	1820	194	43 69 151 107 125
Benzyl alcohol	L N S F M	1833	108	79 108 107 77 51
2-Phenylethanol	L N S F M	1863	122	91 92 122 65 51
Benzyl cyanide	N S F	1875	117	117 90 116 89 51
2-Phenylbut-2-enal	F M	1880	146	146 117 115 118 91
cis-Jasmone	N S F M	1887	164	164 110 79 149 122
β-Ionone	N S F M	1889	194	177 135 93 79 107
4-Methylguaiacol	M	1902	138	123 138 95 67 77
1,2,3-Trimethoxybenzene	M	1909	168	168 153 110 125 95
2-Acetylpyrrole	N F M	1921	109	94 109 66 39 53
5,6-Epoxy-β-ionone	N S F M	1957	208	123 43 135 107 95
Nerolidol	N S F M	2005	222	69 93 55 81 107
Octanoic acid	N S F	2017	144	60 73 43 41 55
(Z)-3-Hexenyl benzoate	L N S F	2032	204	105 82 67 77 123
6,10,14-Trimethylpentadecanone	L N S	2069	268	58 43 71 85 95 109
Bovolide	L N S F M	2069	180	124 55 137 180 41
Theaspirone	L N S F	2120	208	152 110 111 153 96
Nonanoic acid	N S F M	2123	158	60 73 57 43 41
4-Ethylphenol	M	2123	122	107 122 121 77 91
Dihydrobovolide	F M	2131	182	83 55 111 43 182
α-Cadinol	N S	2139	222	43 95 161 121 204
Jasmine lactone	S F	2184	168	99 71 43 55 41
Dihydroactinidiolide	N S F M	2260	180	111 43 137 180 109
Methyl jasmonate	N S F	2269	224	83 151 55 95 67
(E)-Geranic acid	F M	2294	168	69 41 100 123 82
4-Vinylphenol	N S	2325	120	120 91 119 45 65
Indole	N S F M	2365	117	117 90 89 59 63

L: Tea leaves, N: nonfermented tea, S: semi-fermented tea, F: fermented tea.
M: Microbial fermented tea.
KI: Carbowax 20M.

predominantly by fungi such as *Aspergilluis niger*. The pickled tea has a sweet and sour fermentative flavor, and the piled tea has a moldy odor. The microbial fermented teas contain high levels of phenolic compounds and terpenenoids. The main aroma components of pickled tea are (Z)-3-hexenol, linalool and its oxide, methyl salicylate, benzyl alcohol, 2-phenylethanol, acetic acid, and 4-ethylphenol. Phenolic ether compounds such as 1,2,3-trimethoxybenzene having a moldy odor are produced specifically during fungal fermentation (Kawakami and Shibamoto 1990).

In conclusion, a specific tea flavor can be produced by the combination of various manufacturing processes and the selection of a different tea plant variety for each type of tea.

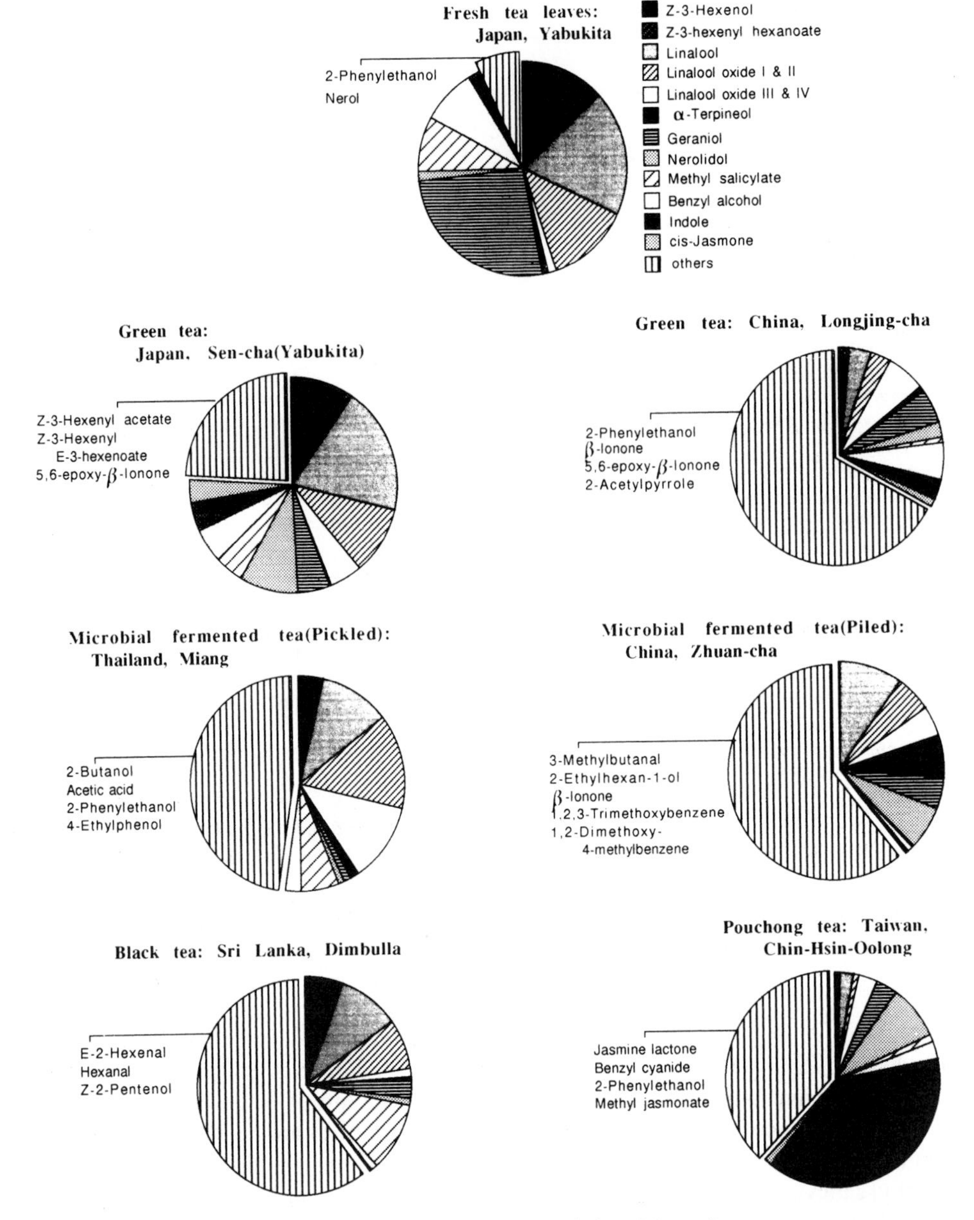

Fig. 5 Comparison of the aroma patterns and the characteristics of the various teas

Table 3. Contents of linalool and Geraniol, and terpene index

	Black tea					Oolong tea		
	B-I	B-II	B-III	B-IV	B-V	O-I	O-II	O-III
(1) Linalool	16.5	18.1	9.5	8.7	6.8	3.0	2.8	2.9
(2) Linalool oxide I and II	10.6	9.0	7.6	11.4	10.0	1.6	1.5	3.1
(3) Linalool oxide III and IV	0.4	1.0	1.6	0.6	6.4	0.7	0.6	2.3
(4) 3,7-Dimethyl-1,5,7- -octatrien-3-ol	0.9	0.2	0.3	1.1	0.9	2.0	2.0	3.1
Geraniol	1.2	1.5	2.9	12.7	29.6	3.4	5.0	12.9
(1)+(2)+(3)+(4) Total	28.4	27.3	19.0	21.8	24.1	7.3	6.9	11.4
terpene index	0.96	0.95	0.87	0.63	0.45	0.68	0.58	0.47

B-I: Assum(var.*assamica*)/Taiwan, B-II: Uva(var.*assamica*)/Sri Lanka, B-III: Dimbulla (var. *assamica*)/Sri Lanka, B-IV: Himaraya(hybrid of var.*assamica* and *sinensis*), B-V: Keemun-(var. *sinensis*)/China,, O-I: Tie-Kuan-Yin(var.*sinensis*)/China, O-II: Sou-Chong (var.*sinensis*)/China, O-III: Huang-Chin-Kuei (var. *sinensis*)/China.

4.2 Specific Components of the Essential Oils

4.2.1 Linalool and Linalool Oxides

Linalool is considered to be present as glycosides in fresh tea leaves. The enzyme hydrolyzes glycoside to linalool during mechanical crushing of the tea leaf cell (Takeo 1981). Linalool is easily oxidized to its oxides by oxydase. During the fermentation process of black tea, and the solar-withering and turnover treatment of pouchong tea, oxidation proceeds rapidly. Therefore, the ratio of linalool oxides to linalool can indicate the extent of oxidation in tea manufacturing.

4.2.2 Linalool and Geraniol

Tea plants are classified into three varieties, i.e., var. *sinensis*, var. *assamica*, and var. *sinensis* forma *macrophylla*. The content ratio of linalool and geraniol, which are the main terpenoids in tea leaves, is specific for different varieties. The ratio of linalool to the total monoterpene alcohol in injured tea shoots can be represented as the Terpene Index (TI) from the following formula:

$$TI = \frac{\text{linalool + linalool der.}}{\text{linalool + linalool der. + geraniol}} .$$

Var. *assamica*, which is cultivated in Sri Lanka, India, and Malaysia, has a TI of nearly 1.0. On the other hand, var. *sinensis* forma *bohea*, cultivated in Taiwan, shows the lowest TI of 0.1. Hybrids of var. *assamica* and var. *sinensis* are distributed widely in Asian countries and have intermediate values between 1.0 and 0.1. The Terpene Index can be correlated with the genetic characteristics and similarity of each tea cultiver, which also indicate the profile of the manufactured tea aroma.

4.2.3 Precursors of Each Component

The precursors of tea aroma components and the mechanisms for aroma formation have not been fully elucidated.

Various ionone derivatives are widely present in all types of tea, but not in fresh tea leaves. All of these ionone-related compounds have been proved to be produced from carotene by studies with many model experiments. Carotene, being included as 50–88 mg/100 g of water-free leaves, is degraded to yield ionone derivatives by heating (Kawashima and Yamanishi 1973; Kawakami 1982), photo oxidation and auto-oxidation (Isoe et al. 1969; Ina and Eto 1972) during the tea manufacturing process, particularly at the stage of fermentation (Sanderson et al. 1971). The identified ionone derivatives in tea essential oil are listed in Fig. 6 (Yamanishi 1981).

The straight-chain alcohols and aldehydes, especially the C_6 alcohols and aldehydes, are almost all derived from fatty acid. The amount of fatty acid present in tea leaves is approximately 100 mg/ 100 g of fresh tea leaves, and comprises 38% of linoleic acid, 25% of linolenic acid, 17% of palmitic acid, and 17% of oleic acid. The biosynthetic pathways for C_6 alcohol and aldehydes from the unsaturated fatty acid in tea leaves have been investigated (Hatanaka et al. 1982; Sekiya et al.

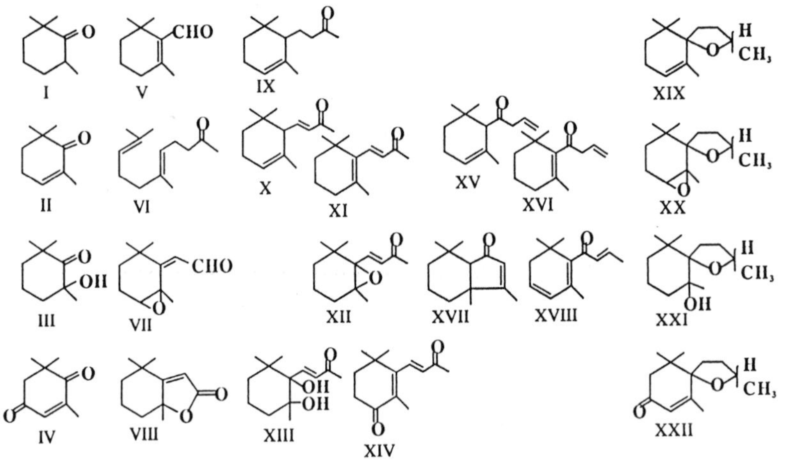

Fig. 6. Ionone series aroma compounds identified in tea. **I** 2,6,6-trimethylcyclohexanone; **II** 2,6,6-trimethylcyclohex-2-en-1-one; **III** 2,6,6-trimethyl-2-hydroxycyyclohexanone; **IV** 2,6,6-trimethylcyclohex-2-en-1,4-dione; **V** β-cyclocitral; **VI** geranylacetone; **VII** 1,1,5-trimethyl-4,5-epoxycyclohexyliden-2-acetaldehyde; **VIII** dihydroactinidiolide; **IX** 7,8-dihydro-α-ionone; **X** α-ionone; **XI** β-ionone; **XII** 5,6-epoxy-β-ionone; **XIII** 5,6-dihydroxy-β-ionone; **XIV** 4-oxo-β-ionone; **XV** α-damascone; **XVI** β-damascone; **XVII** 1,5,5,9-tetramethylbicyclo[4,3,0]non-8-cn-7-one; **XVIII** β-damascenone; **XIX** cis-&trans-theaspirane; **XX** cis- and trans-7,8-epoxytheaspirane; **XXI** cis- and trans-6-hydroxydihydrotheaspirane; **XXII** cis- and trans-theaspirone

1984). The contents of (Z)-3-hexenal, (Z)-3-hexenol, (E)-2-hexanal, and (E)-2-hexenol depend on the activity of each enzyme, i.e., LAH (lipid hydrolase), lipoxigenase, hydroperoxide lyase, and alcohol dehydrogenase, which are influenced by seasonal factors, especially by the cultivating temperature. Degradation products from oleic acid, linoleic acid, and linolenic acid by heating have been reported by Frankel et al. (1981), and auto-oxidation of linoleic acid was also reported by Ullrich and Grosch (1987). The photo-degradation products from linoleic acid and linolenic acid are listed in Table 4.

Acetaldehyde, 2-methylpropanal, 2-methylbutanal, and phenyl acetaldehyde were confirmed to be produced from amino acid by Strecker degradation through an experiment using ^{14}C-labeled amino acids during the fermentation process during black tea manufacture (Sanderson and Graham 1973). These volatile aldehydes are significant in the essential oil of the Japanese green tea, Gyokuro (Fig. 4).

1,2,4-Substituted phenolic compounds such as 4-alkenyl-guaiacol are derived from ferulic acid, and 4-alkylphenols such as 4-ethylphenol are derived from p-coumaric acid by microorganisms during the microbial fermentation process

Table 4. Photo oxidation products[a] of linoleic acid and linolenic acid

	Linoleic acid	Linolenic acid
Aldehyde	Butanal Pentanal Hexanal Heptenal (E)-2-Heptanal (Z/E)-2-Octenal	Propanal Pentanal (E)-2-Butenal Hexanal (Z/E)-2-Pentenal (E)-2-Hexenal (E)-2-Heptenal (E)-2-Octenal (E)-2,(E/Z)-4 Heptadienal (E)-2,(Z)-6-Nonadienal (E)2,(E/Z)-4-Nonadienal
Ketone	2-Heptanone	1-Penten-3-one (E)-3-Penten-2-one (E)3,(E/Z)-5-Octadienone
Alcohol	Butanol Pentanol Hexanol 1-Octen-3-ol	1-Penten-3-ol (Z/E)-2-Pentenol 1-Octen-3-ol
Acid	Acetic acid Butanoic Acid Pentanoic acid Hexanoic acid Heptanoic acid (E)-2-Heptenoic acid Octanoic acid 7-Octenoic acid (E)-2-Octenoic acid Nonanoic acid	Acetic acid Propionic acid Butanoic acid 2-Butenoic acid (Z/E)-Pentenoic acid Hexanoic acid Heptanoic acid (E)-2-Heptenoic acid Octanoic acid
Lactone	4-Hexanolide 2-Hexen-4-olide	4-Hexanolide 2-Hexen-4-olide
Furan	2-Pentylfuran	

[a]solar-degradaded for 8 h.

(Kawakami and Shibamoto 1990). 1,2,3-Polyphenoic compounds are considered to be formed from catechins.

4.2.4 Glycosides of Alcohols

Terpene alcohols and aliphatic alcohols are main components of the essential oil, which have been proved to be formed by terpene synthesis or the enzymatic degradation of fatty acids as described in the preceding section. However, these products are thought to be preserved and circulated as glycosides in the plant body and, during the fermentation process, the free alcohols should be produced by hydrolysis with glycosidase. The essential oil was extracted from a hot-water extract of tea leaves, and the water solution was then treated with β-glycosidase. The typical alcohols, geraniol, nerol, linalool, and benzyl alcohols, were identified by GC. The presence of some glycosides in the water extract was also recognized by HPLC (Yano et al. 1990). The glycosides of terpene alcohols have been found in several plant tissue (Schreier 1986; Winterhalter 1990), and tea leaves also seem to contain various types of glycosides as the precursor of the essential oil components.

4.2.5 Stereochemistry

Due to the restriction of sample quantity, a stereochemical study of each component is rare.

Linalool oxide I (*cis*-furanoid), II (*trans*-furanoid), III (*cis*-pyranoid), and IV (*trans*-pyranoid) are prepared through linalool epoxide, and each structure has been confirmed by a direct comparison with the synthetic one (Felix et al. 1963).

α-Farnesene is a main component of oolong tea, and its structure is related to another main compound, nerolidol, a sesquiterpene alcohol. The (3E,5E)-structure of α-farnesene has been confirmed by the identical KI value with that of an authentic sample derived from farnesol (Nobumoto 1990). The dehydration of nerolidol to α-farnesene did not occur during the separation and isolation process.

Methyl jasmonate has been thought to be one of the main aroma compounds in the essential oil of oolong tea, with jasmin lactone, *cis*-jasmone, and indole (Fig. 7). Recently, Acree et al. (1985) found that the sex pheromone of the oriental fruit moth, *Grapholitha molesta*, was the *cis*-isomer of methyl jasmonate, methyl epijasmonate, which shows a stronger odor than the *trans*-isomer. Kobayashi et al.

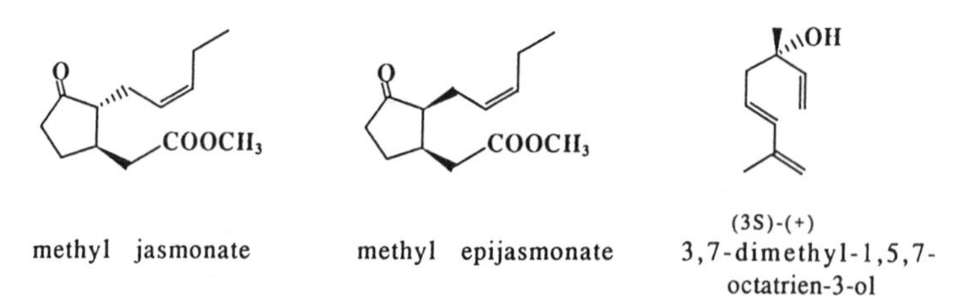

Fig. 7. Some stereoisomers found in tea essential oil

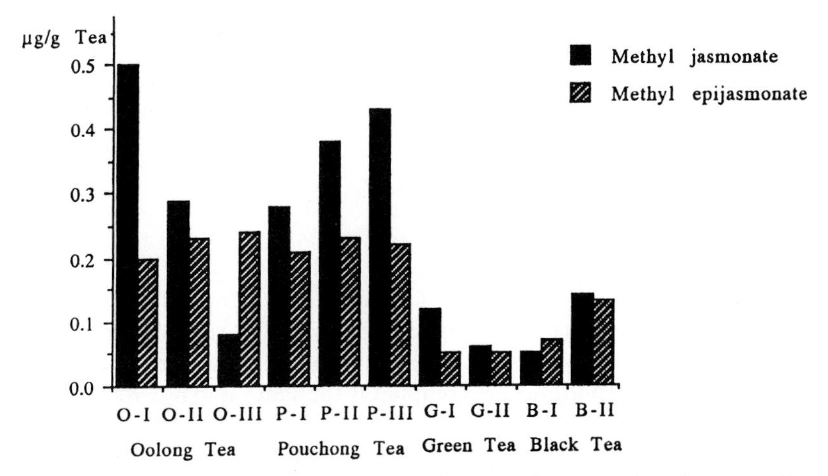

Fig. 8. Concentrations of methyl jasmonate and methyl epijasmonate in various types of tea O-I Huang Chin Kuei, O-II Tie Kuan Yin; O-III Shui Hsin, P-I Chin Hsin Oolong; P-II Chin Hsin Taipon; P-III Wu Yi; G-I Sayama; G-II Shizuoka; B-I Uva; B-II Nilgiri

(1988) reinvestigated the separation and identification conditions of methyl jasmonate from tea and found that the unstable *cis*-isomer was easily transformed to the stable *trans* form during the SDE isolation process or with GC conditions above 180 °C. Avoiding this heat isomerization, the contents of both compounds in several tea samples were quantitatively calculated as shown in Fig. 8. The high content of methyl epi-jasmonate in the semi-fermented teas (oolong and pouchong teas) is clearly demonstrated.

3S(+)–3,7-Dimethyl-1,5,7-octatrien-3-ol is a rare case of its absolute configuration being established in the essential oil of black tea (Nakatani et al. 1969). The same absolute configuration of plant-origin linalool suggests a biosynthetic pathway from linalool to this compound.

5 Pattern Analyses of Gas Chromatograms

High resolution or capillary gas chromatograms show very complicated patterns with 300 or more peaks. It became necessary to compare each gas chromatogram and to draw some information about the composition of the essential oil, the quality of the tea product, and/or a correlation between the organoleptic evaluation and analytical results. Today, personal computers with statistical software have become routine in the chemical laboratory (Warren and Walradt 1984).

If the relative peak areas on a gas chromatogram are treated as variables contributing to the whole character, various statistical calculations and multivariant analyses can be performed. Here, we will discuss only the calculation of similarity among the different essential oil of teas.

The similarity between sample A and B can be calculated by the following equation:

$$S_{AB} = \cos\theta = \frac{\sum_{i=1}^{N} a_i \cdot b_i}{\sqrt{\sum_{i=1}^{N} a_i^2} \cdot \sqrt{\sum_{i=1}^{N} b_i^2}}$$

Here, a_i and b_i are the areas of peak i in samples A and B, respectively. This value fluctuates between 0 and 1, in other words, an angle (θ) between 90° and 0°. $S_{AB} = 1$ means that these two patterns are completely the same, and $S_{AB} = 0$ means there is no similarity.

In the gas chromatograms of essential oils obtained from several oolong and black teas by the SDE method, 53 main peaks were selected, and their peak area percentages were used as the variables for calculating the similarity of each gas chromatogram. The calculated similarity data is summarized in Table 5. From these data, we can easily recognize that the similarity of gas chromatograms among the same type of tea is relatively high.

A multivariant analysis has been developed to qualify the GC data obtained, but a detailed discussion will be omitted here because its application oversteps the bounds of analytical methods for the essential oils.

Table 5. Similarity between six kinds of tea (oolong tea and black tea)[a]

	Oolong tea			Black tea		
	O-I	O-II	O-III	B-IV	B-I	B-II
O-I	1.000	0.982	0.845	0.345	0.297	0.265
O-II		1.000	0.868	0.356	0.267	0.262
O-III			1.000	0.535	0.303	0.275
B-IV				1.000	0.740	0.638
B-I					1.000	0.882
B-II						1.000

[a]cf. Table 3.

References

Acree TE, Nishida R, Fukami H (1985) Odor-thresholds of the stereo isomers of methyl jasmonate. J Agric Chem 33:425–427

Aisaka H, Kosuge M, Yamanishi T (1978) Comparison of the flavors of Chinese "keemun" black tea and Ceylon black tea. Agric Biol Chem 42:2157–2159

Badings HT, DeJong C, Dooper RPM, De Nigs RCM (1985) Rapid analysis of volatile compounds in food products by purge-and-cold-trapping/capillary gas chromatography. In:Adda J(ed) Progress in flavor research 1984. Elsevier, Amsterdam, p 523

Felix D, Melera A, Seibel J, Kovats E (1963) Zur Kenntnis ätherischer Öle. Die Struktur der sogenannten "Linaloolxide". Helv Chim Acta 46:1513–1536

Flankel EN, Neff WE, Selke E (1981) Analysis of autoxidized fats by gas chromatography-mass spectrometry:V11. Volatile thermal decomposition products of pure hydroperoxides from autoxidized and photosensitized oxidized methyl oleate, linoleate and linolenate. Lipids 16:279–285

Hatanaka A, Kajiwara T, Sekiya J, Imoto M, Inoue S (1982) Participation and properties of Lipoxygenase and hydroperoxide lyase in volatile C_6-aldehyde formation from C_{18}-unsaturated fatty acids in isolated tea chloroplasts. Plant Cell Physiol 23:91–99

Ina K, Eto H (1972) Photo-oxidation of β-ionone. Agric Biol Chem 36:1091–1094

Isoe S, Heyron SB, Sakan T (1969) Photo-oxidation of carotenoids. I. The formation of dihydroactinidiolide and β-ionone from β-carotene. Tetrahedron Lett 279–281

Kawakami M (1982) Ionone series compounds from α-carotene by thermal degradation in aqueous medium. Nippon Nogeikagaku Kaishi 56:917–921

Kawakami M, Shibamoto T (1990) Volatile compounds produced from ferulic acid and p-coumaric acid by microbial organisms. Agric Biol Chem (in press)

Kawakami M, Yamanishi T (1981) Aroma characteristics of kabusecha (shaded green tea). Nippon Nogeikagaku Kaishi 55:117–123

Kawakami M, Yamanishi T (1983) Flavor constituents of Longjing tea. Agric Biol Chem 47:2077–2083

Kawakami M, Yamanishi T, Kobayashi A (1986) The application of the pouchong tea process to the leaves from tea plants, var. assamica dominant hybrids. Agric Biol Chem 50:1895–1898

Kawakami M, Chairote G, Kobayashi A (1987a) Flavor Constituents of pickled tea, minang, in Thailand. Agric Biol Chem 51:1683–1687

Kawakami M, Kobayashi A, Yamanishi T (1987b) Flavor constituents of pickled tea: Goishi-cha and Awa-cha. Nippon Nogeikagaku Kaishi 61:345–352

Kawakami M, Kobayashi A, Yamanishi T, Shoujaku S (1987c) Flavor constituents of microbial-fermented teas, Chinese Zhuan-cha and Koku-cha. Nippon Nogeikagaku Kaishi 61:457–465

Kawakami M, Uchida H, Kobayashi A, Yamanishi T (1989) The effects on Awa-cha flavor of pickling and solar drying. Agric Biol Chem 53:271–275

Kawashima K, Yamanishi T (1973) Thermal degradation of beta-carotene Nippon Nogeikagaku Kaishi 47:79–81

Kobayashi A, Tachiyamak, Kawakami M, Yamanishi Y, Juan IM, Chiu W T-F (1985) Effect of solar-withering and turn over treatment during indoor-withering on the formation of pouchong tea aroma. Agric Biol Chem 49:1655–1660

Kobayashi A, Kawamura M, Yamamoto Y, Shimizu K, Kubota K, Yamanishi T (1988) Methyl epijasmonate in the essential oil of tea. Agric Biol Chem 52:2299–2303

Kovats E (1958) Gas-chromatographische Charakterisierung organischer Verbindungen. Teil 1:Retentionsindices aliphatischer Halogenide, Alkonole, Aldehyde und Ketone. Helv Chim Acta 41:1915–1932

Linskens HF, Jackson JF (1986) (eds) Gas chromatography/mass spectrometry. Springer, Berlin Heidelberg New York Tokyo

Mac Lafferty FW, Stanffer DB (eds) (1989) The Wiley/NBS Registry of mass spectral data. John Wiley, New York

Mass Spectrometry Data Centre (1983) Eight-peak index of mass spectra. The Royal Society of Chemistry, Nottingham UK

Nakatani Y, Sato S, Yamanishi T (1969) 3S-(+)-Dimethyl-1,5,7-octatriene-3-ol in the essential oil of black tea. Agric Biol Chem 33:967–968

Nobumoto Y (1990) Study of the aroma of semi-fermented tea. Thesis, Ochanomizu University, Tokyo

Nose M, Nakatani Y, Yamanishi T (1971) Identification and composition of intermediate and high boiling constituents in green tea flavor. Agric Biol Chem 35:261–271

Pouchert CJ (1989) The Aldrich library of FT-IR spectra. Aldrich Chemical, Milwaukee

Sadtler Research Laboratories (1985) The Sadtler standard gas chromatography retention index library, Division of Bio-Rad Laboratories Inc. Philadelphia

Sanderson GW, Graham HN (1973) On the formation of black tea aroma. J Agric Food Chem 21:576–585

Sanderson GW, Co H, Gonzaleg JG (1971) Biochemistry of tea fermentation: the role of carotenes in black tea aroma formation J Food Sci 36:231–236

Schreier P (1986) Biogeneration of plant aromas. In: Birch BB, Lidley MG (eds) Developments in food flavors. Elsevier, London, pp 89

Schultz TH, Flath RA, Mon TR, Eglling SB, Teranishi R (1977) Isolation of volatile components from a model system. J Agric Food Chem 25:446–449

Sekiya J, Kajiwara T, Hatanaka T (1984) Sensonal changes in activities of enzymes responsible for the formation of C_6-aldehydes and C_6-alcohols in tea leaves, and the effects of environmental temperatures on the enzyme activities. Plant Cell Physiol 25:269–280

Smith SL (1984) Coupled systems: capillary GC-MS and capillary GC-FTIR. J Chromatogr Sci 22:143–148

Straten SV, Maarse H (1983) Volatile components in food. Institute CIVO-Analysis TNO, Utrechtseweg

Takei Y, Ishikawa Y, Hirano N, Fuchinoue H, Yamanishi T (1978) The influence of the amount of supplied fertilizer and vinyl-house cultivation on the tea aroma. Nippon Nogeikagaku Kaishi 52:505–512

Takeo T (1981) Production of linalool and geraniol by hydrolytic breakdown of bound forms in disrupted tea shoots. Phytochemistry 20:2145–2147

Takeo T (1982) Variations in aromatic compound content in non-fermented and semi-fermented tea. Nippon Nogeikagaku Kaishi 56:799–801

Tokimoto Y, Ikegami M, YamanishiT, Juan IM, Chieu WT-F (1984) Effects of withering and mass-rolling process on the formation of aroma components in Pouchong type semi-fermented tea. Agric Biol Chem 48:87–91

Tsugita T, Imai T, Doi Y, Kurata H (1979) GC and GC-MS analysis of head-space volatiles by Tenax GC trapping techniques. Agric Biol Chem 43:1351–1354

Ullrich F, Grosch W (1987) Identification of the most intense volatile flavor compounds formed during autoxidation of linoleic acid. Z Lebensm Unters Forsch 184:277–282

Van den Dool H, Kratz PD (1963) A generalization of the retention index system including linear temperature programmed gas-liquid partition chromatography. J Chromatogr 11:463–471

Warren CB, Walradt JP (1984) Computers in flavor and fragrance research. American Chemical Society, Washington D.C.

Winterhalter P (1990) Bound terpenoids in the juice of the purple passion fruit. J Agric Food Chem 38:452–455

Yamaguchi K, Shibamoto T (1981) Volatile constituents of green tea, Gyokuro (*Camellia sinensis* L. var *yabukita*). J Agric Food Chem 29:366–370

Yamanishi T (1978) The aroma of various teas. In: Charalambous G, Inglett GE (eds) Flavor of food and beverages. Academic Press, New York, p 305.

Yamanishi T (1981) Tea, coffee, cocoa and other beverages. In: Teranishi R, Flath RA, Sugisawa H (eds) Flavor research. Marcel Dekker New York, p 231.

Yamanishi T, Nose M, Nakatani Y (1970) Future investigation of flavor constituents in manufactured green tea. Agric Biol Chem 30:599–608

Yamanishi T, Kosuge M, Tokitomo Y, Maeda R (1980) Flavor constituents of Pouchong tea and comparison of the aroma pattern of jasmin tea. Agric Biol Chem 44:2139–2142

Yano M, Okada K, Kubota K, Kobayashi A (1990) Studies on the precursors of monoterpene alcohols in tea leaves. Agric Biol Chem 54:1023–1028

Special Methods for the Essential Oils of the Genus *Thymus*

M.E. CRESPO, J. JIMÉNEZ, and C. NAVARRO

1 Introduction

The genus *Thymus*, a member of the family Lamiaceae, comprises approximately 70–80 species of woody plants and low-lying shrubs. These aromatic species are found in all temperate regions of the northern hemisphere, and are especially common in the Mediterranean area. The large number and diversity of species in this genus have led to its subdivision into sections which, since their inception, have undergone considerable modifications, particularly with respect to the assignment of given species to a section. Currently, the most widely accepted classification is that proposed by Morales (1986), based on the criteria set down by Jalas (1970) and Jalas and Kaleva (1970), who in turn based their work on Velenovski (1906).

Different phytochemical groups have been detected in *Thymus* species, including polyphenol derivatives (Adzet and Martinez 1981; Van den Broucke 1982; Van den Broucke et al. 1982; Ferreres et al. 1985a,b; Voirin et al. 1985; Tomas-Barberan et al. 1985, 1986, 1987; Hernandez et al. 1986; Marhuenda et al. 1987; Adzet et al. 1988, among others) and especially essential oils.

Although this chapter will center on the essential oils of the genus *Thymus*, some general comments on the chemistry of the essential oils of the genus family Lamiaceae may be appropriate. As stated by Hegnauer (1962), these essential oils are mainly monoterpenic (C10), and often contain large amounts of menthol, menthone, pulegol, thymol, and carvacrol. Sesquiterpenes are rarely the major component, and phenyl propane derivatives, although widely present in other families, are only occasionally found in some Lamiaceae essential oils.

The essential oils of the genus *Thymus* are quite complex. On the basis of the results obtained from analyses in our laboratories, as well as from studies published by others on the essential oils of various species of thyme, these can be classified in general terms into two main groups.

The first group contains those species in which aromatic alcohols (thymol and carvacrol) and/or their biosynthetic precursors γ-terpinene and ρ-cymene (Poulose and Croteau 1978a,b) are clearly the predominant component, as in *Th. bornumeueri* (Merichi 1986), *Th. borystenicum* (Sur et al. 1988), *Th. broussonetti* (Richard et al. 1985), *Th. capitatus* (Papageorgiov and Vassilios 1980; Sendra and Cuñat 1980; Papageorgiov and Argyriadou 1981; Solinas et al. 1981; Lawrence 1982; Falchi-Delitala et al. 1983; Ravid and Putievsky 1983), *Th. collinus* (Novruzova and Kasumov 1987; Kasumov 1988), *Th. daciens* (Kisgyorgy et al. 1983), *Th. daghestanicus* (Novruzova and Kasumov 1983), *Th. fedtschenko* (Kasumov 1988), *Th. forminii* (Kasumov 1981a), *Th. herba-barona* (Falchi-Delitala et al. 1983), *Th. karamarianicus* (Kasumov and Farkhadova 1986; Novruzova and

Kasumov 1987), *Th. kotschyanus* (Kulieva et al. 1979; Kasumov and Gadzhieva 1980; Kasumov 1988), *Th. krylovii* (Tikhonov et al. 1988), *Th. maroccanus* (Richard et al. 1985), *Th. marschallianus* (Dembitskii et al. 1981; Sur et al. 1988), *Th. nigricus* (Kasumov 1981b), *Th. orospedanus* (Crespo et al. 1986), *Th. pallidus* (Richard et al. 1985), *Th. pastoralis* (Novruzova and Kasumov 1987), *Th. pectinatus* (Merichi 1986), *Th. piperella* (Adzet et al. 1977a), *Th. rariflorus* (Kasumov 1982; Novruzova and Kasumov 1987), *Th. schimperi* (Lemordant 1986), *Th. serpylloides* (Crespo et al. 1988), *Th. serpyllum* (Popov and Odynets 1977; Mathela et al. 1980; Sur et al. 1988), *Th. spathulifolius* (Merichi 1986), *Th. tiflisensis* (Kasumov 1988), *Th. transcaucasicus* (Kasumov 1981a; Novruzova and Kasumov 1987; Kasumov 1988), *Th zygioides* (Ilisulu and Tanker 1986; Merichi 1986), and *Th. zygis* (Cabo et al. 1974; Richard et al. 1985).

In the essential oils of other members of this genus, however, aromatic ring-containing components are scarce or altogether lacking, whereas other components usually considered to be scarce in thyme species predominate. Examples of this second group include essential oils in which the major components are monoterpene hydrocarbons (MTHC). In *Th. glabrescens* and *Th. balcanus* (Kisgyorgy et al. 1983), open chain hydrocarbons predominate, while in *Th. coriifolius* (Kasumov 1987) α-pinene is the major component and in *Th. granatensis* (Cabo et al. 1986a), myrcene and α-phellandrene predominate.

In this group can also be included the essential oils in which the major component is an oxidated derivative of MTHC, e.g., *Th. eriphorus* (Kasumov 1983), *Th. praecox* (Lundgren and Stenhagen 1982; Stahl 1982, 1984; Stahl-Biskup 1986a), *Th. tosevii* (Katsiotis and Iconomov 1986), and *Th. leptophyllus* (Blazquez et al. 1989), in which free or sterified linalool is the major component, *Th. trautvetteri* (Ismailov et al. 1981), whose major component (11%) is the alcohol geraniol, and *Th. sipyleus* (Merichi 1986), whose major component (33%) is the aldehyde geranial. Borneol is found as the major component of the essential oils of *Th. satureioides* (26%) (Richard et al. 1975, 1985; Miquel et al. 1976), *Th. quinquecostatus* (31%) (Xi-Quing et al. 1980) and *Th. carnosus* (51%) (Marhuenda and Alarcon de la Lastra 1986). In *Th. cilicicus* and *Th. revolutus*, α-terpineol is the major component, making up 33 and 30% respectively of the essential oil (Ilisulu and Tanker 1986; Merichi 1986).

The internal ether 1,8-cineol makes up from 25 to 50% of the essential oil of *Th. membranaceus* (Zarzuelo et al. 1987), *Th. funkii* (Revert 1975), *Th. antoninae* (Revert 1975), *Th. longiflorus* (Cruz et al. 1988), *Th. baeticus (Cruz et al. 1989), and Th. moroderi* (Adzet et al. 1989).

It should be pointed out that certain species of the genus *Thymus* show chemical polymorphism controlled by genetic factors, as part of an adaptational process which allows the plant to survive under different conditions (Brasseur 1983; Vernet et al. 1986). For example, the essential oil of *Th. vulgaris* would normally be assigned to the aromatic alcohol-rich group. However, studies of samples collected from different populations revealed a considerable degree of chemical polymorphism, so that seven chemotypes could be distinguished: 1,8-cineol, geraniol, linalool, α-terpineol, thymol, carvacrol, and a seventh type, which includes essential oils containing free trans-4-tuyanol and free or sterified cis-myrcenol-8, compounds not usually found in thymes (Granger and Passet 1971, 1973; Vernet et al. 1977, 1986; Passet 1979; Vernet and Gouyon 1979).

In *Th. mastichina*, a second chemotype rich in linalool (80%) has been described in Atlantic regions (Adzet et al. 1977a,b; Mateo et al. 1979), in addition to the well-known Mediterranean chemotype rich in 1,8-cineol.

The hypothesis of chemically distinct individuals was confirmed in *Th. nitens* (Granger et al. 1973), in which the most common chemotypes are carvacrol- and geraniol-rich, although a cineol-rich type is also thought to exist.

Thymus piperella (Adzet et al. 1977a), on the other hand, appears to be a chemically homogeneous species whose essential oil, regardless of whether thymol or carvacrol predominate, can be classified as a member of the phenol-rich group.

Lassanyi (1978) established three chemotypes for *Th. pulegioides*: one characterized by the predominance of phenolic compounds (further subdivided by Stahl-Biskup (1986a) into thymol-rich and carvacrol-rich groups), one rich in citral and a third type in which geraniol is the major component.

Two chemotypes have also been distinguished in *Th. hyemalis* (Adzet et al. 1976; Cabo et al. 1980): one rich in phenols and phenolic precursors, and another in which borneol, camphor, cineol, and hydrocarbons predominate.

Although *Th. praecox* essential oil contains linalyl acetate as its major component (Lundgren and Stenhagen 1982; Stahl 1982, 1984), up to eight different chemotypes have been identified on the basis of their sesquiterpene content (Stahl-Biskup 1986a).

In addition to the chemical variants just mentioned, it should be recalled that the essential oils themselves undergo marked seasonal changes in composition, as shown in studies of *Th. hyemalis* (Cabo et al. 1987), *Th. granatensis* (Cabo et al. 1986b) and *Th. carnosus* (Marhuenda and Alarcon de la Lastra 1986), to name some examples.

2 Plant Material

Recalling that the marked variability in the chemical composition of *Thymus* essential oils results mainly from the characteristics of the site where the plants are collected (latitude, altitude, soil type, etc.) (Schratz and Horster 1970; Richard et al. 1975; Mateo et al. 1978; Bellomaria et al. 1981; Cabo et al. 1981, 1986c, 1986d; Kasumov and Akhmedova 1981; Kasumov 1982, among others) and the stage of the growth cycle at the moment of collection (Falchi-Delitala et al. 1981, 1983; Cabo et al. 1986b, 1987; Marhuenda and Alarcon de la Lastra 1986), the conditions under which the samples are obtained should be defined when reporting experimental results. The geographic coordinates of the site and the growth stage should be mentioned. If the purpose of the study is to compare the essential oils of different species or populations of thyme, however, the plants should be collected during the flowering period.

Different criteria have been published regarding the best time of day to collect the plants (Messerschmidt 1964; Muenchow and Pohloudek-Fabini 1964; Tucakov 1964). Weiss and Flück (1970), in their studies of *Th. vulgaris*, concluded that the highest essential oil yields are obtained from plants collected at 15.00 (GMT), when the air reaches the daily maximum, while lowest yields are obtained at 23.00 (GMT). According to these authors, peak content in essential oil parallels the synthesis of terpenes.

Once collected, the plant's identity is confirmed and the specimens are deposited in a herbarium. The woody parts are separated from the plant material and discarded, and the essential oil is extracted (Fehr and Stenzhorn 1979; Koller 1988). Alternatively, if the fresh material cannot be processed immediately, it is dried under controlled temperature conditions and stored at 4 °C away from light until extraction.

3 Extraction

Several procedures from the extraction of essential oil are available, including organic solvent extraction, scarification, enfleurage, pressing, and various types of distillation. The technique to be used should be chosen in accordance with the characteristics of the essential oil; in *Thymus*, distillation in water is the method preferred by most workers (Adzet et al. 1976, 1977b; Papageorgiov and Vassilios 1980; Sendra and Cuñat 1980; Solinas et al. 1981; Stahl 1982; Ravid and Putievsky 1983; Richard et al. 1985; Ilisulu and Tanker 1986; Katsiotis and Iconomov 1986; Blazquez et al. 1989, among others). In our laboratory we use a Clevenger device (Cabo et al. 1986a,b,c; Crespo et al. 1986, 1988; Cruz et al. 1989), and determine the essential oil content with the methods described in Phamacopée Europėene (1975). The isolated oil is dried over anhydrous sodium sulfate and stored at 4–6°C.

The method of extraction used in our laboratory yields from 0.1% (Cabo et al. 1986c) to 2.7% (Cabo et al. submitted) essential oil (v/w) referred to dried plant material depending on the species. These figures are within the range of yields reported by other authors i.e. from 0.1% (Stahl 1982) to 3.6% (Falchi-Delitala et al. 1983) in other thyme species.

4 Analytical Methodology

4.1 Qualitative Analysis

4.1.1 Chromatographic Techniques

4.1.1.1 Thin Layer Chromatography

This simple, rapid technique provides information on the complexity of the essential oil and allows us to identify and quantify some of its components. We use thin layer chromatography to study *Thymus* essential oils, using a variety of conditions regarding the stationary and mobile phases and the method of detection.

As a rule, we recommend Silicagel 60 F254 precoated plates (250-μ thickness) as the stationary phase. Although acid (0.5 N oxalic acid) or alkaline absorbants (0.5 N KOH) (Vernin 1964) have also been said to provide a better separation of the aromatic alcohols, the results obtained with these modifications in our laboratories do not support these claims.

With regard to the mobile phase, mixtures of hexane-heptane/ethyl ether (84:15) or benzene/ethyl acetate (85:15), frequently used to analyze essential oils by TLC (Verderio and Venturini 1965; Karawya et al. 1970), give the best results.

In some specific cases these general use mobile phases can be replaced with others which provide better resolution of certain components. Some authors (Thieme 1967; Thieme and Thi Tam 1968) used benzene/carbon tetrachloride/o-nitrotoluene (1:1:1) to separate thymol and carvacrol. A mixture of cyclohexane/chloroform/acetic acid (4:5:1) was used by Stahl (1965) to identify camphor. This author also recommended butanone (methylethylketone) for the identification of this component.

The most common systems of detection used in TLC are spray reagents:

Vanillin-sulfuric acid is used as a 3% solution (w/v) in ethanol to which is added 4 ml concentrated H_2SO_4 per 100 ml. The spots are revealed by heating to 120 °C (Von Schantz et al. 1962; Attaway et al. 1965; Younos et al. 1972).

Anisaldehyde-sulfuric acid is, after vanillin, the most commonly used reagent in TLC analyses of essential oils. Both mixtures have the advantage of producing a diverse range of colors, which aid in the identification of the components. Heinrich (1973) used a mixture of both aldehydes, although more frequently, different concentrations of anisaldehyde alone have been employed (Schratz and Qédan 1965; Ter Heide 1968; Weiss and Flück 1970).

Phosphomolybdic acid reveals most of the components of thyme essential oils as a blue spot against a yellow background (Wagner et al. 1984).

In addition to the absorbents and mobile phases most effective in separating the major components usually found in thyme essential oils, specific systems of detection are used to detect aromatic alcohols (thymol and carvacrol) and ketones (camphor).

Ultraviolet light (254 nm) replaces chemical treatment for all components which, like thymol and carvacrol, contain at least two double bonds, which eliminate dark spots against a green fluorescent background on the TLC plate (Wagner et al. 1984).

Among the spray reagents used to selectively detect thymol and carvacrol, anisaldehyde-sulfuric acid is effective when used as described by Stahl (1965) as a freshly made mixture of 0.5 ml anisaldehyde, 9 ml 96° ethanol, 0.5 ml H_2SO_4, and 0.1 ml acetic acid. The colors can be enhanced by heating to 120 °C for 2 min. This sensitive reagent gives a red color for thymol and a pink color for carvacrol, and also reveals even trace levels of borneol as an intense green color.

The two phenols can also be identified with a mixture of anisaldehyde-formaldehyde and sulfuric acid, which reveals thymol as a bluish brown color, and carvacrol as reddish brown (Thieme 1967; Thieme and Thi Tam 1968).

Camphor, which cannot reliably be detected with any of the above mentioned general treatments, is usually revealed with Dragendorff's reagent (Stahl 1965). It should be remembered, however, that this reaction, considered exceptional in nonalkaloid compounds (or at least in compounds containing no nitrogen), is produced by some substances which do not obey this rule. Of the components in thyme essential oils, eugenol gives an immediate, long-lasting reaction, camphor and cineol give immediate reactions which gradually fade away after 24 h, and anethole, geraniol, terpineol, and thymol produce positive reactions, which, however, do not appear until 24 h after the addition of the reagent.

Camphor is also identifiable with 2–4 dinitrophenylhydrazine, which detects compounds containing an oxo group: aldehydes (citral, citronellal, geranial, etc.) and ketones (camphor, ionone, fenchone, etc.)

Essential oil components are identified by comparing their Rf and color with patterns run under the same conditions. The essential oil in the pattern is then enriched to check whether the spot becomes larger or more intensely colored.

Figure 1 illustrates the method used to detect thymol and carvacrol, the two most characteristic compounds in thyme essential oils. Although the difference between the Rf of these two aromatic alcohols is slight, they become clearly distinguishable by their color when developed with anisaldehyde-sulfuric acid.

4.1.1.2 Gas-Liquid Chromatography

The greatest sources of difficulty in analyzing the chemical composition of essential oils are the large number of different compounds to be distinguished, and the differences in their relative proportions. Despite their chemical complexity, however, GLC in combination with spectroscopic examination (with mass spectroscopy to give an example) can go a long way toward identifying many substances. Although this technique is not without its own difficulties, it is currently the most effective and widely used method to separate the components of thyme essential oils (Lundgren and Stenhagen 1982; Ravid and Putievsky 1983; Katsiotis and Iconomov 1986; Merichi 1986; Blazquez et al. 1989, among others).

Several factors involved in GLC affect the degree of accuracy and resolution of the mixture under study: the detection system, mobile phase and stationary phase used, and the working temperature. Of the detection systems commonly used, flame ionization (FID) and mass spectroscopy (MS) are the most effective. The first is used in both the identification and the quantification of essential oil components, while the second is used as a backup to confirm the FID findings, as well as to identify compounds which are not identifiable with GLC-FID.

We use nitrogen as the mobile phase in packed columns, and hydrogen and helium in capillary and semicapillary columns. The flow rates for these gases vary from less than 1 ml (hydrogen is used at a rate of 0.2 ml/min) to 30 ml/min in analyses performed in packed columns.

Stationary phases of different polarity are used as an accurate method of confirming the identity of the compounds. In both packed and capillary columns, high polarity (e.g., DEGS at different proportions), and low polarity fillers (e.g., silicone) have been used.

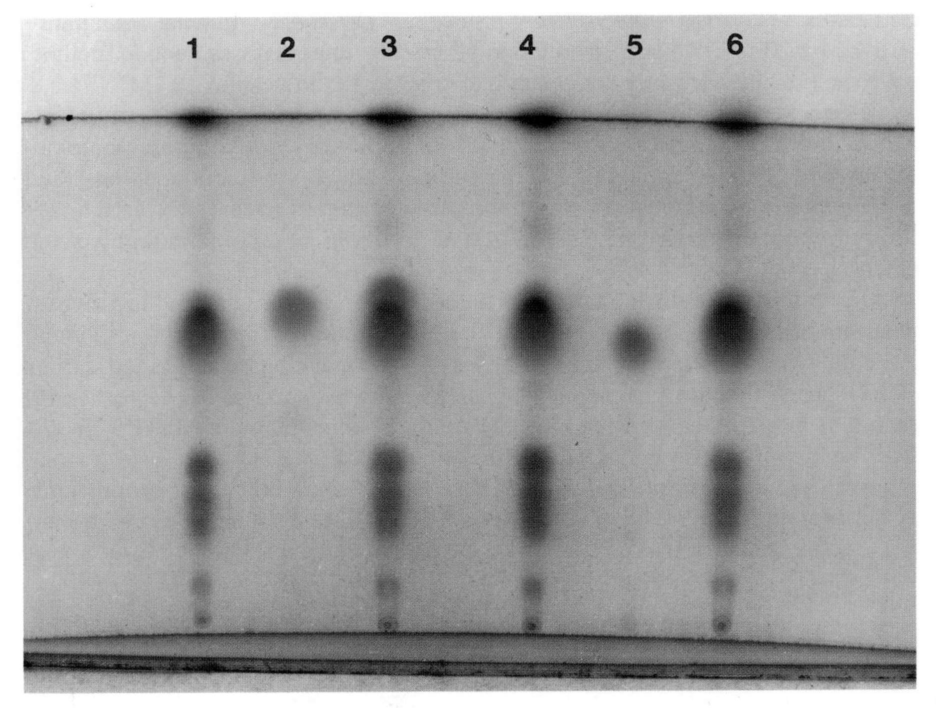

Fig. 1 Thin layer chromatography (one-dimensional ascending) to identify thymol and carvacrol in the essential oil of *Th. serpylloides* ssp. *gadorensis*. Development: simple (15 cm). Silicagel 60 F254 precoated plates were used as the stationary phase, and benzene/carbon tetrachloride/o-nitrotoluene (1:1:1) was used as the mobile phase. The plates were developed with anisaldehyde-sulfuric acid followed by heating to 120 °C for 2 min. A 2-μl sample of the essential oil (10% w/v in ethyl ether) and 2 μl thymol and carvacrol (1% w/v in ethyl ether) were placed on the plate. The order of the samples tested is *1* essential oil; *2* thymol; *3* essential oil + thymol; *4* essential oil; *5* carvacrol; *6* essential oil + carvacrol

The complete chromatogram of the essential oil is obtained by using a specific temperature program to improve the resolution of the different components in the mixture, within the limits imposed by the type of column. For thymes, we generally start the run at 50 °C, which effectively elutes the MTHC, and raise the temperature to 180 °C in order to completely separate the aromatic alcohols in this type of column. Slight variations in this program are used, depending on the species of thyme to be analyzed. In nearly all cases, the temperature is raised in a step-wise fashion, rather than steadily, so that there are pauses at approximately 100 and 150 °C. This gives an acceptable degree of resolution for the oxidated derivatives of MTHC as well as for some esters. Although this method is the most effective for thyme essential oils, some cases require that the column be run under completely isothermal conditions using the relative retention method in order to detect certain compounds in the mixture.

Gas-Liquid Chromatography-Flame Ionization Detection. Initially, qualitative analysis by GLC-FID was performed in all cases by the relative retention method, which gave an idea of which components are present in the sample.

The next step in the process of identifying the components of the essential oils was to develop a method to enrich the peaks. Separate chromatograms were obtained for (1) the essential oils, (2) the pattern and (3) the essential oil enriched in the pattern, added in small amounts. This treatment increases the area under the curve of the peak in comparison to that produced by chromatography of the essential oil only.

To verify, within the limitations of the technique, the presence of the different components, the two methods just described (estimation of relative retentions and enrichment) are used with the two types of stationary phase described earlier (DEGS and silicone) in our analyses of all *Thymus* species. To aid other researchers interested in thyme essential oils, the results obtained with GLC in different members of this genus will be presented below.

We used columns packed with DEGS and silicone in different proportions. As an example of the type of resolution obtained, Fig. 2 shows a chromatogram

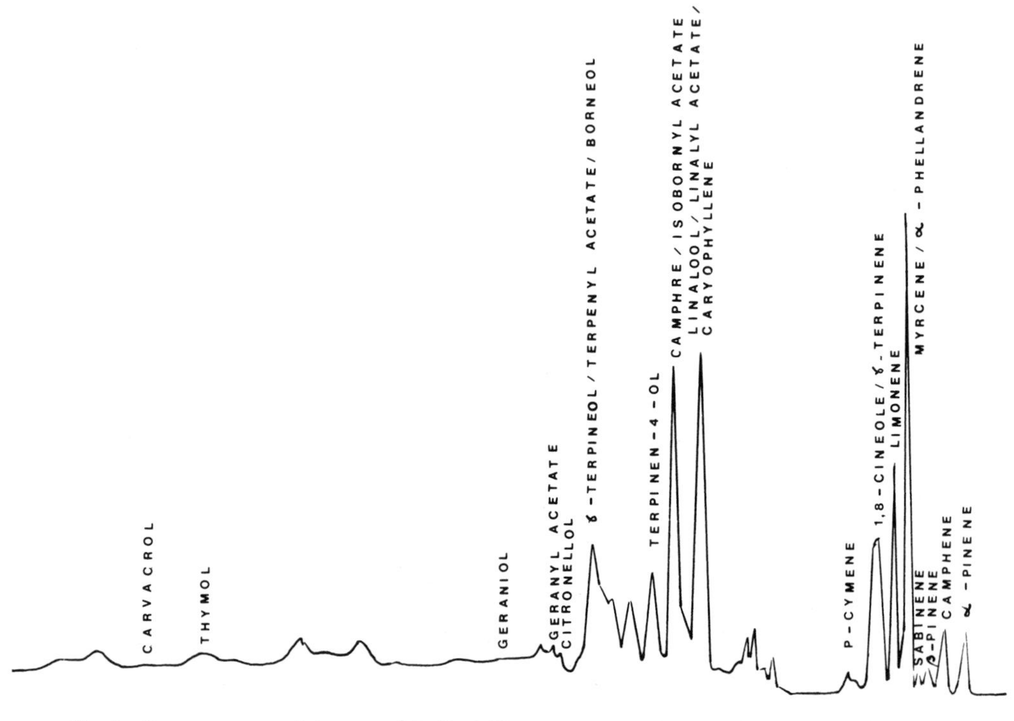

Fig. 2. Chromatogram of the essential oil of *Th. granatensis* Boiss. Working conditions: chromatograph, P-E model F-11, with FID and P-E model 159 recorder. Packed column (2 m x 1.8 mm i.d.) coated with DEGS (10%) over Chromosorb W., 80–100 mesh. Carrier gas: N_2 (flow rate, 30 ml/min). Injection block temperature: 200 °C. Detector temperature: 250 °C. Column temperature program: initial isotherm 70 °C (6 min), 100 °C (10 min), 150 °C (14 min), final isotherm 180 °C (10 min), with a gradient of 30 °C/min

of the essential oil from the flowering apex of *Th. granatensis* Boiss., found endemically in southern Spain. This packed column chromatogram identified most of the essential oil components. However, some of the peaks were difficult to identify. Particularly with regard to the MTHC, it is extremely difficult to detect the presence of myrcene and α-phellandrene, since both have the same retention times under these assay conditions. The same problem arises with the fraction comprising the oxidated derivatives, hence combinations of linalool/lynalyl acetate/caryophyllene, camphor/isobornyl acetate, and terpineol/terpenyl acetate/borneol are not well resolved with these methods.

To obviate as many of these problems as possible, we now use capillary columns filled with different materials; for *Th. granatensis* we use FFAP-A-63 (Carlo Erba). The results are shown in the second chromatogram in Fig. 3. The capillary column resolved a greater number of components than the packed column, as seen for the MTHC γ-terpinene and the internal ether 1,8-cineol, which showed the same retention time with the packed column. Likewise, camphor and isobornyl acetate were separately eluted, in contrast to the results shown in Fig. 2. Caryophyllene was also separated from linalool and linalyl acetate, and borneol

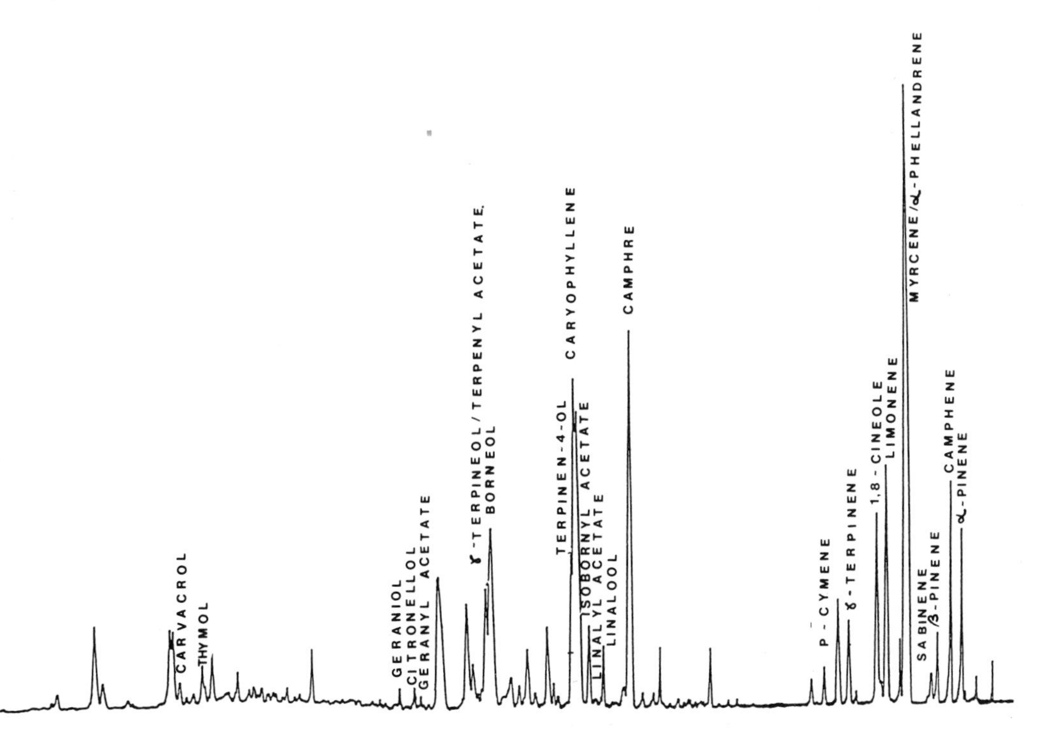

Fig. 3 Chromatogram of the essential oil of *Th. granatensis* Boiss. Working conditions: chromatograph, Carlo Erba model Fractovap 2400 V with FID and a Hewlett-Packard model 3380A integrated recorder. Capillary column FFAP-A-63 (50 m x 0.5 mm i.d.). Carrier gas: H_2 (flow rate 0.2 ml/min). Injection block temperature: 200 °C. Detector temperature: 250 °C. Column temperature program: initial isotherm 60 °C (18 min), 120 °C (23 min), final isotherm 180 °C (15 min), with a gradient of 30 °C/min

was distinguished from terpineol and terpenyl acetate. However, myrcene was not effectively separated from α-phellandrene, nor was terpineol distinguishable from its acetate.

The appearance of capillary columns for GLC was followed some time later by the use of semi-capillary columns. We have tested the advantages and disadvantages of the two types of columns in the chromatography of thyme essential oils, using for these studies another species endemic to southern Spain *Th. serpylloides* ssp. *gadorensis*. Fig. 4 illustrates the intermediate degree of resolution provided by the semi-capillary columns, which is superior to the packed column

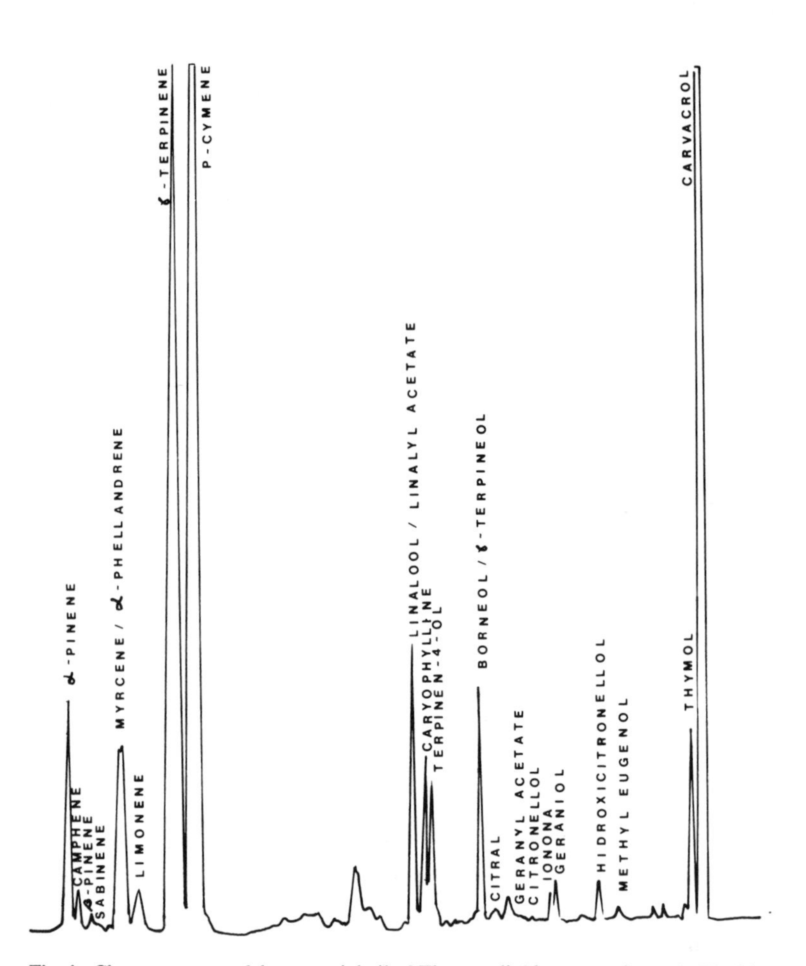

Fig. 4. Chromatogram of the essential oil of *Th. serpylloides* ssp. *gadorensis*. Working conditions: P-E model 8310B chromatograph with FID and a P-E model GP-100 integrated recorder. Semicapillary column (12 m x 0.53 mm i.d.) coated with Carbowax 20M. Carrier gas H_2 (flow rate 10 ml/min). Injection block temperature: 200 °C. Detector temperature: 250 °C. Column temperature program: initial isotherm 45 °C (1 min), 50 °C (0 min), 150°C (0 min), final isotherm 180 °C (5 min, with a gradient of 5 °C/min

but not as precise as the capillary column. With respect to MTHC, the results are similar to those obtained with capillary columns, whereas the oxidated derivatives of MTHC are not as well resolved, e.g., borneol is not separated from α-terpineol, nor is linalool separated from linalyl acetate, in contrast to the results obtained with capillary columns.

We conclude that although GLC-FID is a necessary element in the analysis of thyme essential oils, it is also necessary to resort to other systems of detection, such as mass spectrometry, a method which is currently widely accepted by most authors who have analyzed essential oils of the genus *Thymus* (Sendra and Cuñat 1979; Han and Kim 1980; Papageorgiov and Vassilios 1980; Lundfren and Stenhagen 1982; Stahl 1982; Katsiotis and Iconomov 1986; Stahl-Biskup 1986b and others).

Gas-Liquid Chromatography-Mass Spectrometry. Given the capability of mass spectrometry to provide significant information on the structural characteristics of molecules, its use as a detector in combination with gas chromatography has led to one of the most effective methods currently in use in the study of the composition of complex mixtures, including thyme essential oils. A further advantage of this technique is that even small amounts of the sample are sufficient to provide reliable results.

In applying this technique to our analyses of thyme essential oils we use the same conditions as those described above for gas chromatography in capillary columns and FID. For mass spectrometry, the components are identified by comparison with the mass spectra of authentic samples analyzed under the same conditions, with spectra published in the literature (Yamaguchi 1970; Masada 1976; Scheffer 1978; Hiltunen et al. 1982 and others) and with spectra in the mass spectrometry library.

With GLC-MS we have identified some components of thyme essential oils which could not be distinguished with GLC-FID, either because they were present only at trace levels or because no patterns were available for these compounds. In this connection, GLC-MS has been used to detect a series of monoterpenic (bornylene, Navarro et al. 1990, β-ocymene and terpinolene, Cabo et al. 1986a) and sesquiterpenic hydrocarbons (δ-cadinene and α-humulene, Cabo et al. 1986a, bisabolene, Cruz et al. 1988, β-germacrene, Navarro et al. 1990), oxidated derivatives including β-ionone (Cabo et al. 1986a), and aromatic alcohol-related components such as carvacryl acetate (Cabo et al. 1986a) 1,1 dimethylethylphenol, and 4-methyl-2-(1,1-dimethylphenol) (Navarro et al. 1990).

GLC-MS has also led to the identification of some peaks which could not be precisely classified with GLC-FID. Figure 3 shows a chromatogram of *Th. granatensis* Boiss., in which one peak could not be unequivocally assigned to either myrcene or α-phellandrene with FID, but was shown to be myrcene with MS. Further confirmation is obtained by comparing our mass spectral data with data in the literature. Although the peak for all MTHC, including myrcene and α-phellandrene, was relatively high at m/e = 136= M^+, each MTHC yields different results upon fractionation and analysis by the relative amounts method. In α-phellandrene, m/e = 93 is especially abundant; this compound also makes up a large proportion of myrcene. However, in α-phellandrene, m/e = 69 and m/e=11 are relatively scarce, whereas they make up 73.2 and 75.6% respectively in myrcene. These data lead us to conclude that the combined use of GLC and MS is highly suitable to the study of the composition of thyme essential oils, as it not only verifies the results obtained with GLC-FID, but is also capable of identifying components which FID overlooks. To this we should add that less well-eluted peaks whose resolution with FID leaves something to be desired can often be identified with confidence thanks to the possibility of obtaining the spectra of any given portion of the relatively unresolved peak. This naturally allows us to determine whether the peak represents a single component, or whether on the other hand it contains all the components suggested by GLC-FID.

Headspace Gas Chromatography. Although we have no practical experience with this method, we refer the reader to the studies of Hiltunen et al. (1985), Wampler et al. (1985), and Holm et al. (1988). These workers have defined the optimal conditions for the analysis of compounds in essential oils, in both liquid (previously extracted essential oil) and solid samples (powder of various thyme species).

4.1.1.3 High Performance Liquid Chromatography

Since the development of HPLC, some studies have described the preparatory separation of essential oils (Kubeczka 1985) although this method has not been as widely accepted as GLC with FID or MS for the analysis of essential oils. This may be due mainly to the structural characteristics of the molecules in essential oils, which in most cases lack chromophoric groups detectable at wavelength of 254 nm. In the genus *Thymus* , however, HPLC has been successfully used to identify two characteristic components, namely thymol and carvacrol (Falchi-Delitala et al. 1981; Solinas et al. 1981).

4.1.2 Spectroscopic Techniques

After being isolated with any of the methods described below, the components of thyme essential oils can be identified with the help of spectroscopic techniques. Components have thus far been characterized by their IR and MNR spectra using ^{13}C and H^+ (Adzet et al. 1976, 1977b, Stahl 1984; Ravid and Putievsky 1985). Mass spectroscopy has also been used to confirm the identity of various components, as noted above in the section on GLC-MS.

All the methods described in this section can be used for whole thyme essential oils or for previously fractionated and decomplexed oils. Fractionation can be achieved by the formation of soluble derivatives (Kubeczka 1985), fractionated distillation, preparative layer chromatography, and column chromatography, among others. Of these techniques, the latter is the best suited to the fractionation of thyme essential oils. In general, we recommend Silicagel as the stationary phase, and mixtures of apolar solvents with an elution gradient as the mobile phase (e.g., hexane, hexane/dichloromethane mixtures and dichloromethane) (Han and Kim 1980; Blazquez et al. 1989).

4.2 Quantitative Analysis

4.2.1 Physicochemical Indices

Certain physicochemical indices are of use in the quantification of some groups of thyme essential oil components, such as the esters (saponification index), alcohols (acetyl index), aldehydes and ketones (bisulphite mixture method) (Guenther 1948), and phenols (solubility in alkaline solutions) (Jenkins et al. 1951). Of these methods, we should underline the importance with respect to *Thymus* of phenol quantification: essential oils which contain these compounds decrease in volume upon being mixed with sodium hydroxide or potassium hydroxide solutions as a result of the high solubility of phenolic components in these solutions, while the nonphenolic portion remains undissolved. We have used this method in different species of thyme, and obtained results which concur fully with those obtained with other quantitative techniques to be described below. The quantification of phenols has the advantage over other techniques of simplicity, although on the other hand it involves the destruction of a considerable volume of essential oil (1 ml), and has the further disadvantage of quantifying all aromatic alcohols present in the sample together as a single layer.

4.2.2 Chromatographic Techniques

4.2.2.1 Thin Layer Chromatography-Photodensitometry

As noted above, TLC is considered one of the basic techniques in the study of thyme essential oils. Its use is not limited to qualitative analyses; in combination with photodensitometry, it is highly useful for the quantification of components identified previously with TLC.

Quantification by photodensitometry is based on the fact that when light hits a pigmented spot, part of the light is absorbed, the intensity of absorption being related to the color and the surface of the spot. The amount of light absorbed is measured as the difference between the amount of light reflected and incident light. A precise relationship exists between the area under the curve observed in the integrator and the amount of test substance present in the sample (Shellard 1968). This method has been used to quantify various components in *Thymus* essential oils, including geraniol, citral, terpinen-4-ol (Cabo et al. submitted), cineol, and γ-terpineol (Cabo et al. 1988).

In general, working conditions are as follows.

1. Stationary phase: silicagel precoated plates
2. Mobile phase: benzene/ether (85:15)
3. Developers: 4% sulfuric vanillin, 2,4 dinitrophenylhydrazine
4. Samples: 5% dilution of the essential oil in ethanol
5. Patterns: dilutions such that the concentration of the component being studied in the sample is somewhere between the highest and lowest concentrations used in the patterns.

6. Sample volume: 2 µl
7. Photodensitometry: Shimadzu model CS-920 High Speed TLC Chromato-Scanner
8. Wavelength: 460 nm for plates developed with sulfuric vanillin, 410 nm for plates developed with 2,4 dinitrophenylhydrazine.

The results obtained are highly reproducible, and agree well with the data obtained with other techniques.

4.2.2.2 Gas-Liquid Chromatography

Gas-Liquid Chromatography-Flame Ionization Detection. Of the methods used most widely in quantitative analyses of essential oils in general and those of the genus *Thymus* in particular, GLC is the most effective in providing a rapid evaluation of all components present in the essential oil, even when no preliminary identification has been performed. The quantitative calculations are based on finding the area under the peaks eluted or the height of the peaks themselves, and relating the magnitudes of these values to the amount or concentration of the components by one of the following methods:

Internal normalization of the areas. This method provides an approximate estimate of the different compounds eluted by comparison of the areas under the peaks. The surface of the peaks are then related to the sum total of the areas. This method does not take into account response factors or calibration factors, which are dependent on the response of the detector used. Consequently, the values obtained for each peak eluted do not necessarily correspond to the absolute values of the amount of a given component in a complex mixture such as an essential oil. In our opinion, this does not detract from the method, as the resulting data are highly useful in comparative studies (growth cycle, geographical variations, etc). The working conditions are as described earlier for chromatography aimed at the identification of pure components. Calculations are performed with the help of a Hewlett-Packard 3380A integrator-recorder from a minimum of six measurements per sample.

Internal pattern method. This method allows the quantification, in absolute terms, of the different components in a given essential oil. Calibrations with internal patterns are performed by mixing known amounts of the essential oil with a compound chemically similar to but not contained in the sample, chosen so that the peak for this substance, under the assay conditions used, does not overlap that of any compound in the sample, but is eluted as close as possible to the components to be determined.

The calculations are performed either numerically to determine the calibration factor and the percentage amount of each component, or graphically. With this method we have quantified in absolute terms the main components of a number of important essential oils, e.g., cineol, ρ-cymene, thymol, and carvacrol in *Th. hyemalis* (Cabo et al. 1980), camphor, myrcene, limonene, cineol, γ-terpinene, and terpinen-4-ol in *Th. granatensis* (Cabo et al. 1986a), thymol and carvacrol in *Th.*

serpylloides ssp. *gadorensis* (Crespo et al. 1988), camphor and cineol in *Th. longiflorus* (Cruz et al. 1988), and citral, geraniol, and terpinen-4-ol in *Th. baeticus* (Cabo et al. submitted). Working conditions are the same as described above for qualitative analyses. As internal patterns we use n-hexanol to quantify MTHC and its oxidated derivatives, and eugenol to quantify aromatic alcohols.

In all cases the results are highly reproducible, as shown by statistical analyses. In addition, the values obtained for phenols agree well with the data yielded by calculating the solubility of the essential oil in alkaline solutions, while the results for citral, geraniol, and terpinen-4-ol come close to the values obtained by both the internal standard method and by TLC-photodensitometry.

Head Space Gas Chromatography. The quantitative study of certain components (MTHC) in thyme essential oils has also been performed (Hiltunen et al. 1985) with HSGC. These results, however, do not agree completely with the data obtained by quantitative GLC. HSGC is nevertheless useful to trace quantitative variations in composition throughout the growth cycle as well as to investigate geographical variations within a given species of thyme because reliable results can be obtained with very small samples.

4.2.2.3 High Performance Liquid Chromatography

As noted in the section on qualitative analysis, HPLC has been used to identify and quantify thyme essential oil components, especially phenolics such as thymol and carvacrol. Both the internal normalization of areas and internal pattern method, described under gas-liquid chromatography, can be used.

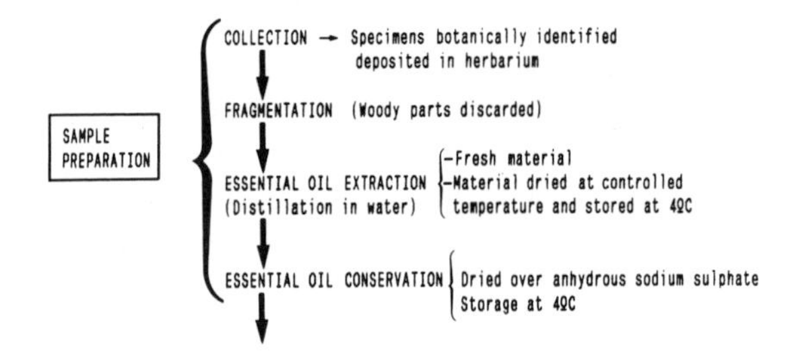

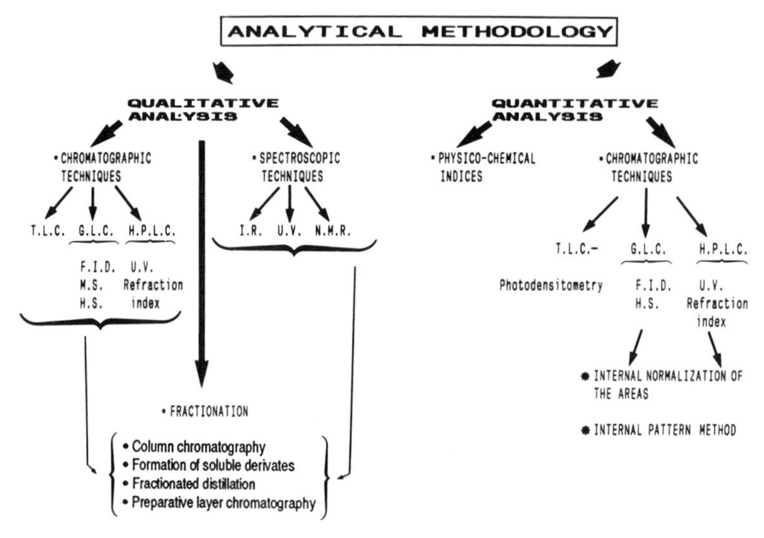

Fig. 5. Protocol recommended for the identification and/or quantification of essential oil components in the genus *Thymus*

5 Conclusions

On the basis of our experience in the analysis of essential oils from various species of thyme, and that of other workers reviewed in this chapter, we recommend the protocol illustrated in Fig. 5 for studies aimed at analyzing the components of the essential oils from this genus.

References

Adzet T, Martinez F (1981) Flavonoids in the leaves of *Thymus*: a chemotaxonomic survey. Biochem Syst Ecol 9:293–295
Adzet T, Granger R, Passet J, San Martin R (1976) Chimiotypes de *Thymus hyemalis* Lange. Plant Med Phytother 10:6–15
Adzet T, Granger R, Passet J, San Martin R (1977a) Le polymorphisme chimique dans le genre *Thymus*: sa signification taxonomique. Bio Sys Ecol 5:269–272
Adzet T, Granger R, Passet J, San Martin R (1977b) Chimiotypes de *Thymus mastichina* L. Plant Med Phytother 11:275–280
Adzet T, Vila R, Caniguerai S (1988) Chromatographic analysis of polyphenols of some Iberian *Thymus*. J Ethnopharmacol 24:147–154
Adzet T, Vila R, Batllori X, Ibañez C (1989) The essential oil of *Thymus moroderi* Pau ex Martinez (Labiatae). Flavour Fragance J 4:63–66
Attaway JA, Barabas LJ, Wolford RW (1965) Analysis of terpene hydrocarbons by thin-layer chromatography. Anal Chem 37:1289–1290
Bellomaria B, Hruska K, Valentini G (1981) Essential oils of *Thymus longicaulis* C. Presl from different localities of central Italy. G Bot Ital 115:17–27
Blazquez MA, Zafra-Polo MC, Villar A (1989) The volatile oil of *Thymus leptophyllus* growing in Spain. Planta Med 55:198
Brasseur T (1983) Études botaniques, phytochimiques et pharmacologiques consacrées au thym. J Pharm Belg 38:261–271
Cabo J, Jimenez J, Revert A (1974) Estudio cuali y cuantitativo del aceite esencial de *Thymus zygis*. Pharm Med 10:329–342.
Cabo J, Bravo L, Jimenez J, Navarro C (1980) *Thymus hyemalis*. II. Étude quali-et quantitative de son huile essentielle par CG. Planta Med 39:270
Cabo J, Jimenez J, Revert A, Bravo L (1981) Influencia de factores ecologicos (altitud) en el contenido y composición de la esencia de dos muestras de *Thymus zygis* L. recolectadas en diversas localidades. Ars Pharm 22:187–194
Cabo J, Cabo MM, Crespo ME, Jimenez J, Navarro C (1986a) *Thymus granatensis* Boiss. I. Étude qualitative et quantitative de son huile essentielle. Plant Med Phytother 20:18–24
Cabo J, Cabo MM, Crespo ME, Jimenez J, Navarro C (1986b) *Thymus granatensis* Boiss. II. Étude de son cycle évolutif. Plant Med Phytother 20:129–134
Cabo J, Cabo MM, Crespo ME, Jimenez J, Navarro C, Zarzuelo A (1986c) *Thymus granatensis* Boiss. III. Étude comparative de différents échantillons d'origines géographiques divers. Plant Med Phytother 20:135–147
Cabo J, Crespo ME, Jimenez J, Navarro C (1986d) A study of the essences from *Thymus hyemalis* collected in three different localities. Fitoterapia 57:117–119
Cabo J, Crespo ME, Jimenez J, Navarro C, Risco S (1987) Seasonal variation of essential oil yield and composition of *Thymus hyemalis*. Planta Med 53:380–382
Cabo J, Cabo MM, Cruz T, Jimenez J (1988) Estudio mediante cromatografía en capa fina de la esencia de *Thymus longiflorus* Boiss. Ars Pharm 29:77–84
Cabo MM, Cabo J, Castillo MJ, Cruz T, Jimenez J (1990) Étude de l'essence de *Thymus baeticus* Boiss. Plant Med Phytother (in press)
Crespo ME, Cabo J, Jimenez J, Navarro C, Zarzuelo A (1986) Composition of the essential oil in *Thymus orospedanus*. J Nat Prod 49:558–560
Crespo ME, Gomis E, Jimenez J, Navarro C (1988) The essential oil of *Thymus serpylloides* ssp. gadorensis. Planta Med 2:161–162
Cruz T, Jimenez J, Navarro C, Cabo J, Cabo MM (1988) Sur l'huile essentille de *Thymus longiflorus* Boiss. Plant Med Phytother 22:225–230
Cruz T, Jimenez J, Zarzuelo A, Cabo MM (1989) The spasmolytic activity of *Thymus baeticus* Boiss. in rats. Phytother Res 3:106–108
Dembitskii AD, Yurina RA, Krotova GI (1981) Composition of *Thymus marschallianus* essential oil. Khim Prir Soedin 4:522
Falchi-Delitala L, Solinas V, Gessa C (1981) Ricerche sulle variazioni dei componenti fenolici dell'olio essenziale di *Thymus capitatus* Hoffmgg. e Lk durante il ciclo vegetativo mediante HPLC. Riv Ital EPOS 2

Falchi-Delitala L, Solinas V, Gessa C (1983) Variazioni stagionali quantitative e qualitative di olio essenziale e dei suoi fenoli in *Thymus capitatus* Hofmgg. et Lk ed in *Thymus herba-barona* Loisel. Fitoterapia 54:87–96

Fehr D, Stenzhorn G (1979) Studies on the shelf life of peppermint leaves, rosemary leaves and thyme. Pharm Ztg 124:2342–2349

Ferreres F, Tomas-Barberan FA, Tomas F (1985a) 5,6,4'-trihydroxy-7,8-dimethoxy flavone from *Thymus membranaceus*. Phytochemistry 24:1869–1871

Ferreres F, Tomas F, Tomas-Barberan FA, Hernandez L (1985b) Free flavone aglycones from *Thymus membranaceus* Boiss. subsp. *membranaceus*. Plant Med Phytother 19:89–97

Granger R, Passet J (1971) Types chimiques de l'espiece *Thymus vulgaris* L. C R Acad Sci Paris, Ser D, 273:2350–2353

Granger R, Passet J (1973) *Thymus vulgaris* spontané de France: races chimiques et chemotaxonomie. Phytochemistry 12:1683–1691

Granger R, Passet J, Teulade-Arbousset G, Auriol P (1973) Types chimiques de *Thymus nitens* Lamotte endemique cévénol. Plant Med Phytother 7:225–233

Guenther E (1948) The essential oils, vol.1. Van Nostrand, New York, pp 27–305

Han DS, Kim KW (1980) Studies on the essential oil components of *Thymus magnus* Nakai. Saengyak Hakhoe Chi 11:1–6

Hegnauer R (1962–1966) Chemotaxonomie der Pflanzen vol.4. Birkhäuser, Basel, p 292

Heinrich G (1973) Über das ätherische Öl von *Monarda fistulosa* und den Einbau von markiertem CO_2 in dessen Komponenten. Planta Med 23:201–212

Hernandez LM, Tomas Barberan A, Tomas-Lorente F (1986) A chemotaxonomic study of free flavone aglycons from some iberian *Thymus* species. Biochem Syst Ecol 15:61–67

Hiltunen R, Raisanen S, Von Schantz M (1982) The use of mass fragmentography in the analysis of essential oils. In: Kubeczka KH (ed) Ätherische Öle. Ergeb Int Arbeitstag, Thieme, Stuttgart, pp 33–41

Hiltunen R, Vuorela H, Laakso I (1985) Quantitative headspace gas chromatography in the analysis of volatile oils and aromatic plants. In: Baerheim Suendsen A, Scheffer JJC (eds) Essential oils and aromatic plants. Nijhoff/Junk, Dordrecht, pp 23–41

Holm Y, Aho E, Hiltunen R (1988) Analysis of the essential oil from *Thymus vulgaris* by headspace gas chromatography. Acta Pharm Fenn 97:13–19

Ilisulu F, Tanker M (1986) The volatile oils of some endemic *Thymus* species growing in southern Anatolia. Planta Med 5:340

Ismailov NM, Kasumov F Yu, Akhmedova Sh A (1981) Essential oil of *Thymus trautvetteri*. Dokl Akad Nauk Az SSR 37:64–66

Jalas J (1970) Notes on *Thymus* L. (Labiatae) in Europe. I. Supraespecific classification and nomenclature. Bot J Linn Soc 64:199–215

Jalas J, Kaleva K (1970) Supraspezifische Gliederung und Verbreitungstypen in der Gattung *Thymus* L. (Labiatae). Feddes Repert 81:93–106

Jenkins GL, Dumez AG, Christian JE, Hager GP (1951) Quimica farmaceutica cuantitativa. Atlante S. A., Mejico, pp 383–385

Karawya MS, Balbaa SI, Hifnawy MSM (1970) Essential oils of certain labiaceous (Labiate) plants of Egypt. Am Perfum Cosmet 85:23–28

Kasumov F Yu (1981a) Composition of essential oils of *Thymus fominii* and *Thymus transcaucasicus*. Khim Prir Soedin 5:665–666

Kasumov F Yu (1981b) Components of thyme essential oils. Khim Prir Soedin 4:522

Kasumov F Yu (1982) Essential oil of *Thymus rariflorus*. Maslo-Zhir Prom-st 7:36–37

Kasumov F Yu (1983) Essential oils of *Thymus transcaucasicus* Ronn. and *Thymus eriphorus* Ronn. Maslo-Zhir Prom-st 1:29

Kasumov F Yu (1987) Essential oil composition of *Thymus* species. Khim Prir Soedin 5:761–762

Kasumov F Yu (1988) Composition of essential oil from *Thymus* species in the Armenian flore. Khim Prir Soedin 1:134–136

Kasumov F Yu, Akhmedova Sh A (1981) *Thymus trautvetteri* essential oil. Maslo-Zhir Prom-st 3:30–31

Kasumov F Yu, Farkhadova MT (1986) The composition of *Thymus karamarianicus* essential oil. Khim Prir Soedin 5:642–643

Kasumov F Yu, Gadzhieva TG (1980) Components of *Thymus kotschyanus*. Khim Prir Soedin 5:728

Katsiotis S, Iconomov N (1986) Contribution to the study of the essential oil from *Thymus tosevii* growing wild in Greece. Planta Med 4:334–336

Kisgyorgy Z, Csedo K, Horster H, Gergely J, Racz G (1983) Essential oil of the more important indigenous *Thymus* species occuring in the composition of *Serpylli herba*. Rev Med (Tirgu-Mures, Rom) 29:124–130

Koller WD (1988) Problems with the flavor of herbs and spices. Dev Food Sci 17:123–132

Kubeczka KH (1985) Progress in isolation techiques for essential oils constituents. In: Baerheim Svendsen A, Scheffer JJC (eds) Essential oils and aromatic plants. Nijhoff/Junk, Dordrecht, pp 107–126

Kulieva ZT, Guseinov D Ya, Kasumov F Yu, Akhundov RA (1979) Studies on the chemical composition and some pharmacological and toxicological properties of the essential oil of *Thymus kotschyanus*. Dokl Akad Nauk Az SSR 35:87–91

Lassanyi Z (1978) Cytophotometric measurement of phenol components of volatile oils present in the gland squame of *Thymus* species. Acta Pharm Hung 48:168–171

Lawrence BM (1982) Progress in essential oils. Perfumer and Flavorist 7:35–40

Lemordant D (1986) Identification of a sample of a commercial thyme from Ethiopia Int J Crude Drug Res 24:107–119

Lundgren L, Stenhagen G (1982) Leaf volatiles of *Thymus vulgaris, Th. serpyllum, Th. praecox, Th. pulegioides* and *Th. citriodorus* (Labiatae). Nord J Bot 2:445–452

Marhuenda E, Alarcon de la Lastra C (1986) Composition of essential oil of *Thymus carnosus* and its variation. Fitoterapia 57:448–450

Marhuenda E, Alarcon de la Lastra C, Garcia MD, Cert A (1987) Flavones isolated from *Thymus carnosus* Boiss. Ann Pharm Fr 45:467–470

Masada Y (1976) Analysis of essential oils by gas chromatography and mass spectrometry. John Wiley, New York

Mateo C, Morera MP, Sanz J, Calderon J, Hernandez A (1978) Analytical study on essential oils derived from Spanish plants. I. Species of the *Thymus* genus. Riv Ital EPPOS 60:621–627

Mateo C, Morera MP, Sanz J, Hernandez A (1979) Estudio analitico de aceites esenciales procedentes de plantas españolas. I. Especies del género *Thymus*. Riv Ital EPPOS 61:135–136

Mathela CS, Agarwall I, Taskinen J (1980) Composition of essential oil of *Thymus serpyllum* Linn. J Indian Chem Soc 57:1249–1250

Merichi F I (1986) Evaluation of the thymol contents of endemic *Thymus* species growing in Turkey. Doga: Tip Eczacilik 10:187–200

Messerschmidt W (1964) Gas and thin-layer chromatographic investigations of the volatile essential oils of several *Thymus* species. I. Influence of various factors on the formation and variation of the essential oil. Planta Med 12:501–511

Miquel JD, Richard HMJ, Sandret FG (1976) Volatile constituents of Moroccan Thyme oil. J Agric Food Chem 24:833–834

Morales R (1986) Taxonomia de los generos *Thymus* (excluida la sección serpyllum) y *Thymbra* en la peninsula Ibérica. Ruizia 3:7–11

Muenchow P, Pohloudek-Fabini R (1964) The microanalysis of essential oils. XV. The thymol content of *Thymus vulgaris*. Pharmazie 19:655–661

Navarro MC, Arrebola ML, Socorro O, Jimenez J (1990) Estudio botánico farmacoquimico de *Thymus hyemalis* Lange. Act Soc Brot (in press)

Novruzova ZA, Kasumov F Yu (1987) Anatomy of Caucasican representatives of the genus *Thymus* L. (Lamiaceae) in relation to the essential oil composition. Izv Akad Nauk Az SSR, Ser Biol Nauk 6:18–24

Papageorgiov VP, Argyriadou N (1981) Trace constituents in the essential oil of *Thymus capitatus* Phytochemistry 20:2295–2297

Papageorgiov VP, Vassilios P (1980) GLC-MS computer analysis of the essential oil of *Thymus capitatus*. Planta Med 40:29–33

Passet J (1979) Chemical variability within thyme, its manifestations and its significance. Parfums Cosmet Arômes 28:39–42

Pharmacopée Européene (1975) Maisonneuve, Strasbourg, vol 3, p 69

Popov VI, Odynets AI (1977) Study of the chemical composition of the essential oil of Ukrainian thyme grown in Belorussia. Mater S'ezda Farm B SSR 3:166–181

Poulose AJ, Croteau R (1978a) γ-Terpinene synthetase: key enzyme in the biosynthesis of aromatic monoterpenes. Arch Biochem Biophys 191:400–411

Poulose AJ, Croteau R (1978b) Biosynthesis of aromatic monoterpenes. Conversion of γ–terpinene to p-cymene and thymol in *Thymus vulgaris* L. Arch Biochem Biophys 187:307–314

Ravid U, Putievsky E (1983) Constituents of essential oil from *Majorana syriaca, Coridothymus capitatus* and *Satureja thymbra*. Planta Med 49:248–249

Ravid U, Putievsky E (1985) Essential oils of israeli wild species of Labiatae. In:Baerheim Suendsen A, Scheffer JJC (eds) Essential oils and aromatic plants. Nijhoff/Junk, Dordrecht, pp 155–162

Revert A (1975) Estudio comparativo de diversas especies del género *Thymus* L. desde el punto de vista de sus esencias. Doctoral Thesis, Granada

Richard NMJ, Miquel JD, Sandret FG (1975) The essential oils of the thyme of Morocco and the thyme of Provence. Parfums Cosmet Aromes 6:69–78

Richard H, Benjilali B, Banquour N, Baritaux O (1985) Étude de diverses huiles essentielles de thyme du Maroc. Lebensm-Wiss Technol 18:105–110

Scheffer JJC (1978) Analysis of essential oils by combined liquid-solid and gas-liquid chromatography. Drukkerrij J. H. Pasmans, 'S-Gravenhage

Schratz E, Horster H (1970) Composition of the essential oils of *Thymus vulgaris* and *Thymus marschallianus* in relation of leaf age and season of the year. Planta Med 19:160–1976

Schratz E, Qedan S (1965) Composition of the essential oil of *Thymus serpyllum* complex. I. Detection of the oil components by thin-layer chromatography. Pharmazie 20:710–713

Sendra JM, Cuñat P (1979) Constituyentes del aceite esencial de orégano español. I. Hidrocarburos y compuestos oxigenados ligeros. Rev Agroquim Tecnol Aliment ESP 19:102–118

Sendra JM, Cuñat P (1980) Volatile phenolic constituents of Spanish origanum (*Coridothymus capitatus*) essential oil. Phytochemistry 19:89–92

Shellard EJ (1968) Quantitative paper and thin layer chromatography. Academic Press, London. pp 70

Solinas V, Gessa C, Falchi-Delitala L (1981) High-performance liquid chromatographic analysis of carvacrol and thymol in the essential oil of *Thymus capitatus*. J Chromatogr 219:332–337

Stahl E (ed) (1965) Thin-layer chromatography, a laboratory handbook. Springer, Berlin Heidelberg New York

Stahl E (1982) The essential oil from *Thymus praecox* ssp. *articus*. World Crops Prod Util Deser 7:203–206

Stahl E (1984) The essential oil from *Thymus praecox* ssp. *articus* growing in Iceland. Planta Med 50:157–160

Stahl-Biskup E (1986a) The essential oil from Norwegian *Thymus* species. I. *Thymus praecox* ssp. *articus* Planta Med 1:36–38

Stahl-Biskup E (1986b) The essential oil from Norwegian *Thymus* species. II. *Thymus pulegioides*. Planta Med 3:233–235

Sur SV, Tulyupa FM, Tolok A Ya (1988) Composition of oils from thyme herbs. Khim-Farm Zh 12:1361–1366

Ter Heide R (1968) Terpenes. II. Characterization of monoterpene esters by gas and thin-layer chromatography. Fresenius'Z Anal Chem 236:215–227

Thieme H (1967) Thin-layer chromatographic separation of thymol and carvacrol. Pharmazie 22:722–723

Thieme H, Thi Tam N (1968) Method for spectrophotometric determination of methylchavicol in estragon oils. Pharmazie 23:339–340

Tikhonov VN, Khan UA, Kalinkina GI (1988) Composition of the essential oil of *Thymus krylovii*. Khim Prir Soedin 6:886–887

Tomas-Barberan FA, Hernandez L, Ferreres F, Tomas F (1985) Highly methylated 6-hydroxyflavones and other flavonoids from *Thymus piperella*. Planta Med 5:452–454

Tomas-Barberan FA, Hernandez L, Tomas F (1986) A chemotaxonomic study of flavonoids in *Thymbra capitata*. Phytochemistry 25:561–562

Tomas-Barberan FA, Husain SZ, Gil MI (1987) The distribution of methylated flavones in the Lamiaceae. Biochem Syst Ecol 16:43–46

Tucakov J (1964) Influence des facteurs exogénes sur le rendement et la qualité de l'huile essentiale de *Thymus vulgaris* L. Fr Parf 7:277–283

Van den Broucke CO (1982) New pharmacologically important flavonoids of *Thymus vulgaris*. World Crops Prod Util Descr 7:271–276
Van de Broucke CO, Dommisse Ra, Esmans EL, Lemli JA (1982) Three methylated flavones from *Thymus vulgaris*. Phytochemistry 21:2581–2583
Velenovski J (1906) Vorstudien zu einer Monographie der Gattung *Thymus* L. Beih Bot Zentralbl 19:271–287
Verderio E, Venturini D (1965) Thin layer chromatography of mandarin essential oil. Riv Ital EPPOS 47:430–434
Vernet Ph, Gouyon PH (1979) Le polymorphisme chimique de *Thymus vulgaris*. Parfums Cosmet Arômes 30:31–45
Vernet Ph, Guillerm JL, Gouyon PH (1977) Le polymorphisme chimique de *Thymus vulgaris* L. Oecol Plant 12:159–190
Vernet Ph, Gouyon PH, Valdeyron G (1986) Genetic control of the oil content in *Thymus vulgaris* L.: a case of polymorphism in a biosynthetic chain. Genetica 69:227–231
Vernin G (1964) Les techniques analytiques modernes et leurs applications en parfumerie, vol I. Chauvet Scillans, pp 106
Viorin B, Viricel MR, Favre-Bonvin J, Van den Broucke CO, Lemli J (1985) 5,6,4'-trihydroxy -7,3'-dimethoxyflavone and other methoxylated flavonoids isolated from *Thymus satureioides*. Planta Med 6:523–525
Von Schantz M, Lopmeri A, Stroemer E, Salonen R, Brunni S (1962) Separation and identification of the constituents of some essential oils by thin layer chromatography. Farm Aikakauslehti 71:52–88
Wagner H, Bladt S, Zgainski EM (1984) Plant drug analysis. A thin layer chromatography atlas. Springer, Berlin, Heidelberg New York, Tokyo, p 8
Wampler TP, Bowe WA, Levy EJ (1985) Splitless capillary GC analysis of herbs and spices using cryofocusing. Am Lab 17:76–81
Weiss B, Flück H (1970) Variability of content and composition of the essential oil in *Thymus vulgaris* leaves and plants. Pharm Acta Helv 45:169–183
Xi-Qing P, Kuei-Sheng H, Fu-Tang Ch, Te-Len Y (1980) A note on the study of the essential oil from Bailixiang (*Thymus quinquecostatus*) for medicinal use. Chung Ts'ao Yao 11:101–102
Yamaguchi K (1970) Spectral data of natural products, vol.2. Elsevier, Amsterdam, pp 389–393
Younos Ch, Mortier F, Pelt JM (1972) Contribution a l'étude chimique et pharmacologique des essences de Labiées d'Afghanistan. I. Méthodes analytiques utilisées pour la dosage et le fractionnement des huiles essentielles. Plant Med Phytother 6:171–177
Zarzuelo A, Navarro C, Crespo ME, Ocete MA, Jimenez J, Cabo J (1987) Spasmolytic activity of *Thymus membranaceus* essential oil. Phytother Res 1:114–116

Chemical Races Within the Genus *Mentha* L.

S. Kokkini

1 Introduction

The genus *Mentha* belongs to the family Lamiaceae; it occurs in all five continents, although its native occurrence in the New World is restricted to a single species in the North. It is infrequent in the Tropics, and in Australia there are a number of species, with unclear relationships to the rest of the genus. Although the genus consists of approximately 25 species and rather fewer hybrids, one can find more than 900 binomials listed in the Index Kewensis.

This high number of different taxonomic rank names attributed by the taxonomists during the past 200 years to the mint plants reflects a great morphological variation. The occurrence of hermaphrodite and male-sterile (female) plants in the wild populations, the frequent hybridization, particularly between the members of the subgenus *Menthastrum*, as well as the long history of cultivation since the times of antiquity, have led to the complex variation patterns characterizing most wild populations (Harley 1963; Harley and Brighton 1977; Kokkini 1983).

Most mint species possess a vigorous rhizome system as a means of spread and dispersal, and therefore wild populations often consist of a few or even only one genotype. The importance of vegetative reproduction can be seen in the prominence of certain clones which are widely cultivated and occasionally become naturalized.

Apart from their high morphological variability, most mint species are characterized by a great chemical diversity, with respect to their essential oil constituents, rarely encountered in other temperate zone species. The great differences in essential oil composition found in the members of this genus afford the experimentalist with a source of strains high in linalool, menthol, menthone, carvone, pulegone, or other commercially valuable compounds (Table 1).

The most important mint oils from the economic standpoint are peppermint oil, cornmint oil, and spearmint oil. Less important, produced in smaller quantities, are pennyroyal oil, *M. citrata* oil, and bergamot-mint oil (Lawrence 1985).

The essential oils of mints are accumulated in secretory structures on the aerial plant part. These structures consist of two types of glandular trichomes (1) peltate, with eight secretory cells, a stalk cell, and a basal cell (Amelunxen 1965), and (2) capitate, with one secretory cell, a stalk cell, and a basal cell (Amelunxen 1964). Because of their size and number, the peltate trichomes contain the bulk of the essential oil (Maffei and Sacco 1987; Maffei et al. 1989). Recently, Gershenzon et al. (1989) demonstrated that essential oils of mint plants are not only accumulated but also biosynthesized in the glandular trichomes.

Table 1. Species and hybrids of the genus *Mentha* with essential oils particularly rich in commercially interesting compounds

Compound	Taxon (Ref*)	% Oil content	% Composition
A. Alcohols			
Menthol	*M. arvensis* L. var. *piperascens* Malinv. (2)	0.9–1.8	70–80
Neomenthol	*M. sacchalinensis* (Briq.) Kudo[a](2)	0.05–0.10	58
Linalool	*M. longifolia* (L.)L. (2)	0.14–0.24	83–88
trans-Sabinene hydrate	*M. candicans* Miller[b] (1)		85
B. Esters			
Menthyl acetate	*M.* × *dumetorum* Schultes (2)	0.10–0.25	10–30
	M. × *rotundifolia* (L.) Huds. (3)	2.20	51
Linalyl acetate	*M. citrata* Ehrh. (2)	0.15	42–78
cis-Carvyl acetate	*M. suaveolens* Ehrh. (2)	0.18	23
Neoisodihydro-carvyl acetate	*M. spicata* L. (2)	0.30–0.40	20–23
α-Terpinyl acetate	*M.* × *verticillata* L. (2)	0.1	75
C. Ketones			
Carvone	*M. spicata* L. (2)	0.30–0.50	40–75
Menthone	M. arvensis L. var. cens Malinv. (2)	0.40–0.80	70–75
cis-Dihydrocarvone	*M.* × *rotundifolia* (L.) Huds. (2)	0.14–0.20	45–69
Piperitone	*M. pulegium* L. (2)	0.15–0.28	70–88
Pipertitenone	*M. longifolia* (L.)L. (2)	0.14–0.20	20–26
cis-Isopulegone	*M. arvensis* L. (2)	0.20–0.30	41–78
D. Oxides			
Menthofuran	*M. aquatica* L. var. *hypeuria* Briq. (5)	0.49	59
Piperitenone oxide	*M. longifolia* (L.) L. (4)	0.11	77
	M. spicata L. (2)	0.30–0.50	65–73
	M. spicata L. (6)	0.80–2.20	64–70
trans-Piperitone oxide	*M. suaveolens* Ehrh. (2)	0.20–0.50	20–60

*Data from (1) Karasawa and Shimizu (1978); (2) Lawrence (1978, 1989); (3) Kokkini and Papageorgiou (1988b); (4) Maffei (1988a); (5) Sacco and Maffei (1988); (6) Kokkini and Vokou (1989).
[a] A synonym of *M. arvensis* L. var. *piperascens* Malinvaud for "Sakahlin mint", as it is known in the Soviet Union.
[b] Probably a variety of *M. longifolia* L.

2 Biosynthesis of *Mentha* Essential Oils

The use of crossing or hybridization techniques to study the genetic control of mint monoterpene biosynthesis has received a considerable amount of attention over the past 30 years. During this period, Murray and his various co-workers have published extensively on the nature of some of the genes involved in controlling the formation or nonformation of various *Mentha* oil constituents. Using Mendelian analysis of the inheritance of major oil constituents in the progeny of artificial crosses, the genetic basis for the production of most mint monoterpenes was determined. The very first experiments by Murray were based solely on a few main components, which were mainly characterized by their odor, but later the use of gas chromatography gave more comprehensive analyses, including quantitative data (Hefendehl and Murray 1976; Lincoln and Murray 1978; Murray et al. 1980, and literature therein).

In an attempt to explain oil biogenesis in terms of genetic control of specific reactions and gene interactions, Lincoln et al. (1986) proposed a hypothetical biosynthetic pathway (Fig. 1). There is general agreement for most of the monoterpenoid interconversions presented in this scheme, although some of its aspects, for example the roles of geranyl, neryl, and linalyl pyrophosphates as initial precursors (Cori 1983), are in dispute. Some biosynthetic steps have been studied in more detail using enzymatic assays (Kjonaas et al. 1985; Croteau and Venkatachalan 1986).

Based on both short-term tracer studies and long-term periodic analysis, there is considerable evidence that mint monoterpenes are metabolically active and subject to rapid turnover in plants (Burbott and Loomis 1969). Thus, l-menthone, the major monoterpene component of peppermint young leaves, is metabolized during the plant development to l-menthol, which accumulates in the essential oils, and to d-neomenthol. The latter is glucosylated and transported to the rhizome, whereupon the β-D-glucoside is hydrolyzed, the aglycone oxidized back to l-menthone, and this ketone converted to 1–3, 4-menthone-lactone (Croteau and Virendar 1985).

3 Chemical Races

The existence of different chemical races or chemotypes, based on qualitative differences within a species, is a common feature in many genera of the Lamiaceae family. In the wide variety of published papers concerning the composition of *Mentha* essential oils, one can easily discern great intraspecific differences, which may result in the definition of particular chemotypes. Although different approaches have been proposed for the recognition and acceptance of infraspecific chemical races, it is generally accepted that the definition of them may concern the presence or absence of a particular biosynthetic pathway and not the presence or absence of a particular single compound (Tétényi 1973; Harborne and Turner 1984).

3.1 Oils Rich in Acyclic Compounds

3.1.1 Oils Rich in Geraniol and/or Its Acetate

A rare mint chemotype was reported first for an Austrian population of *M. arvensis* (geraniol 12.6%, geranyl acetate 16.6%) (Malingré 1971). Lawrence

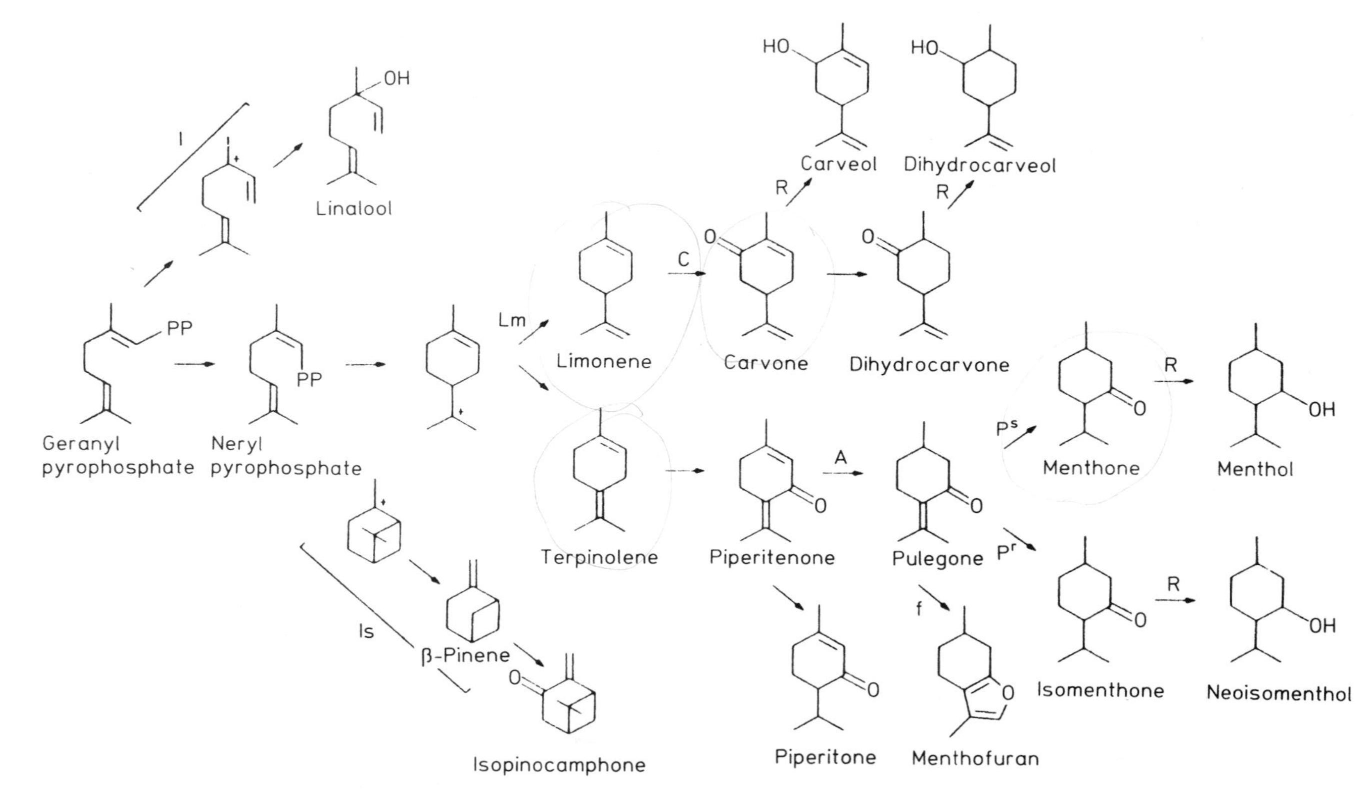

Fig. 1. Hypothesized biosynthetic pathway for some pricipal monoterpenoids of *Mentha* species and proposed gene actions (Lincoln et al. 1986)

(1978) found in a *M.* × *verticillata* strain originating from the USSR (Sukhumi) a considerable amount geranyl acetate (14.1%) but only a small amount of geranol (0.3%).

According to Katsuhara (1966), geraniol is produced from geranyl pyrophosphate. Esterification of geraniol to the corresponding acetate is controlled by a dominant gene (or genes) *E*; this gene is responsible for the formation of monoterpene acetates from all the monoterpene alcohols occurring in the different mint species (Hefendehl and Murray 1976; Lawrence 1978).

3.1.2 Oils Rich in Linalool and/or its Acetate

Linalool and linalyl acetate have been reported as main constituents in several *Mentha* species essential oils, such as *M. arvensis* (Gill et al. 1973), *M. citrata* (Todd and Murray 1968), *M. longifolia* (Góra et al. 1975; Lawrence 1978), *M. spicata* (Kokkini and Vokou 1989), as well as in the hybrid *M.* × *gentilis* (Lawrence and Morton 1972).

Murray and Lincoln (1970) demonstrated that the dominant gene *I* allowed the accumulation of linalool, which may or not be converted into linalyl acetate, and prevents or nearly prevents the formation of the cyclic ketones and their derivatives (Fig. 1).

3.2 Oils Rich in 2-Substituted Compounds

3.2.1 Oils Rich in Carvone with More or Less *cis*- and *trans*-dihydrocarvone and Their Related Alcohols and Acetates

Carvone, dihydrocarvone, and their related compounds carveol, carvyl acetate, dihydrocarveol, and dihydrocarvyl acetate have been found as main components in a number of "spearmint oils" (Smith et al. 1963; Lawrence 1978; Maffei et al. 1986). The two main types of spearmint oil produced commercially, mainly in the United States, are the Scotch spearmint and the Native spearmint oil (Lawrence 1985). Scotch spearmint oil is obtained from the hybrid *M.* × *gentilis* nm. *cardiaca* Gray (*M. arvensis* × *spicata*). The Native type is produced either from the species *M. spicata* (2n = 48) or from the sterile hybrid *M. longifolia x spicata* (2n = 36). Apart from these plants, cultivated for their carvone-rich essential oils, there are also other native *Mentha* species or hybrids producing similar oils with main constituents C-2 compounds, e.g., *M. longifolia*, *M. suaveolens*, *M.* × *villosa*, and *M.* × *villoso-nervata* (Table 2), (Lawrence 1978; Kokkini 1983; Kokkini and Papageorgiou 1987, 1988a; Kokkini and Vokou 1989).

Ravid et al. (1987) have isolated optically pure (R–)-(–) carvone from the essential oils of *M. spicata* and *M. longifolia*. It should be noted that carvone occurs in nature in two more forms, the (S–)-(+) and (RS). The (S–)-(+) form, giving what we call caraway odor, is the major constituent of *Carum carvi* L. (caraway) and *Anethum graveolens* L. (dill) essential oils.

A recent study of the headspace volatiles of living and picked American spearmint (Mookherjee et al. 1989) has shown that dramatic chemical changes occur

Table 2. Chemical races occuring within the commonest species and hybrids of the genus *Mentha*

Taxa	Chemotypes[a]								
	I	II	III	IV	V	VI	VII	VIII	IX
Subgenus *Puleqium* (Miller) Lamk. et DC									
Section *Eupleqia* Briq.									
1. *M. puleguim L.*				X				X	
Subgenus *Menthastrum* Cossom et Germain									
Section *Spicatae* L.									
2. *M. suaveolens* Ehrh.			X		X				
3. *M. longifolia* (L.)L.		X	X		X				
4. *M. spicata* L.		X	X	X	X			X	
Section *Capitatae* L.									
5. *M. aquatica* L.						X			
M. citrata L.		X							
Section *Verticillatae* L.									
6. *M. arvensis* L.	X	X					X	X	X
Hybrids									
(2 × 3) *M.* × *rotundifolia* (L.) Hudson			X		X				X
(2 × 4) *M.* × *villosa* Hudson			X		X				
(2 × 5) *M.* × *maximilianea* F.W. Schultz						X			X
(2 × 6) *M.* × *muellerana* F.W. Schultz			X						
(3 × 4) *M.* × *villoso-nervata* Opiz			X		X				
(3 × 5) *M.* × *dumetorum* J.A. Schultes			X		X	X			
(3 × 6) *M.* × *dalmatica* Tausch			X						
(4 × 5) *M.* × *piperita* L.						X			X
(4 × 6) *M.* × *gentilis* L.		X	X		X				
(5 × 6) *M.* × *verticillata* L.	X					X			

[a] The main components of each chemotype are I: geranyl/geranyl acetate, II: linalool/linalyl acetate, III: carvone/dihydrocarvone, IV: piperitenone/piperitone, V: piperitenone epoxide/piperitone epoxide, VI: menthofuran, VII: pulegone/*cis*- and *trans*-isopulegone, VIII: pulegone/menthone/-isomenthone, IX: menthone/isomenthone/isomers of menthol.

following the harvesting of plant material. It was found that harvested spearmint kept at room temperature for 24 h and reduced in weight by 50%, possessed 70% carvone and 2% limonene in its headspace, while the relative amounts of the headspace of living spearmint were 24% and 18%. The essential oils of *Mentha* species have either C-2 (as carvone and related compounds) or C-3 (as piperitenone, piperitone, pulegone, menthone) oxygenated compounds, but not both, at least if one does not take trace constituents into account. Genetic studies have shown that the formation of C-2 oxygenated compounds is controlled by the dominant gene *C*; the recessive *cc* genotype allows the formation of C-3 oxygenated compounds (Murray 1960b; Hendriks 1974; Hefendehl and Murray 1976; Hendriks et al. 1976). Furthermore, it has been postulated that there is a close coupling phase linkage between the dominant genes *C* and L_m, the latter gene almost entirely preventing the formation of terpinolene, and in consequence the development of the 3-oxygenated compounds causes the accumulation of limonene (Fig. 1). Therefore, individuals with high limonene content occur only when the dominant L_m gene is separated from both the dominant gene *I*, that interrupts oil biogenesis at a stage earlier than limonene formation, and the dominant *C* gene as a result of rare quadrivalent pairing and crossing-over between homologous chromosomes (Lincoln et al. 1971; Hefendehl and Murray 1973; Murray and Hefendehl 1973; Hefendehl and Murray 1976).

In 1972, Hefendehl and Murray showed that a dominant gene *R* is responsible for the conversion of carvone to carveol and dihydrocarvone to dihydrocarveol; the same gene is also responsible for the conversion of menthone to menthol (Fig. 1), (Murray 1960a).

There is no sufficient evidence for the biosynthetic origin of *cis*- and *trans*-isomers of dihydrocarvone. Lawrence (1978) postulated that two stereospecific reductases controlled by the dominant genes A_s and A_r are responsible for the reduction of (–)-carvone to (+)-*cis*-dihydrocarvone and (+)-*trans*-dihydrocarvone, respectively. In addition, it is assumed that the reduction of carvone to *cis*-dihydrocarvone, always found in higher proportion, is of the same type as that described for the conversion of piperitenone to (+)-pulegone (Hendriks et al. 1976; Lawrence 1978).

3.3 Oils Rich in 3-Substituted Compounds

3.3.1 Oils Rich in Piperitenone and/or Piperitone

Piperitenone and/or piperitone are very common constituents of the different mint essential oils, rich in 3-substituted compounds, though in fairly low amounts. However, in a few cases these two compounds have been found in high amounts in mint oils. Essential oils particularly rich in piperitone have been reported from *M. pulegium* plants growing in Austria (Zwaning and Smith 1971) and Italy (Lawrence 1978). Piperitenone was found in relatively high amounts in *M. longifolia* plants grown in Israel (Lawrence 1978), (Table 1).

Terpinolene is considered to be the precursor of piperitenone (Burbott et al. 1974), a key component for the formation of C-3 oxygenated compounds (Fig. 1). Biogenetic data have demonstrated that the dominant gene *A* is responsible for the reduction of piperitenone to pulegone, whereas the recessive gene a, preventing this reduction is responsible for the accumulation of piperitone. Furthermore, crosses between *M. spicata* × *M. longifolia* and *M. spicata* × *M. suaveolens* revealed that the dominant *C* and *A* genes are independently inherited (Murray 1960a,b).

According to Lawrence (1978), (+)-piperitone is the stereoisomer commonly occurring in mint essential oils. This enantiomer was found to participate by 76–85% in *M.* × *piperita* cv. Black Mitcham essential oil; it is considered that peppermint produces both enantiomers of piperitone (presumably by means of separate, stereospecific enzymes) but accumulates (+)-piperitone by reducing (–)-piperitone, at least in part, to (+)-isomenthone (Burbott et al. 1983).

3.3.2 Oils Rich in Piperitenone Epoxide and/or Piperitone Epoxide

This chemotype is often encountered in the wild populations of the three species of the section Spicatae. In particular, piperitenone oxide accompanied or not by low quantities of *cis*- and *trans*-piperitone oxide has been found in high amounts in the essential oils of (1) *M. longifolia* s.lat. plants from the Netherlands (Shimizu and Ikeda 1961), India (Melkani et al. 1989), Italy (Sacco and Scannerini 1968; Maffei 1988a), and Syria (Lawrence 1978); (2) *M. suaveolens* plants from the United Kingdom (Lawrence 1978), Germany (Hendriks 1971), Greece (Kokkini 1983), the Netherlands (Hefendehl and Nagell 1975), and Italy (Handa et al. 1964; Sacco and Scannerini 1968; Lawrence 1978); and (3) *M. spicata* plants from Greece (Kokkini and Vokou 1989), Belgium, Denmark, Sweden, Czechoslovakia, India (Shimizu and Ikeda 1962), and Italy (Shimizu and Ikeda 1962; Lawrence 1978).

Cis- and *trans*-piperitone oxide have been reported as main components of the essential oils of (1) *M. longifolia* s.lat. plants from Belgium, Italy (Shimizu and Ikeda 1961), Germany (Baquar and Reese 1965; Hefendehl and Nagell 1975), Greece (Kokkini and Papageorgiou (1988a); Egypt, France, Poland, SW Africa, and India (Lawrence 1978); (2) *M. spicata* plants from Belgium (Shimizu and Ikeda 1962), Greece (Kokkini and Vokou 1989), and Poland (Lawrence 1978); and (3) *M. suaveolens* plants from the Netherlands (Hendriks 1971; Hendriks et al. 1976; Lawrence 1978), France, and Morocco (Lawrence 1978). Essential oils, rich in piperitenone oxide and/or *cis*- and *trans*-piperitone oxide, have also been found in the wild-growing hybrids between the three Spicatae species (Lawrence 1978; Kokkini 1983; Kokkini and Papageorgiou 1987,1988b).

Hefendehl and Murray (1976) suggested that the epoxidation of both piperitenone and piperitone is controlled by a recessive gene *o*. To explain the occurrence of two isomeric forms of piperitone oxide in *Mentha* essential oils, Lawrence (1978) suggested that (1) the recessive gene *o* is responsible for the formation of piperitenone oxide and *trans*-piperitone oxide from piperitenone and piperitone respectively, and (2) the reduction of piperitenone to (+)-piperitone and the conversion of (+)-piperitenone oxide to (–)-*cis*-piperitone oxide are further controlled by the dominant genes P_s and P_r, respectively.

3.3.3 Oils Rich in Menthofuran

Menthofuran has been reported as the major constituent of *M. aquatica* essential oils by a number of authors. It occurs also in relatively high amounts in the essential oils of *M. aquatica* hybrids (Table 2), (Lawrence 1978; Kokkini 1983; Sacco and Maffei 1988, and references therein).

Crosses between *M. arvensis*, possessing 70–75% menthol and less than 0.1% menthofuran, and *M. aquatica*, possessing 60–80% menthofuran and less than 1% menthol, have shown that the oxidation of pulegone to menthofuran is controlled by a single incompletely dominant gene *F*, the reduction of pulegone to menthone, however, by the gene *P* (Fig. 1). The *FF* genotype allows the formation of less than 0.1% menthofuran, the *Ff* genotype of less than 25% and the *ff* genotype of 60–80%, with the oxidation of pulegone to menthofuran taking precedence over the reduction of pulegone to menthone or isomenthone (Murray and Hefendehl

1972). Lawrence (1978), confirming these findings, considered pulegone as the only precursor of menthofuran, and proposed the possible mechanism of how this unusual reaction came about.

Although *M. aquatica* seems to be a stable species with respect to its qualitative oil composition, characterized by the high content of menthofuran, Shimizu et al. (1966) reported isopinocamphone-rich strains from Belgium, Italy, and Portugal. There is no other reference for any wild-growing isopinocamphone-rich mint species. Lincoln et al. (1986) studied the oil composition of hybrids resulting from a number of crossing experiments between a *M. citrata* linalool/linalyl acetate-rich strain and *M. aquatica* menthofuran-rich strain. They demonstrated that if the gene I_s is separated from the linked I gene, responsible for linalool accumulation, and is substituted into *M. aquatica* of the recessive *ii* genotype, the oils of the resulting hybrids will have high amounts of β-pinene and isopinocamphone (Fig. 1), as well as a number of related monoterpenoids, seldom observed in *Mentha* oils.

3.3.4 Oils Rich in Pulegone, *cis*- and/or *trans*-Isopulegone

Pulegone, *cis*- and *trans*-isopulegone participating in various amounts, have been reported as the main components in the essential oils of *M. arvensis* plants growing wild in North America (Gill et al. 1973; Lawrence 1978).

Katsuhara (1966) and Hendriks and Van Os (1976) postulated that isopulegone was biosynthesized from isopiperitenone. Lawrence (1978) considered that pulegone can be stereospecifically isomerized to *cis*-isopulegone (isomerization controlled by the dominant H_r gene) and *trans*-isopulegone (isomerization controlled by the dominant H_s).

3.3.5 Oils Rich in Pulegone, Menthone, and Isomenthone

Pulegone, menthone, and isomenthone in various amounts, characterize the essential oils of some North American *M. arvensis* (Gill et al. 1973) and North American or European *M. spicata* plants (Shimizu and Ikeda 1962; Lawrence 1978; Kokkini and Vokou 1989). High content of pulegone characterizes the *M. pulegium* plants exploited for their oils (European pennyroyal oil), (Handa et at. 1964; Lawrence 1978).

Katsuhara (1966) first suggested the (+)-pulegone is the precursor of both (+)-isomenthone and (–)-menthone; this conclusion was based solely upon the stereochemical structure of the compounds. Cell-free enzyme preparations obtained from Mitcham peppermint were able to convert (+)-pulegone to menthone and to isomenthone (Burbott et al. 1974). It has been found also that cell cultures, derived from different *Mentha* chemotypes, transform (+)-pulegone and (–)-menthone into (+)-isomenthone and (+)-neomenthol, respectively (Aviv and Galun 1978; Aviv et al. 1981).

Data obtained from a number of crossing experiments provide evidence that the reduction of piperitenone to the ketone pulegone is controlled by the dominant allele A, while for the conversion of (+)-pulegone to (–)-menthone or (–)-isomenthone a single locus having multiple alleles is involved; the allele the P^s

allele is not completely dominant over the P^r allele and both are dominant over the recessive allele *p*, that largely prevents menthone development (Fig. 1). The quantitative amounts of the two isomers in a diploid species are believed to be controlled by the six combinations of the three alleles; graded effects are obtained in the more complex genotypes in double diploid and octaploid species (Murray et al. 1980).

3.3.6 Oil Rich in Menthone, Isomenthone, Isomers of Menthol, and Their Acetates

Cornmint oil and peppermint oil obtained from *M. arvensis* var. *piperascens* and different cultivated varieties of *M.* × *piperita* plants respectively are characterized by the preponderance of menthone, isomenthone, and mainly of the different isomers of menthol. In particular, cornmint oil is valued as a source of l-menthol. (55–82%) of the total oil) (Lawrence 1978; Duriyaprapan et al. 1986; Tyagi and Naqvi 1987, and references therein), which is generally obtained by simple freeze crystallization. The dementholized oil is also an important trading commodity as an inexpensive, rather harsh peppermint-like oil.

M. × *piperita*, a sterile hybrid of the species *M. aquatica* and *M. spicata*, is, on commercial grounds, probably the most important commercial essential oil-bearing plant in the world today from the standpoint of the number of acres grown for distillation. The chemical composition of peppermint essential oil has been the subject of many investigations; the plant is one of the most widely studied of all aromatic plants. Three detailed studies by Lawrence et al. (1972), Sheldon et al. (1972), and Fleisher et al. (1974) have identified in total 138 constituents in peppermint oil from cultivated plants in the United States. Most of them are monoterpene and sesquiterpene hydrocarbons and monoterpene alcohols.

Published information on peppermint oil, obtained from cultivated plants of various origin, indicates that it is characterized by the presence of oxygenated p-menthane monoterpenes such as l-menthol (20–54%), l-menthone (5–43%), menthofuran (1–8%), menthyl acetate (1–29%), and their relatives like isomenthol, isomenthone, neomenthol, neoisomenthol, isomenthyl acetate, neomenthyl acetate, and neoisomenthyl acetate (Lawrence 1978; Chialva et al. 1982; Wilkins et al. 1982; Kjonaas and Croteau 1983; Katsiotis et al. 1985; Gilly et al. 1986; Srinivas 1986; Gasic et al. 1987; Maffei and Sacco 1987; Udyanskaya et al. 1987; Maffei 1988b). Wild-growing *M.* × *piperita* plants have higher content of menthofuran (6–24%) than the commercial peppermint oil (Kokkini 1983).

Remarkable differences have been found in the quantitative composition of peppermint oil during plant development. The apical young leaves contain mainly monoterpenes in a more oxidized condition (i.e., menthone), and the basal old leaves more reduced and esterified compounds (i.e., menthol and menthyl acetate). The decrease in menthone with leaf age and the increase in menthol, neomenthol, and menthyl acetate may be the result of direct biosynthetic conversion as the pathway to menthols proceeds via menthone (Fig. 1), (Lawrence 1978; Clark and Menary 1980; Martinkus and Croteau 1981; Croteau 1984). Analyses of essential oils extracted from the peltate trichomes of peppermint leaves have shown that menthone and menthol, exhibiting an opposite trend during leaf development, were found to be present in larger amounts on adaxial epidermises, whereas neomenthol and isomenthol had higher percentages on abaxial epidermises (Maffei et al. 1989). Furthermore, it should be noted that the essential oils obtained from inflorescences and leaves of *M.* × *piperita* plants differ significantly in their menthofuran content; flower oils contain 18–22% menthofuran, but leaf oils contain only 1–6% menthofuran (Lawrence 1978; Tzimourtas et al. 1980; Maffei and Sacco 1987).

Lawrence (1978) has postulated that two kinds of reductase enzymes control the development of the four menthol isomers occurring in mint oils: (1) the R_r enzyme is responsible for the conversion of (−)-menthone to (−)-menthol and (+)-isomenthone to (+)-neo-isomenthol, and (2) the R_s enzyme converts (−)-menthone to (+)-neo-menthol and (+)-isomenthone to (+)-isomenthol.

The compounds that characterize the above-mentioned chemotypes are those commonly occurring as main components of mint essential oil (Table 2); their formation reflects differences in biosynthetic pathways. A few other compounds have also been reported sporadically as main mint oil components , e.g., *trans*-sabinene hydrate in *M. candicans* (Karasawa and Shimizu 1978). Because of their unclear biosynthetic origin, it is difficult to accept that they characterize distinct chemotypes. Nevertheless, it should be taken into account that despite the numerous works for the commercially exploited mint essential oils, inclusive chemosystematic treatments of the native *Mentha* plants are unfortunately very few. Comparison of the essential oil composition of cultivated mint plants and those growing wild shows that in some species great differences occur in both qualitative and quantitative composition. A characteristic example is the case of *M. arvensis* s. lat., which is widely distributed in temperate Eurasia and North America. Although commercially exploited *M. arvensis* plants are always rich in l-menthol, wild populations are very variable. Different chemotypes were distinguished within the species characterized by the high contribution of the following compounds in the essential oils: (1) 3-octanol and its acetate (Sacco and Shimizu 1965), (2) 3-octanone with myrcene and/or *cis*- and *trans*-ocimene (Malingré 1971; Lawrence 1978), (3) geraniol and its acetate with more or less 3-octanone (Van Os and Smith 1970; Malingré 1971), (4a) linalool with more or less 3-octanone or γ-terpinene (Malingré 1971; Lawrence 1978), (4b) linalool with *cis*- and *trans*-ocimene (Gill et al. 1973; Lawrence 1978), (5) pulegone, menthone, and isomenthone (Sacco and Shimizu 1965; Gill et al. 1973), (6) *trans*-sabinene hydrate with more or less 3-octanone and terpinen-4-ol (Lawrence 1978), (7) 1,8-cineole and *cis*- and *trans*-ocimene (Gill et al. 1973; Lawrence 1978), (8) *cis*- and *trans*-isopulegone (Gill et al. 1973; Lawrence 1978), (9) pulegone, 1,8 cineole, and α-terpineol (Lawrence 1978), and (10) β-pinene and *trans*-piperitone oxide (Lawrence 1978); this chemotype needs confirmation, since the taxonomic status of the plants examined is uncertain. The chemotypes 1,2,3,4a,6,9, and 10 have been reported from European plants (2n = 72), the chemotypes 4b,7, and 8 from North American plants (2n = 96), while the chemotype 5 has been found in both European and North American plants.

From the taxonomic point of view, the study of *Mentha* essential oils has been mainly useful as an aid in (1) defining the species or species relationships; for example, the two chemotypes characterized by the main components carvone/dihydrocarvone and piperitenone/piperitone oxide occur exclusively in the close relative species of the section Spicatae (Table 2), (2) for detecting hybridization in natural populations, e.g., the hybrids of *M. aquatica* are always characterized by a more or less high amount of menthofuran (Table 2), and (3) in confirming the existence of geographical races; for example, the essential oils of the cytologically and morphologically distinct North American and European populations of *M. arvensis* present differences in both their qualitative and quantitative composition (Gill et al. 1973; Lawrence 1978).

4 Conclusions

Great inter- and intraspecific variability characterizes the essential oil composition of the genus *Mentha*. Research published to date suggests strongly that the qualitative production of mint essential oils is clearly controlled by simple genetic systems. Biogenetic data demonstrate that three different dominant genes are involved in the formation of the main ketones occurring in mint essential oils. A dominant gene *C* is responsible for the formation of C-2 oxygenated compounds (like carvone, dihydrocarvone), whereas the recessive *cc* genotype allows the formation of C-3 oxygenated compounds (like piperitenone, piperitone, pulegone, and menthone). The formation of pulegone through the reduction of piperitenone is controlled by the dominant gene *A*. An incompletely dominant gene *F* is responsible for the conversion of pulegone to menthofuran, while pulegone may also be converted into either menthone (P^s allele) or isomenthone (P^r allele) when the dominant gene *P* is present. Furthermore, a dominant gene *R* is responsible for the formation of alcohols from the corresponding ketones, and finally, the formation of monoterpene acetates from all monoterpene alcohols is controlled by the dominant gene *E*.

Three dominant genes, viz. *I*, causing accumulation of linalool, I_s, causing formation of considerable amount of β-pinene and isopinocamphene, and I_m, causing accumulation of limonene, can interrupt the terpene biogenesis at earlier stages than the formation of monocyclic C-2 and C-3 oxygenated compounds. These "blocking" genes, detected in unexpected segregations in hybrids resulting from experimental interspecific crossings, do not appear to occur as common segregates in natural populations.

Chemical races have been detected in almost every *Mentha* species or hybrid that has been studied extensively. Plants belonging to four distinct chemotypes, viz. rich in menthone/menthol, carvone/dihydrocarvone, linalool/linalyl acetate, and pulegone are more or less widely exploited all over the world. However, mint plants could be the raw material for more than the presently exploited oils and compounds, as the different species and hybrids contain relatively high amounts of commercially valuable and/or potentially interesting constituents.

References

Amelunxen F (1964) Elektronenmikroskopische Untersuchungen an den Drüsenhaaren von *Mentha piperita* L. Planta Med 12:124–139

Amelunxen F (1965) Elektronenmikroskopische Untersuchungen an den Drüsenhaaren von *Mentha piperita* L. Planta Med 13:457–473

Aviv D, Galun E (1978) Biotransformation of monoterpenes by *Mentha* cell lines: conversion of pulegone to isomenthone. Planta Med 33:70–77

Aviv D, Krochmal E, Dantes A, Galun E (1981) Biotransformation of monoterpenes by *Mentha* cell lines: conversion of menthone to neomenthol. Planta Med 42:236–243

Baquar SR, Reese G (1965) Cytotaxonomische und gaschromatographische Untersuchungen an norddeutschen *Mentha*-Formen. Pharmazie 20:159–168

Burbott AJ, Loomis WD (1969) Evidence for metabolic turnover of monoterpenes in peppermint. Plant Physiol 44:173–179

Burbott AJ, Croteau R, Shine WE, Loomis WD (1974) Biosynthesis of cyclic monoterpenes by cell-free extracts of *Mentha piperita* L. Paper no. 17, VIth International Essential Oil Congress, San Francisco

Burbott AJ, Hennessey JP, Johnson WC, Loomis WD (1983) Configuration of piperitone from oil of *Mentha piperita*. Phytochemistry 22:2227–2230

Chialva F, Gabri G, Liddle PAP, Ulian F (1982) Qualitative evaluation of aromatic herbs by direct headspace $(GC)^2$ analysis. Applications of the method and comparison with the traditional analysis of essential oils. In: Margaris N, Koedam A, Vokou D (eds) Aromatic plants: basic and applied aspects. Martinus Nijhoff. The Hague, pp 183–195

Clark RJ, Menary RC (1980) Environmental effects on peppermint (*Mentha piperita* L.). II. Effects of temperature on photosynthesis, photorespiration and dark respiration in peppermint with reference to oil composition. Aust J Plant Physiol 7:685–692

Cori OM (1983) Enzymic aspects of the biosynthesis of monoterpenes in plants. Phytochemistry 22:331–341

Croteau R (1984) Biosynthesis and catabolism of monoterpenes. In: Nes WD, Fuller G, Tsai LS (eds) Isoprenoids in Plants: biochemistry and function. Marcel Dekker, New York, pp 31–65

Croteau R, Venkatachalam KV (1986) Metabolism of monoterpenes: demonstration that (+)-*cis*-isopulegone, not piperitenone, is the key intermediate in the conversion of (—)-isopiperitenone to (+)-pulegone in peppermint (*Mentha piperita*). Arch Biochem Biophys 249:306–315

Croteau R, Virendar KS (1985) Metabolism of monoterpenes: evidence for the function of monoterpene catabolism in peppermint (*Mentha piperita*) rhizomes. Plant Physiol 77:801–806

Duriyaprapan S, Britten EJ, Basford KE (1986) The effect of temperature on growth, oil yield and oil quality of Japanese mint. Ann Bot 58:729–736

Fleischer J, Hopp R, Kaminsky HJ, Bauer K, Mack H, Köpsel M (1974) A new way to produce optically active menthols and its importance for substituting peppermint oil. Paper no. 142, VIth International Essential Oil Congress, San Francisco

Gasic O, Mimika-Dukic N, Adamovic D, Borojevic K (1987) Variability of content and composition of essential oil in different genotypes of peppermint. Biochem Syst Ecol 15:335–340

Gershenzon J, Maffei M, Croteau R (1989) Biochemical and histochemical localization of monoterpene biosynthesis in the glandular trichomes of spearmint (*Mentha spicata*). Plant Physiol 89:1351–1357

Gill LS, Lawrence BM, Morton JK (1973) Variation in *Mentha arvensis* L. (Labiatae). I. The North American populations. Bot J Linn Soc 67:213–232

Gilly G, Garnero J, Racine P (1986) Menthes poivrées–composition chimique analyse chromatographie. Parfum Cosmet Arom 71:79–86

Góra J, Druri M, Kaminska J, Kalemba D (1975) Chemical composition of essential oil from *Mentha spicata* subsp. *longifolia* (L.) Tacik. Herva Pol 20:357–365

Handa KL, Smith DM, Nigam IC, Levi L (1964) Essential oils and their constituents XXIII. Chemotaxonomy of the genus *Mentha*. J Pharm Sci 53:1407–1409

Harborne JB, Turner BL (eds) (1984) Plant chemosystematics. Academic Press, London

Harley RM (1963) Taxonomic studies in the genus *Mentha*. Thesis, Oxford University, Oxford

Harley RM, Brighton CA (1977) Chromosome numbers in the genus *Mentha* L. Bot J Linn Soc 74:71–96

Hefendehl FW, Murray MJ (1972) Changes in monoterpene composition in *Mentha aquatica* produced by gene substitution. Phytochemistry 11:189–195

Hefendehl FW, Murray MJ (1973) Monoterpene composition of a chemotype of *Mentha piperita* having high limonene. Planta Med 23:101–109

Hefendehl FW, Murray MJ (1976) Genetic aspects of the biosynthesis of natural odors. Lloydia 39:39–52

Hefendehl FW, Nagell A (1975) Unterschiede in der Zusammensetzung der ätherischen Öle von *Mentha rotundifolia*, *Mentha longifolia* und des F_1 Hybriden beider Arten. Parfum Kosmet 56:189–193

Hendriks H (1971) Chemotaxonomisch onderzoek van *Mentha rotundifolia* (L.) Hudson. Pharm Weekbl 106:158–165

Hendriks H (1974) De vluchtige olie van enkele chemotypen van *Mentha suaveolens* Ehrh. en van hybriden met *Mentha longifolia* (L.) Hudson. Thesis, Rijksuniversiteit, Groningen

Hendriks H, Van Os FHL (1976) Essential oil of two chemotypes of *Mentha suaveolens* during ontogenesis. Phytochemistry 15:1127–1130

Hendriks H, Van Os FHL, Feenstra VJ (1976) Crossing experiments between some chemotypes of *Mentha longifolia* and *Mentha suaveolens*. Planta Med 30:154–162

Karasawa D, Shimizu S (1978) *Mentha candicans*, a new chemical strain of section Spicatae, containing *trans*-sabinene hydrate as the principal component of essential oil. Agric Biol Chem 42:433–437

Katsiotis S, Ktistis G, Ikonomou GN (1985) Investigation of the simultaneous influence of three different parameters on the essential oil yield and quality of *Mentha piperita*. Pharm Acta Helv 60:228–231

Katsuhara J (1966) An aspect of biogenesis of terpenoids in *Mentha* species. Koryo 83:51–62

Kjonaas R, Croteau R (1983) Demonstration that limonene is the first cycle intermediate in the biosynthesis of oxygenated p-menthane monoterpenes in *Mentha piperita* and other *Mentha* species. Arch Biochem Biophys 220:79–89

Kjonaas R, Venkatachalam KV, Croteau R (1985) Metabolism of monoterpenes: oxidation of isopiperitenol to isopiperitenone, and subsequent isomerization to piperitenone by soluble enzyme preparations from peppermint (*Mentha piperita*) leaves. Arch Biochem Biophys 238:49–60

Kokkini S (1983) Taxonomic studies in the genus *Mentha* L. in Greece. Thesis, University of Thessaloniki, Thessaloniki

Kokkini S, Papageorgiou VP (1987) Constituents of essential oils from *Mentha* × *villoso-nervata* Opiz. growing wild in Greece. Flav Frag J 2:119–121

Kokkini S, Papageorgiou VP (1988a) Constituents of essential oils from *Mentha longifolia* growing wild in Greece. Planta Med 54:59–60

Kokkini S, Papageorgiou VP (1988b) Constituents of essential oils from *Mentha* × *rotundifolia* growing wild in Greece. Planta Med 54:166–167

Kokkini S, Vokou D (1989) *Mentha spicata* (Lamiaceae) chemotypes growing wild in Greece. Econ Bot 43:192–202

Lawrence BM (1978) A study of the monoterpene interrelationship in the genus *Mentha* with special reference to the origin of pulegone and menthofuran. Thesis, Groningen State University, Groningen

Lawrence BM (1985) A review of the world production of essential oils (1984). Perfum Flav 13:2–16

Lawrence BM (1989) Labiatae oils — Mother Nature's chemical factory. Presented in XIth International Congress of Essential Oils, Fragrances and Flavors, New Delhi

Lawrence BM, Morton JK (1972) 3-Dodecanone in *Mentha* × *gentilis*. Phytochemistry 11:2639–2640

Lawrence BM, Hogg JW, Terhune SJ (1972) Essential oils and their constituents. X. Some new trace constituents in the oil of *Mentha piperita* L. Flav Ind 467–472

Lincoln DE, Murray MJ (1978) Monogenic basis for reduction of (+)-pulegone to (-)-menthone in *Mentha* oil biogenesis. Phytochemistry 17:1727–1730

Lincoln DE, Marble PM, Cramer FJ, Murray MJ (1971) Genetic basis for high limonene-cineole content of exceptional *Mentha citrata* hybrids. Theor Appl Genet 41:365–370

Lincoln DE, Murray MJ, Lawrence BM (1986) Chemical composition and genetic basis for the isopinocamphone chemotype of *Mentha citrata* hybrids. Phytochemistry 25:1857–1863

Maffei M (1988a) A chemotype of *Mentha longifolia* (L.) Hudson particularly rich in piperitenone oxide. Flav Frag J 3:23–26

Maffei M (1988b) Environmental factors affecting the oil composition of some *Mentha* species grown in North West Italy. Flav Frag J 3:79–84

Maffei M, Sacco N (1987) Chemical and morphometrical comparison between two peppermint nothomorphs. Planta Med 53:214–216

Maffei M, Codignola A, Fieschi M (1986) Essential oil from *Mentha spicata* L. (spearmint) cultivated in Italy. Flav Frag J 1:105–109

Maffei M, Chialva F, Sacco T (1989) Glandular trichomes and essential oils in developing peppermint leaves. I. Variation of peltate trichome number and terpene distribution within leaves. New Phytol 111:707–716

Malingre TM (1971) Chemotaxonomisch onderzock van *Mentha arvensis* L. Pharm Weekbl 106:165–171

Martinkus C, Croteau R (1981) Metabolism of monoterpenes. Evidence for compartmentation of 1-menthone metabolism in peppermint (*Mentha piperita*) leaves. Plant Physiol 68:99–106

Melkani AB, Shah GC, Parihar R (1989) *Mentha longifolia* subsp. *himalaiensis*: A new species as a source of aroma chemicals. In: Bhattacharyya SC, Sen N, Sethi KL (eds) Proc. 11th International Congress of Essential Oils, Fragrances and Flavours, vol 4, Oxford & IBH Publishing, New Delhi, pp 177–180

Mookherjee BD, Trenkle RW, Wilson RA (1989) Live vs. dead. Part II. A comparative analysis of the headspace volatiles of some important fragrance and flavor raw materials. J Ess Oil Res 2:85–90

Murray MJ (1960a) The genetic basis for the conversion of menthone to menthol in Japanese mint. Genetics 45:925–929

Murray MJ (1960b) The genetic basis for a third ketone group in *Mentha spicata* L. Genetics 45:931–937

Murray MJ, Hefendehl FW (1973) Changes in monoterpene composition of *Mentha aquatica* produced by gene substitution from a high limonene strain of *M. citrata*. Phytochemistry 12:1875–1880

Murray MJ, Lincoln DE (1970) The genetic basis of acyclic constituents in *Mentha citrata* Ehrh. Genetics 65:457–471

Murray MJ, Lincoln DE, Hefendehl FW (1980) Chemogenetic evidence supporting multiple allele control of the biosynthesis of (-)-menthone and (+)-isomenthone stereoisomers in *Mentha* species. Phytochemistry 19:2103–2110

Ravid U, Bassat M, Putievsky E, Weinstein V, Ikan R (1987) Isolation and determination of optically pure carvone enantiomers from caraway (*Carum carvi* L.), dill (*Anethum graveolens* L.), spearmint (*Mentha spicata* L.) and *Mentha longifolia* (L.) Huds. Flav Frag J 2:95–97

Sacco T, Maffei M (1988) *Mentha aquatica* var. *hypeuria* Briq., a selection particularly rich in menthofuran. In: Lawrence BM, Mookherjee BD, Willis BJ (eds) Flavors and fragrances: a world perspective. Elsevier Science, Amsterdam, pp 141–145

Sacco T, Nano G (1971) Contributo allo studio botanico e chimico del genere *Mentha* gruppo arvensis. Riv. Ital. Essenze profumi 54:325–327

Sacco T, Scannerini S (1968) On the cytotaxonomy of selected mints of the section Spicatae. I. Mints with prevalent piperitenone oxide in the oil: *Mentha rotundifolia* (L.) Huds. var. *bullata* Briq., *Mentha longifolia* (L.) Huds. var. *grandis* Briq., *Mentha longifolia* (L.) Huds. var. *typica* Fiori. Allionia 14:177–192

Sacco T, Shimizu S (1965) Botanical and chemical studies on a mint of the arvensis group (*Mentha arvensis* L. var. *praecox* (Sole 1798) Fiori), self-sown in Piedmont, Italy. Perfum Essent Oil Rec 56:211–213

Sheldon RM, Walters LA, Druell WE (1972) The synthesis of flavour chemicals and oils from domestic raw materials. Paper no. 38, 164th Am Chem Soc National Meeting, New York

Shimizu S, Ikeda N (1961) Isolation of (+)-piperitenone oxide or (–)-piperitone oxide from *Mentha spicata* L. with 24 chromosomes in the somatic cells. Perfum Essent Oil Rec 52:708–713

Shimizu S, Ikeda N (1962) The essential oils of *Mentha spicata* L. with 36 or 48 somatic chromosomes. Agric Biol Chem 26:543–545

Shimizu S, Karasawa D, Ikeda N (1966) A new mint (variety of *Mentha aquatica* L.) containing (-)-isopinocamphone as a major constituent of essential oil. Agric Biol Chem 30:200–201

Smith DM, Skakum W, Levi L (1963) Determination of botanical and geographical origin of spearmint oil by gas chromatographic and ultraviolet analysis. J Agric Food Chem 11:268–276

Srinivas SR (ed) (1986) Atlas of essential oils. Bronx, New York

Tétényi P (1973) Homology of biosynthetic routes: The base in chemotaxonomy. In: Bendz G, Santesson J (eds) Chemistry in botanical classification. Academic Press, New York, pp 67–78

Todd WA, Murray MJ (1968) New essential oils from hybridization of *Mentha citrata* Ehrh. Perfum Essent Oil Rec 59:97–102

Tyagi BR, Naqvi AA (1987) Relevance of chromosome number variation to yield and quality of essential oil in *Mentha arvensis* L. Cytologia 52:377–385

Tzimourtas CA, Papageorgiou VP, Sagredos AN, Alexiades CA (1980) Phytochemical study of *Mentha piperita* cultivated in Greece. Part I: Essential oils. Chim Chron, New Ser 9:13–27

Udyanskaya IL, Dzhabarov DN, Rudenko BA (1987) Gas chromatographic comparative study of essential oils of the Prilukskaya-6 and Krasnodar variety peppermints. Nauh, Tr VNII Farmatsii 25:178–181

Van Os FHL, Smith D (1970) De vluchtige olie van *Mentha arvensis* L. subsp. *austriaca* (Jacquin) Briquet. Pharm Weekbl 105:1273–1276

While JGH, Iskandar SH, Barnes MF (1987) Peppermint: effect of time of harvest on yield and quality of oil. N Z J Exp Agric 15:73–79

Wilkins CL, Giss GN, White RL, Brissey GM, Onyiriuka EC (1982) Mixture analysis by gas chromatography/Fourier transform infrared spectometry/mass spectometry. Anal Chem 54:2260–2264

Zwaning JH, Smith D (1971) Composition of the essential oil of Austrian *Mentha pulegium*. Phytochemistry 10:1951–1953

Special Methods for the Essential Oil of Ginger

T.A. VAN BEEK

1 Introduction

1.1 Description and Use

Ginger oil is produced by steam or hydrodistillation of ground rhizomes of *Zingiber officinale* Roscoe (Zingiberaceae). It is valued for its pleasant, aromatic, more or less lemony odor. Ginger oil finds much use in the food and drink industry, e.g., ginger ale, ginger beer, and various cookies and desserts. The oil is further used in small quantities in the cosmetic, pharmaceutical, and perfume industry. It has still not been clarified which compounds are responsible for the characteristic ginger aroma. The various investigations have all come to different conclusions and in some respects contradict each other. The citral (= geranial and neral combined) content is responsible for the lemony note.

1.2 Chemical Composition

Because of the economic interest and its widespread cultivation, ginger oil is well studied. The various investigations have been summarized by Lawrence (1984,1988) and van Beek et al. (1987). Not mentioned in the above reviews are the recent investigations by Ibrahim and Zakaria (1987), Erler et al. (1988), Ekundayo et al. (1988), and Miyazawa and Kameoka (1988).

Upon a first examination of the published literature there appears to be a considerable variation in the chemical constituents reported in the 20 investigations carried out on ginger oil so far. However, a closer study reveals that a combination of some compounds can be considered as fairly characteristic for ginger oil. Major deviations of the general chemical picture are due to incorrect identifications. These have been caused by poor separations and very similar mass spectral data of many sesquiterpenes. In a few cases there has been some nomenclatural confusion.

Much of the remaining chemical variation may be explained by the storage period of the rhizomes (Sakamura 1987), drying conditions (MacLeod and Pieris 1984; McGaw et al. 1984; Ekundayo et al. 1988), distillation time (van Beek et al. 1987), local conditions, and chemical varieties.

The combined occurrence of the following monoterpene hydrocarbons (MTHC) is characteristic for ginger: α-pinene, myrcene, limonene, camphene, and β-phellandrene, with the latter two occurring in the highest concentrations. There

Structure 1 and 2

neral geranial

is relatively little variation in this group. More variation is reported for the oxygenated monoterpenes (OXMT). Important compounds are 1,8-cineol, linalool, neral 1, borneol, geranial 2, geranyl acetate, citronellol, and geraniol. Most of the difference in odor between the various oils is probably due to this class of compounds. For instance, the citral content is reported to vary between extremes of 0.22% (Lin and Hua 1987) for a Chinese oil and 64% (Smith and Robinson 1981) for a Fijian oil.

Except for some citral-rich Japanese and Fijian ginger oils, quantitatively the most important group of compounds are the sesquiterpene hydrocarbons (SQHC). α-Zingiberene 3 always occurs in the highest concentration, usually 10–30% of the oil. Many authors simply report this compound as zingiberene. This is, however, incorrect, as also β-zingiberene 4 has been tentatively reported. Other major SQHC are β-bisabolene 5, (E,E)-α-farnesene 6, β-sesquiphellandrene 7 and *ar*-curcumene 8. Oxygenated sesquiterpenes (OXSQ) normally occur in small amounts (0.4–2%). β-Sesquiphellandrol is a characteristic constituent of ginger oil.

Finally, some phenylpropanoids, e.g., methyl isoeugenol, and aliphatics e.g., 2-undecanone, occur as minor compounds in ginger oil. These ketones are decomposition products formed from the non-volatile hot-tasting gingerols during steam distillation.

Structure 3-8

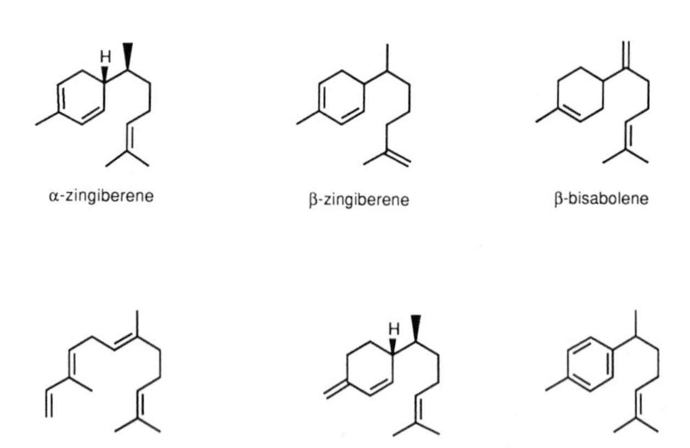

1.3 Isolation, Separation and Quality Evaluation of Ginger Oil

Several old and modern methods exist for the isolation, separation, quality control, and identification of ginger oil and its constituents. The older ("traditional" or "classical") techniques will be briefly mentioned in Section 2. Many of the newer methods will be mentioned in Section 3.1, and several of them will be highlighted in Section 3.2 and following. Examples of their application in the field of ginger oil will be given. Of course, all the techniques used for ginger oil can be applied to other essential oils as well.

2 Traditional Methods of Extraction, Separation, and Control

2.1 Traditional Isolation Methods

Commercial isolation of ginger oil is almost exclusively carried out by means of steam distillation. In the laboratory, hydrodistillation is practiced as well. Solvent extraction with, for instance, acetone gives the so-called ginger oleoresin, which contains both the essential oil and the hot-tasting nonvolatile compounds. This process is also carried out on a commercial scale.

2.2 Traditional Separation Methods

Separation of ginger oil components can be carried out by means of fractionated vacuum distillation, chemical methods, column chromatography on silica gel, and preparative GC. Especially the first method is still of value if it is carried out on modern equipment, e.g., on a spinning band or Spaltrohr column with ca.100 theoretical plates. Advantages are a separation mechanism which is different from all types of column chromatography, high capacity, and no solvent consumption.

2.3 Traditional Quality Control

The single most important characteristic of ginger oil remains its odor. If the odor is unpleasant it will have a low market value no matter how good all other physical and chemical data may be. Odor evaluation is a specialized task, usually carried out by trained professionals.

Apart from the odor, several other easily measurable parameters such as density, refraction index, optical rotation, and solubility in alcohol are usually determined to detect any gross adulterations with synthetic materials or cheap essential oils. Adulteration of ginger oil is rare. Physical and chemical data of ginger oil can be found in the review of Govindarajan (1982).

3 Modern Methods of Extraction, Separation, and Identification

3.1 Introduction

In the last 30 years many more techniques have been introduced for the study of essential oils. A number of them have been summarized in Table 1. Due to space limitations it is not possible to discuss identification techniques 2, 4, 6, and 7. Techniques 2, 4, and 6 are now routine methods used mainly in the structure elucidation of compounds isolated from ginger. UV spectroscopy can also be used as detection technique in HPLC, see Section 3.3.4.2. Color reactions can be used after a separation by TLC, see Section 3.3.1.

Table 1. Modern methods for the isolation, separation, and identification of ginger oil and its constituents

Isolation	Separation	Identification
1. Headspace methods	1. TLC	1. Chromatographic retention
2. Direct vaporization	2. Gas chromatography	2. UV spectroscopy
3. Liquid CO_2 extraction	3. Small-scale column chromatography	3. IR spectroscopy
	4. HPLC	4. PNMR spectroscopy
		5. CNMR spectroscopy
		6. Mass spectroscopy
		7. Color reaction

3.2 Modern Methods of Isolation

3.2.1 Headspace Methods

With headspace methods the volatiles in the surrounding atmosphere of, for example, a ginger rhizome are directly collected and studied with appropriate methods, for instance GC-MS. Thus no steam distillation is necessary and the risk of artifact formation, e.g., hydrolysis of labile esters or rearrangements of certain hydrocarbons, will be greatly diminished.

Two different versions of the headspace technique exist:

1. equilibrium collection
2. dynamic collection

The latter method is more sensitive and is thus more useful for plants with contain many higher boiling constituents, such as ginger. It has been used by De Pooter et al. (1985) in the analysis of ginger volatiles. In a 250 ml flask 1.5 g of ginger powder was flushed for 15 min with ca. 4 l of air at 17 °C. The volatiles were trapped on Tenax GC and thermally desorbed. All the compounds were "injected" with a two-step procedure on a capillary column. The GC pattern was similar to that of the corresponding essential oil prepared by hydrodistillation. This stresses the usefulness of the headspace method, as the analysis can be carried out in considerably less time.

From the similarity of the two GC's one could further conclude that artifact formation due to thermal stress during distillation is not significant for ginger oil.

3.2.2 Direct Vaporization

Chen et al. (1987) have constructed a special vaporizer directly connected to a GC. In the vaporizor chamber 3–15 mg pulverized ginger powder was heated for 60 s at 250 °C. After instant vaporization, the volatiles are swept by the carrier gas onto the capillary GC column and analyzed. Variations in retention times and area % values were less than 5%. There was a good correspondence between the results obtained with this method and the results with an essential oil prepared from the same material. The advantage is, of course, a much faster GC or GC-MS analysis of the plant material. The authors noticed little decomposition, as longer heating times or different temperatures gave the same area % values.

3.2.3 Isolation by Means of Liquid Carbon Dioxide

Chen and Ho (1988) have extracted freeze-dried ginger powder with liquid carbon dioxide at ca. 40 atm for 48 h. In this way they obtained an extract which consisted of one third volatiles. The rest was made up by pungent components and some unidentified components. The authors tentatively detected a lot of unknown volatile sesquiterpenes esters in their extracts which have not been found before in ginger oils prepared by steam distillation. They also found very low concentrations of straight chain aldehydes and 2-alkanones, which are thermal degradative products of the nonvolatile gingerols. The advantage of this isolation method is that little or no decomposition takes place.

It should be remarked that although the three isolation methods discussed above may be interesting from an analytical point of view, they are of no importance for the commercial preparation of ginger oil.

3.3 Modern Methods of Separation

3.3.1 Thin Layer Chromatography (TLC)

TLC is of limited value in the analysis of complex essential oils, especially if they contain large amounts of sesquiterpene hydrocarbons like ginger oil. Usually silica gel is used, which is not very well suited for the separation of hydrocarbons.

Govindarajan and Raghuveer (1973) used three-directional TLC to circumvent some of these shortcomings of silica gel. After the usual two-directional development with (1) hexane : ethyl acetate = 85:15 and (2) benzene : ethyl acetate = 1:1 as solvents, the plate was dried and then developed a third time in the reverse direction of the second elution with pure hexane at –15 °C. This solvent-temperature system separates the terpenic hydrocarbons without affecting the pattern of the separated oxygenated compounds and without the addition of silver nitrate to the stationary phase.

The plates were sprayed with 5% vanillin in sulfuric acid as chromogenic reagent. Ca. 35 spots could be detected. Significant differences, both qualitative and quantitative, could be observed between different oils, although no spots were actually assigned. 2,4-Dinitrophenyl hydrazine was also used for the selective detection of carbonylic compounds, e.g., citral.

3.3.2 Gas Chromatography (GC)

3.3.2.1 Analytical GC

GC is the single most powerful technique for the separation of essential oils. The current generation 0.25 mm i.d. capillary columns routinely generate 4000 plates/m which means over 200 000 plates for 60 m columns. However, often there is still an overlap of peaks in some areas of the chromatogram for complex essential oils like ginger oil.

There is not a best type of stationary phase for complex oils because a better separation of one pair of compounds is likely to be counteracted by the overlapping of another pair of compounds. For ginger oil both a methyl silicone phase (OV-1, DB-1, CP-Sil5) and a polyethylene glycol phase (Carbowax, DB-Wax) give equally good, although very different, results. Chromatograms of a ginger oil with a high content of citral ("citral" type) on 60 m DB-1 and DB-Wax columns are given in Fig. 1a and 1b respectively.

On the apolar DB-1 column there is a good separation of MTHC (t_R 10 to 13 min), OXMT (t_R 15–26 min), SQHC (t_R 25–32 min), and OXSQ (t_R 31–38 min). The separation within these groups is in some cases less satisfactory, e.g., limonene and β-phellandrene overlap. Unfortunately, the monoterpene ether 1,8-cineol (eucalyptol) coincides with these two compounds. Separation is essentially in order of boiling point, so alcohols elute before aldehydes and these elute before esters. The separation of the five main SQHC is good.

On the polar DB-Wax column there is considerable overlap between OXMT and SQHC, which complicates the already difficult identification of these compounds even more. However, the separation within each group is a little better than on the DB-1 column: peaks are more spread out. Limonene, β-phellandrene, and 1,8-cineol are now separated. Separation of the five main SQHC is different but not better than on the DB-1 column (see also Sect. 3.4.1).

A 50% methyl silicone/50% phenyl silicone column (OV-17, DB-17) is a poor choice for the analysis of ginger SQHC. (Z,Z)-α-farnesene 6, β-bisabolene 5, and *ar*-curcumene 8 more or less coincide on this column. The use of packed columns, whatever the stationary phase, should now be strongly discouraged. No packed column has sufficient separation power for the analysis of ginger oil.

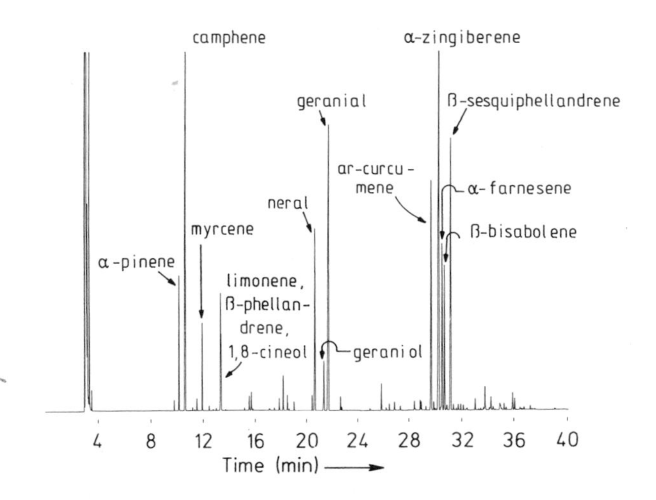

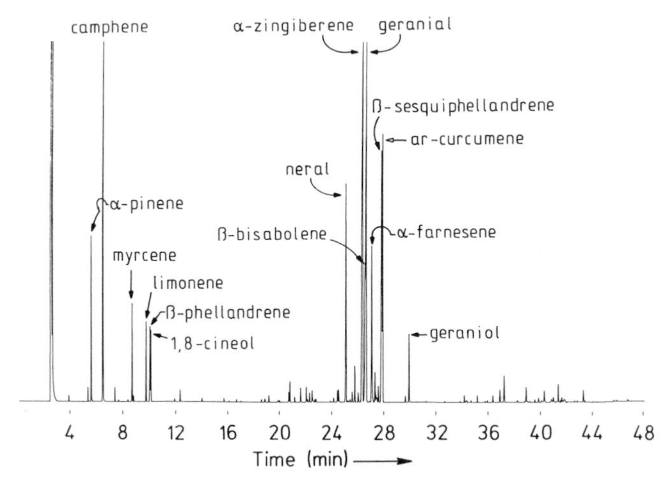

Fig. 1. a GC of an Indonesian ginger oil with high citral content ("citral type") on DB-1. **b** Same oil on DB-Wax. Experimental conditions: dimensions 60 m×0.25 mm, film thickness 0.25 μm, temperature program 50 °C (0 min) to 238 °C (1 min) with 4 °C/min, carrier gas hydrogen 35 cm/s, split 1:100, FI detection

3.3.2.2 Preparative GC

GC has been used for preparative purposes as well. Regrettably, preparative packed columns possess even less separation power than normal packed columns. Recently, wide-bore capillary columns of 0.53 mm i.d. have been introduced ("megabore"columns) and when coated with thick films of 3–5 μm of stationary phase they can be used for preparative purposes as well. They combine reasonable separation power (up to 30 000 plates) with excellent inertness (no tailing or decomposition) and reasonable loadability.

A 30-m 3-μm DB-1 megabore column was used with success for the final separation of some ginger SQHC (van Beek and Lelyveld 1990). One fraction obtained with preparative HPLC (see Sect. 3.3.4.1) still consisted of 69% α-zingiberene 3, 19%-β-bisabolene 5, 9% β-sesquiphellandrene 7, and 3% minor compounds. By injecting a 20–50% solution of the mixture in pentane on this column and trapping the compounds in small glass tubes at room temperature, a 5-mg quantity of β-bisabolene 5 sufficiently pure for PNMR and CNMR could be obtained. Ca. 50 injections were necessary, which took 4 h. Although each separation actually took 8 min, every 4 min one injection was possible without adverse effects on the separation of the previous injection (See Fig. 2). α-Zingiberene 3 and β-sesquiphellandrene 7 were purified from a fraction which consisted solely of these two compounds.

3.3.3 Prefractionation by Small-Scale Column Chromatography

As mentioned above, some overlap between OXMT and SQHC exists on polar columns. This makes GC-MS studies and identifications based on retention times less reliable. One possibility to solve this problem is to fractionate an essential oil on silica gel in a hydrocarbon and an oxygenated fraction. Formerly this was carried out on larger amounts of oil, but nowadays one drop of oil will suffice.

Ginger oil can be fractionated by dissolving 20 μl of dry oil in 0.5 ml pentane (no turbidity should be visible) and bringing it on a solid phase extraction column filled with 0.5 g silica gel of 25–40 μm particle size. The column is eluted with 5 ml of pentane and then eluted with 2 × 5 ml pentane : ether = 2:1 or pentane : acetone = 4:1. The first 5 ml will contain all hydrocarbons and the second 5 ml only oxygenated compounds. An exception are terpenes containing a furan ring. These

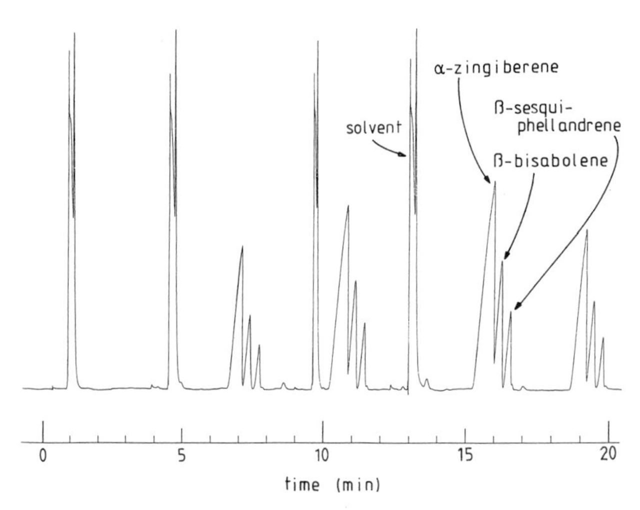

Fig. 2. Preparative GC separation of a ginger SQHC fraction on DB-1 megabore column. Experimental conditions: dimensions 30 m×0.53 mm, film thickness 3 μm, temperature program 175 °C isothermal, carrier gas hydrogen, packed injector, TC detection. Injection volume 1 μl

can occur in both fractions. Examples of furans in ginger oil are perillene and rosefuran. Finally, highly polar compounds such as lauric acid will end up in the last fraction.

In Fig. 3 the four chromatograms belonging to a fractionation of a "mixed type" of ginger oil are shown. This type of oil has relative high concentrations of aldehydes (citral), esters (geranyl acetate), and alcohols (geraniol) beside the always present MTHC and SQHC. The large amount of overlap on the DB-Wax column without a prefractionation can be easily imagined by comparing the two chromatograms at the bottom. Also on DB-1 some overlap between MTHC and oxygenated nonterpenoids and 1,8-cineol (peaks between 10 and 16 min in the chromatogram at the right hand top) would have taken place.

3.3.4 High-Pressure Liquid Chromatography (HPLC)

HPLC can be used (1) for the preparative separation of essential oil compounds and (2) for the quantitative determination of important compounds. An example of both possibilities will be given for ginger oil.

3.3.4.1 Preparative Separation

In the literature much confusion exists about the identity of the major SQHC of ginger oil. To establish unequivocally the identity of the five major SQHC of ginger it was necessary to isolate at least 5 mg of each SQHC in a purity of 85% or higher for PNMR and CNMR measurements.

The first step was the HPLC separation of a ginger oil rich in SQHC. This Indian oil with >70% SQHC is called a "hydrocarbon"-type oil. As stationary phase C-18 modified silica gel was chosen. Unmodified silica gel is unsuitable because of the lack of any polar groups in SQHC. In four runs a total of 130 mg of oil was injected on a 25 × 1 cm column filled with 5 μm C-18 reversed phase material. As solvent MeCN – H_2O = 88:12 was used at 4 ml/min. UV detection at 215 and 245 nm was used to detect both SQHC with unconjugated and conjugated double bonds. A chromatogram of a semi-preparative run with 4 mg of oil is given in Fig. 4. Under these conditions at least 25 different peaks are clearly distinguishable. Some of the later eluting peaks have been assigned. After the four preparative runs *ar*-curcumene 8 was obtained in >99% purity and (E,E)-α-farnesene 6 in 84% purity, which was sufficient for its identification by PNMR and CNMR. β-Sesquiphellandrene 7, α-zingiberene 3, and β-bisabolene 5 could be separated under analytical conditions but not under these preparative conditions. Two fractions were collected which consisted of (1) 53% α-zingiberene and 40% β-sesquiphellandrene and (2) 69% α-zingiberene, 19% β-bisabolene, and 9% β-sesquiphellandrene. These fractions were further separated into the pure compounds by means of preparative capillary GC (see Sect 3.3.2.2).

3.3.4.2 Quantification of Selected Compounds

By adding a suitable detector to the HPLC, one or more of the compounds of interest may be quantified. A sensitive and selective detector is the UV detector.

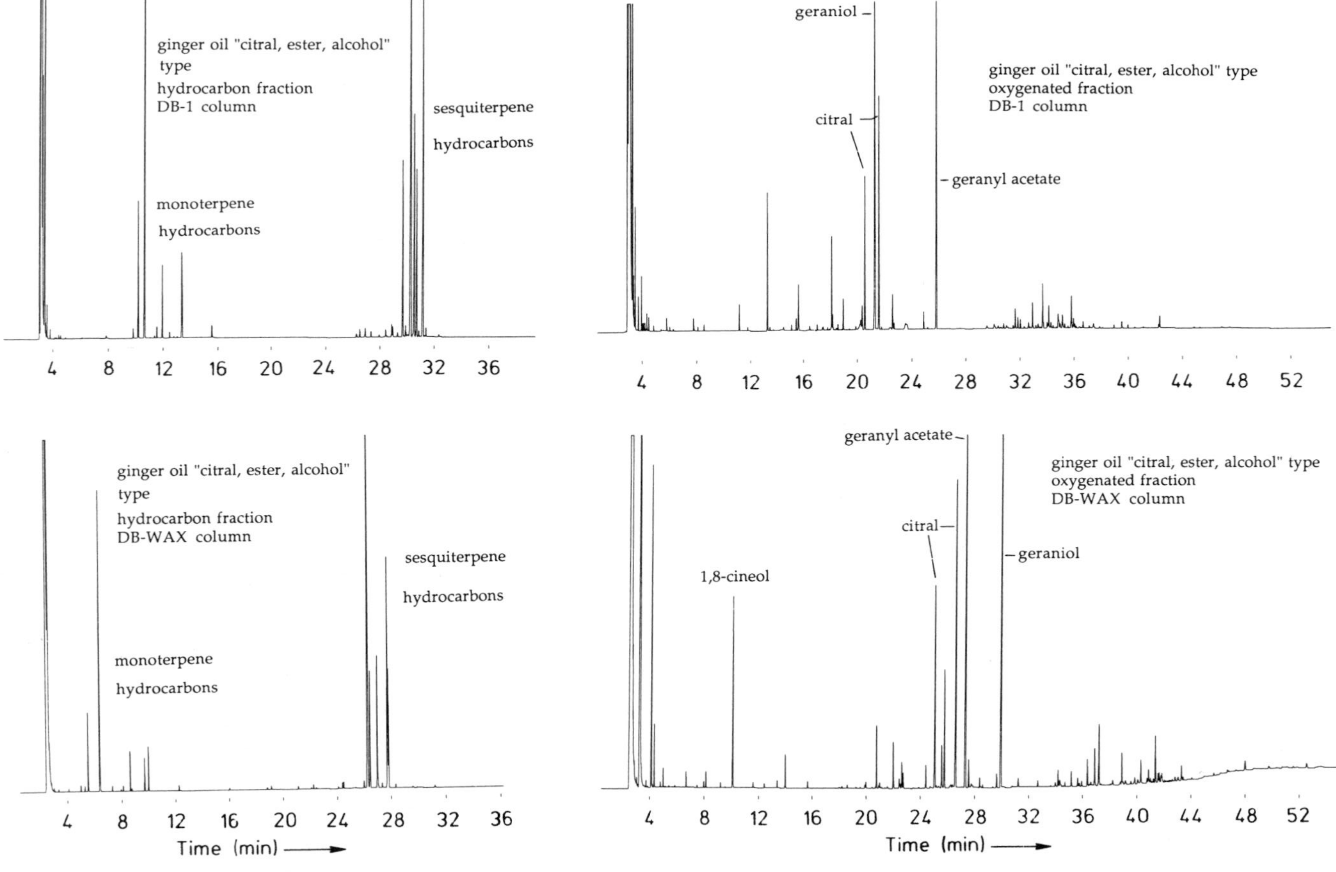

Fig. 3. Chromatograms of the hydrocarbon and oxygenated fraction of a "mixed type" ginger oil on both DB-1 and DB-Wax. Experimental conditions as in Fig. 1

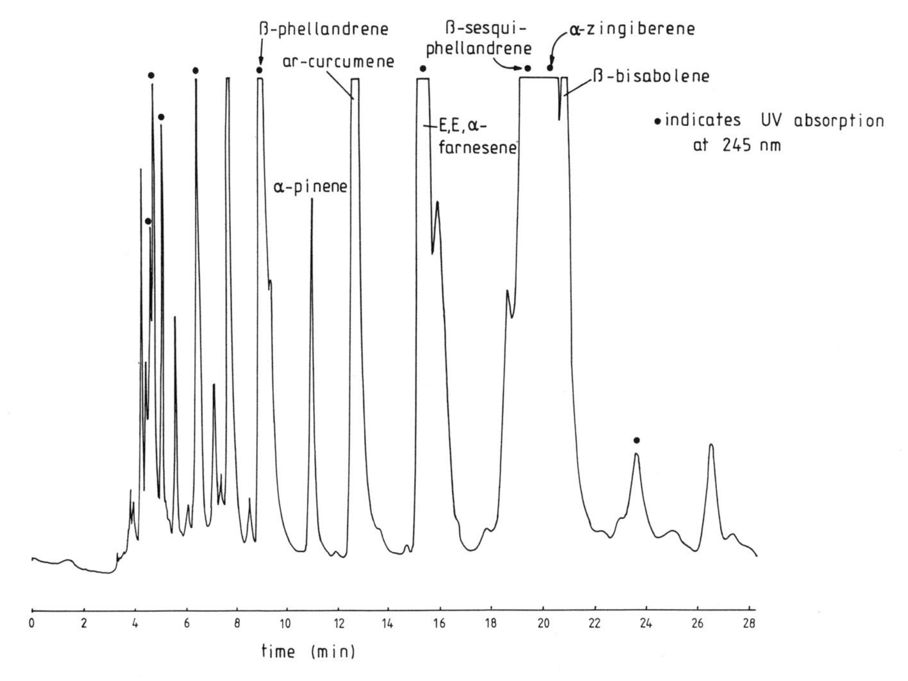

Fig. 4. Semi-preparative HPLC of a "hydrocarbon"-type ginger oil. Experimental conditions: load 4 mg oil dissolved in 4 µl 100% EtOH, column dimensions 25×1 cm filled with Microsorb 5 µm C-18 silica gel, solvent MeCN: H_2O = 9:1, 4 ml/min, detection UV 215 nm

An example is given in Fig. 5, where a wavelength was selected which corresponds with the UV maximum of geranial and neral (together citral). With this reversed-phase system the citral content of any ginger oil can be measured in minutes. With a minimum detectable quantity of ca. 1 ng citral even very low concentrations of citral can be measured. The citral content is important for the lemony note of ginger oil. In this gradient run only four other peaks are visible, corresponding with myrcene, β-phellandrene, (E,E)-α-farnesene, β-sesquiphellandrene, and α-zingiberene. The latter two coincide.

If an isocratic system of MeCN: H_2O = 6:4 had been used, many runs could have been carried out without any interference of the hydrocarbons because of their low k' value in this system. At the end of the day flushing with 100% MeCN removed all the hydrocarbons at the same time.

3.4 Modern Methods of Identification

3.4.1 Identification by Retention Times

Together with GC-MS, identification by retention times or preferably retention indices (Kovats indices) is widely practiced in the essential oil field. Regrettably,

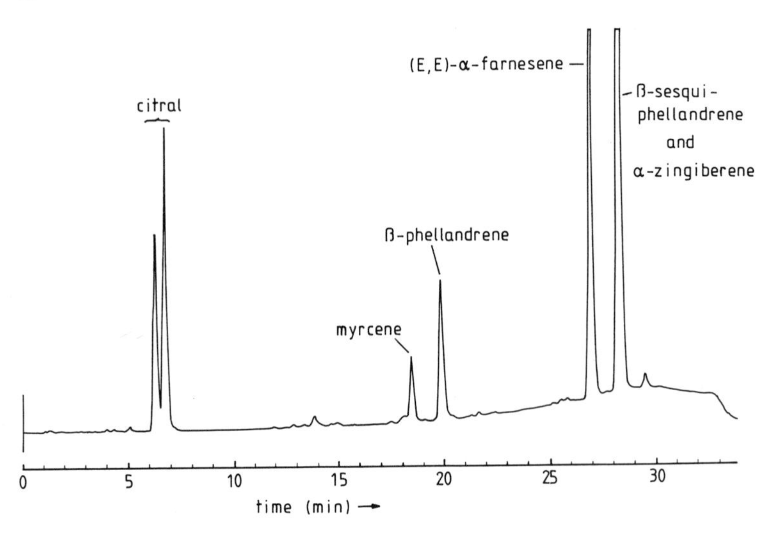

Fig. 5. Analytical HPLC of a "citral"-type ginger oil. Experimental conditions: column dimensions: 15×0.46 cm filled with Microsorb 5 μm C-18 silica gel, solvent MeCN : H_2O = 6:1 to MeCN: H_2O = 95:5 in 30 min, 1 ml/min, detection UV 236 nm

however, for both techniques the old axiom "what you cannot separate you cannot identify or quantify" is true. Most of the misidentifications or nonidentifications of ginger oil compounds have been caused by insufficient separations resulting in unreliable retention times, mixed mass spectra, or obscuring of minor compounds by major compounds. Therefore one should buy the best GC columns available. Remember that the most expensive MS cannot restore the poor performance of a GC column costing perhaps less than 0.1% of the MS apparatus.

Even the best GC columns, however, will not be able to separate all compounds present in ginger oil (see also Sect. 3.3.2.2). A better approach therefore is the use of two GC columns of different polarity, e.g., an apolar and a polar one. Probably no two compounds will have the same retention on two different columns. This is nicely illustrated by the five major SQHC of ginger. Even these highly apolar compounds, which possess a double bond as only functionality, show considerable differences in their order of elution, see Table 2. Compounds with conjugated, exocyclic double bonds, and especially aromatics are more retained on polar columns than others lacking these characteristics. Thus compounds, which can be distinguished only with difficulty with MS, e.g., β-bisabolene, (E,E)-α-farnesene, and β-sesquiphellandrene can be reliably identified by their retention indices if the right columns are used.

An even better approach is to mount the two columns in one oven and to install a 1:1 splitter just after the injection port. In that case one can also make use of peak areas which should be approximately the same for one compound on both columns if no peak overlap occurs. This procedure should reduce the number of false positive identifications or obscured compounds especially with more complex essential oils. Even this method may fail, as a minor compound can co-elute with one major compound on column 1 and another on column 2. Examples of

Table 2. Retention indices of major SQHC in ginger oil

Compound	RI DB-1	Peak no.	RI DB-Wax	Peak no.	Difference in RI
α-Zingiberene 3	1489	2	1730	1	241
β-Bisabolene 5	1502	4	1736	2	234
(E,E)-α-Farnesene 6	1496	3	1755	3	259
β-Sesquiphellandrene 7	1518	5	1781	4	273
ar-Curcumene 8	1473	1	1783	5	310

such compounds, which remained unobserved during normal GC analysis, are given in Section 3.4.3. Only fractionation by column chromatography or HPLC, changing of the GC experimental conditions, CNMR, or true two-dimensional GC can solve such problems. The last technique, in which a heart-cut of column 1 is brought onto column 2 and further separated, is probably the best for minor components, but has so far not been used for ginger oil.

GC can also be used for the qualitative identification of ginger oils and for the detection of any volatile adulterants. In a recent study by the author (van Beek and Lelyveld 1990) 48 ginger oils of nine different countries were investigated by GC on two different columns. In spite of many inevitable differences between the growing, storage, drying, and distillation conditions, all 48 oils were qualitatively fingerprint identical. There were only quantitative differences in composition. Another species from the same genus, *Z. cassumunar*, was totally different both in chemical composition and odor. So intraspecies differences appear to be small when compared with differences between various *Z.* species.

3.4.2 Infrared (IR) Spectroscopy

IR spectroscopy can be used off-line or on-line (GC-IR) for the structure elucidation of separated compounds, but also for the analysis of the total oil. An IR spectrum of the total oil can be recorded as a film between NaCl windows, as a solution in $CHCl_3$ or simply as a film on a KBr disc. The IR spectra of three different types of ginger oil recorded with the latter method are given in Fig. 6. The three oils can be easily distinguished by the intensity of the peaks at 3470 cm^{-1} (alcohols), 1743 cm^{-1} (esters), and 1680 cm^{-1} (conjugated aldehydes). The rest of the spectrum is roughly similar for the three oils.

Thus IR spectroscopy can serve as a quick identification method, mark additions of large quantities of extraneous materials, and distinguish between various types of ginger oils, e.g., "hydrocarbon", "citral", or "mixed" types.

3.4.3 Carbon-13 Nuclear Magnetic Resonance Spectroscopy (CNMR)

The application of CNMR spectroscopy in the analysis of essential oils has been pioneered by the work of Kubeczka (Kubeczka and Formácek 1984 and references cited therein). The biggest advantage of this method is that the oils are

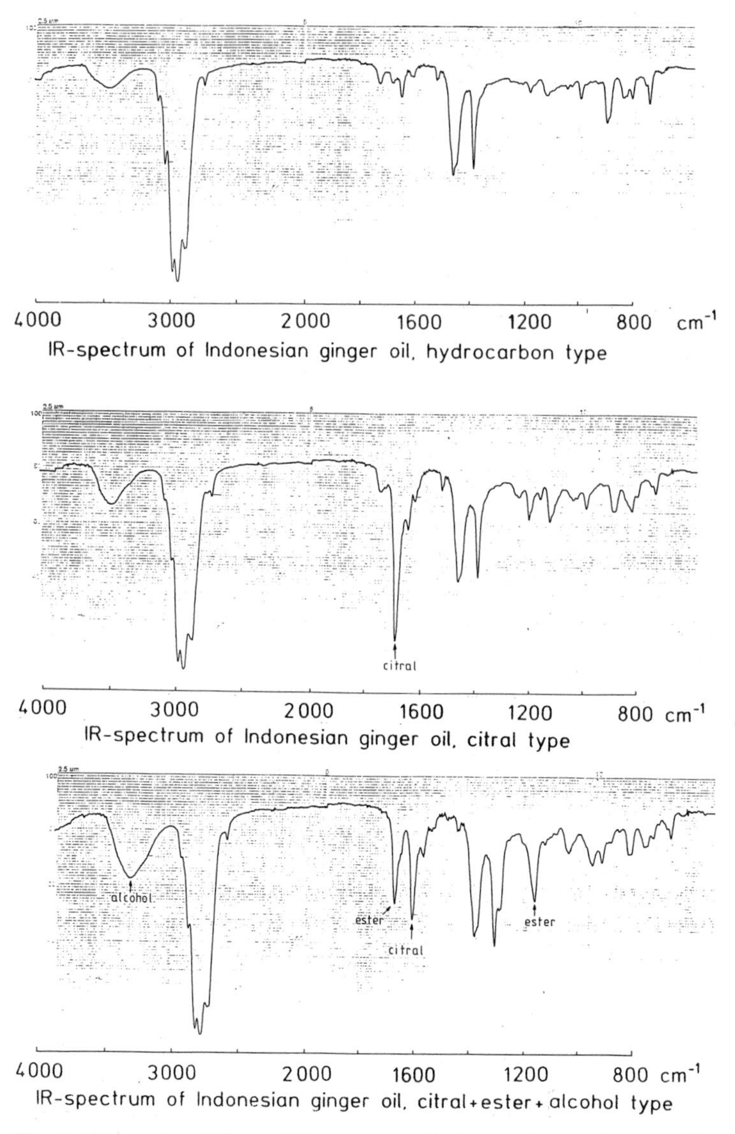

Fig. 6. IR spectra of three different types of ginger oil recorded as a film on KBr discs. **Top** "hydrocarbon"-type oil; **middle** "citral"-type oil; **bottom** "mixed" type of oil. See also text

measured directly at room temperature. Thus no thermal degradation can take place and no artifacts can be formed or introduced as a result of the analysis. Further, all substances present in the oil which contain carbon atoms are measured, i.e., also nonvolatile substances added to adulterate oils.

The theoretical maximum number of peaks which can be observed is high. In the spectrum, the peaks, each of which corresponds with a carbon atom, are spread out between 5 and 215 ppm, which with a resolution of 0.05 ppm means that theoretically 4000 peaks can be observed. Unfortunately, in CNMR spectra of essential oils, not all the peaks are evenly spread out over the entire spectral range. Especially the areas between 20 and 40 ppm and 120 and 135 ppm contain a disproportionate number of peaks. Another problem is that smaller peaks can be obscured by adjacent larger peaks. In spite of this almost all compounds have between one and three easily observable resonances which together are highly characteristic for that particular compound. As in GC, the peak heights or peak areas are with the exception of peaks of quaternary carbons approximately the same for all resonances of one compound. This can help the interpretation. The total set of shifts for one compound varies little from one lab to another, with deviations usually smaller than 0.5 ppm. The largest compilation of spectra can be found in the book by Formácek and Kubeczka (1982).

Disadvantages of CNMR are the necessity of a routine NMR spectrometer and the relative insensitivity. Compounds occurring below 0.2% are not easily measurable even after 16 h of scanning.

In Fig. 7 a CNMR spectrum of a ginger oil is given. After a short examination, it is clear that this is the CNMR spectrum of "mixed" type of oil because large signals are observed for neral and geranial (together citral), geranyl acetate, and geraniol. The largest constituent, viz. α-zingiberene, can be recognized by its 13 nonquaternary signals. Camphene, bornyl acetate, borneol, 1,8-cineol, β-bisabolene, and (E,E)-α-farnesene are also detectable. Without any prior knowledge, the combination of these compounds alone would suggest the presence of a ginger oil.

Although initially it may seem difficult to assign all peaks in the spectrum, this is, however, possible after some study. In Fig. 8 the aromatic region of a "hydrocarbon"-type ginger oil is expanded and almost all peaks have been assigned.

One should keep in mind, however, that normally a CNMR spectrum is recorded after some GC and GC-MS studies have been carried out. CNMR spectra should be used to confirm the presence of compounds found in GC and GC-MS and generally not be interpreted without prior knowledge. This is possible, but much more difficult and time-consuming. CNMR is an extra identification technique in addition to the regular techniques based on retention indices and MS.

In the spectrum of Fig. 8 only a few larger peaks at 104.2, 105.8, 111.3, 134.0, 150.8 and 153.0 ppm remained unassigned. According to their peak heights, they must belong to compounds occurring in concentrations of 1–2%. As no unknown compounds occurring in such high concentrations had been detected in our initial GC-MS studies, they must coincide with other SQHC occurring in even higher concentrations. After more careful GC-MS studies on another type of column and different carrier gas velocities, two more compounds became visible. On the basis

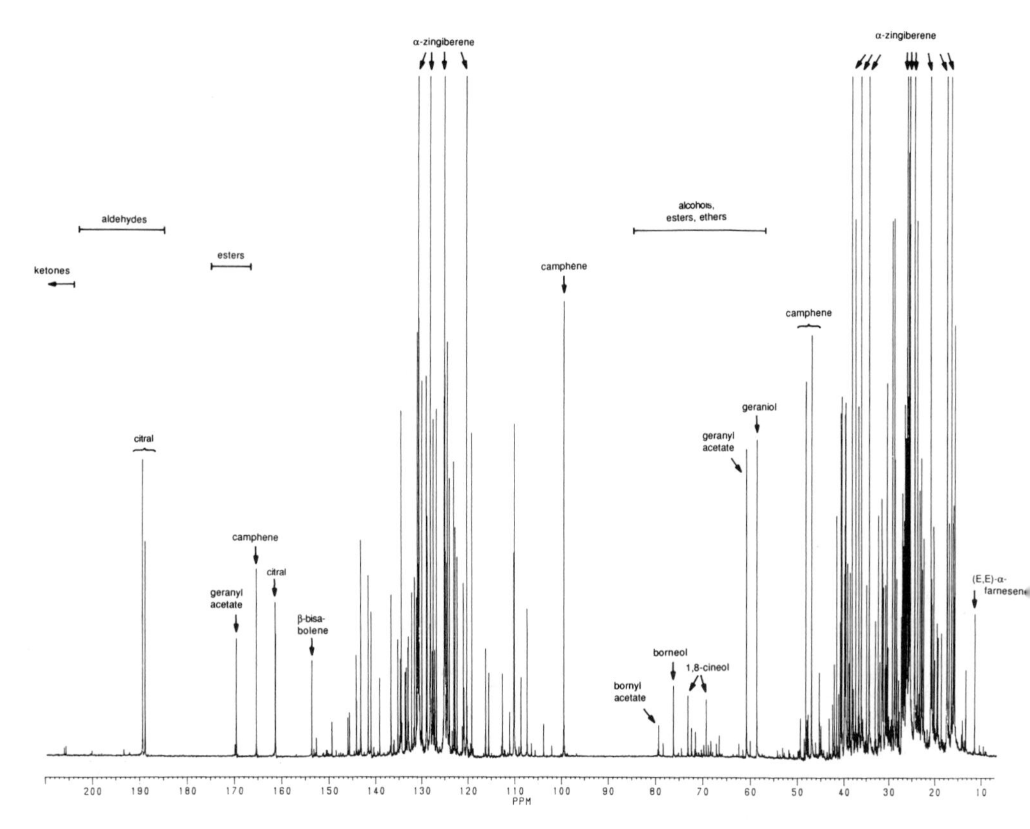

Fig. 7. CNMR of a "mixed" type of ginger oil in C_6D_6 containing MTHC, citral, geraniol, geranyl acetate, 1,8-cineol, and SQHC

of their MS data they were tentatively identified as selina-4(14), 11-diene (7-*epi*-β-selinene) 9, and γ_2-cadinene 10. The shifts at 104.2, 121.2, 134.0, and 153.0 could belong to γ_2-cadinene. In Fig. 8 the peak at 121.2 ppm is assigned to limonene, but actually the peak is a little too big for limonene only. Biougne et al. (1987) give shifts of 104.7, 120.3, 132.7, and 152.6 ppm in $CDCl_3$ for this compound. The differences could be the result of the fact that our spectrum was recorded in C_6D_6. The smaller signals at 105.8, 111.3, and 150.8 ppm could belong to selina-4(14), 11-diene. The values are characteristic for double bonds of the sort present in this compound.

Fig. 8. Expanded aromatic region of the CNMR spectrum of "hydro-carbon"-type ginger oil in C_6D_6. Assignments: *ac ar*-curcumene; *α*f (E,E)-α-farnesene; *αp* α-pinene; *αt* α-terpineol; *αz* α-zingiberene; *βb* β-bisabolene; *βe* β-elemene; *βp* β-phellandrene; *βs* β-sesquiphellandrene; *ca* camphene; *el* elemol; *ga* geranial; *gd* germacrene-D; *li* limonene; *ll* linalool; *mh* 6-methyl-5-heptene-2-one; *my* myrcene; *ne* neral; *φd* C_6D_6

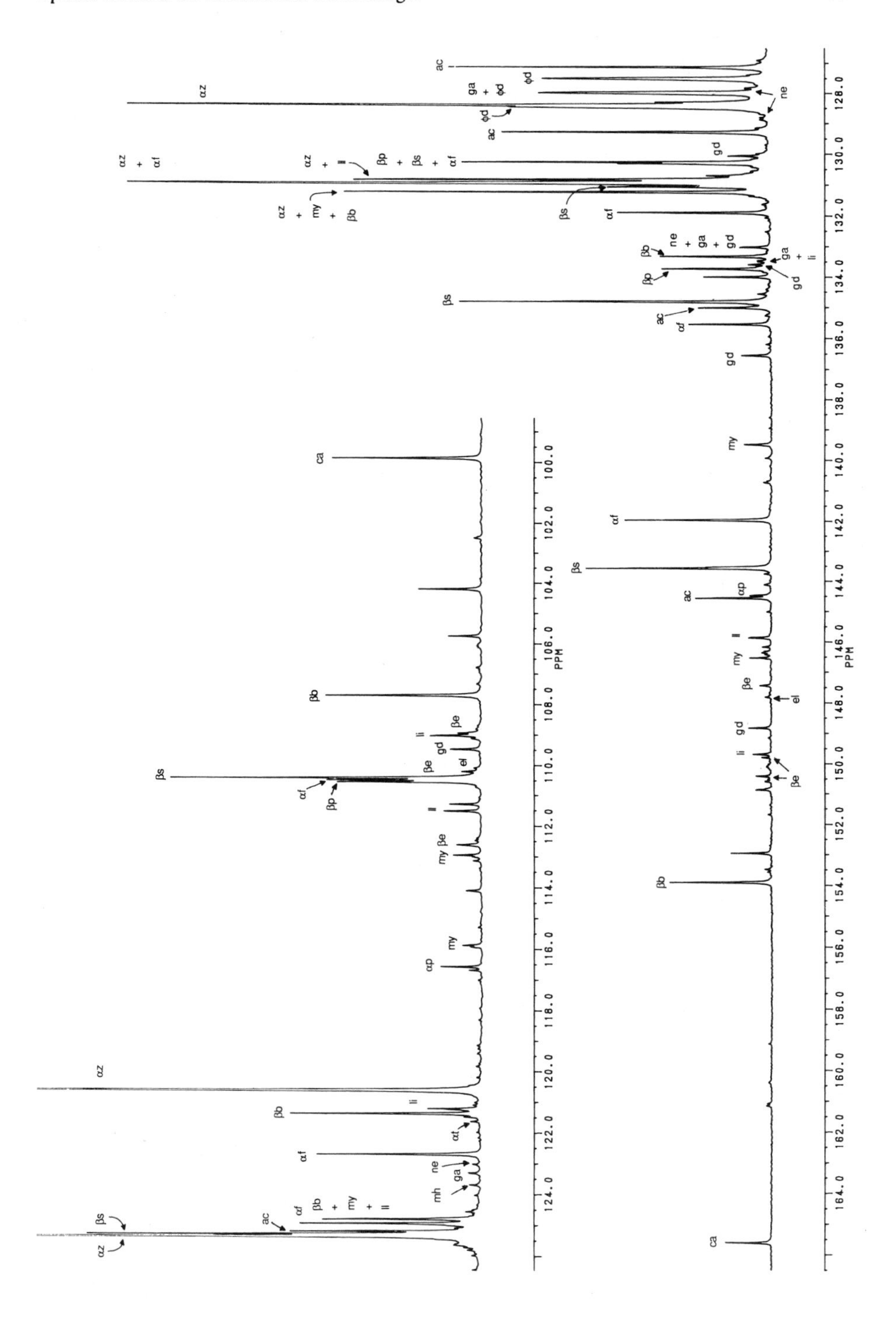
100.0
102.0
104.0
106.0
PPM
108.0
110.0
112.0
114.0
116.0
118.0
120.0
122.0
124.0
128.0
130.0
132.0
134.0
136.0
138.0
140.0
142.0
144.0
146.0
PPM
148.0
150.0
152.0
154.0
156.0
158.0
160.0
162.0
164.0

The above case nicely illustrates the strength and weakness of CNMR. It can detect facts which remain unobserved by routine GC-MS studies, but it is difficult to deduce what kind of structure is exactly present.

Structure 9-10

selina-4(14),11-diene

γ_2-cadinene

References

van Beek TA, Lelyveld GP (1990) Isolation and identification of the five major sesquiterpene hydrocarbons of ginger. Phytochem Anal (in press)

van Beek TA, Posthumus MA, Lelyveld GP, Phiet HV, Yen BT (1987) Investigation of the essential oil of Vietnamese ginger. Phytochemistry 26:3005–3010

Biougne J, Chalchat J-C, Garry R-P, Michet A (1987) Caractérisation des cadinènes dans les huiles essentielles. Parfums Cosmét Arom 73:59–61

Chen C-C, Ho C-T (1988) Gas chromatographic analysis of volatile components of ginger oil (*Zingiber officinale* Roscoe) extracted with liquid carbon dioxide. J Agric Food Chem 36:322–328

Chen Y, Li Z, Xue D, Qi L (1987) Determination of volatile constituents of Chinese medicinal herbs by direct vaporization capillary gas chromatography/mass spectrometry. Anal Chem 59:744–748

De Pooter HL, Coolsaet BA, Dirinck PJ, Schamp NM (1985) GLC of the headspace after concentration on tenax GC and of the essential oils of apples, fresh celery, fresh lovage, honeysuckle and ginger powder. In: Baerheim Svendsen A, Scheffer JJC (eds) Essential oils and aromatic plants. Nijhoff/Junk, Dordrecht, p 1

Ekundayo O, Laakso I, Hiltunen R (1988) Composition of ginger (*Zingiber officinale* Roscoe) volatile oils from Nigeria. Flav Fragr J 3:85–90

Erler J, Vostrowsky O, Strobel H, Knobloch K (1988) Über ätherische Öle des Ingwer, *Zingiber officinalis* Roscoe. Z Lebensm Unters Forsch 186:231–234

Formácek V, Kubeczka K-H (1982) Essential oils analysis by capillary gas chromatography and carbon-13 NMR spectroscopy. Wiley, Chichester

Govindarajan VS (1982) Ginger — chemistry, technology, and quality evaluation: part 2. CRC Crit Rev Food Sci Nutr 17:189–258

Govindarajan VS, Raghuveer KG (1973) Evaluation of spice oils and oleoresins. I. A three-dimensional procedure for thin layer chromatography of essential oils and oleoresins. Lab Pract 22:414–416

Ibrahim H, Zakaria MB (1987) Essential oils from three Malaysian Zingiberaceae species. Malays J Sci 9:73–76

Kubeczka K-H, Formácek V (1984) Application of direct carbon-13 NMR spectroscopy in the analysis of volatiles. In:Schreier P (ed) Analysis of volatiles: methods and applications. de Gruyter, Berlin, p 219

Lawrence BM (1984) Major tropical spices — ginger (*Zingiber officinale* Rosc.). Perf Flav 9:2–40

Lawrence BM (1988) Progress in essential oils, ginger oil. Perf Flav 13:69–74

Lin Z-K, Hua Y-F (1987) Chemical constituents of the essential oil from *Zingiber officinale* Rose, of Sichuan. Youji Huaxue 6:444–448

MacLeod AJ, Pieris NM (1984) Volatile aroma constituents of Sri Lankan ginger. Phytochemistry 23:353–359

McGaw DR, Chang Yen I, Dyal V (1984) The effect of drying conditions on the yield and composition of the essential oil of West Indian ginger. In: Toei R (ed) Proc 4th Int Drying Symp 2:612–615
Miyazawa M, Kameoka H (1988) Volatile flavor components of Zingiberis rhizoma (*Zingiber officinale* Roscoe). Agric Biol Chem 52:2961–2963
Sakamura F (1987) Changes in volatile constituents of *Zingiber officinale* rhizomes during storage and cultivation. Phytochemistry 26:2207–2212
Smith RM, Robinson JM (1981) The essential oil of ginger from Fiji. Phytochemistry 20:203–206

GC-MS (EI, PCI, NCI, SIM) SPECMA Bank Analysis of Volatile Sulfur Compounds In Garlic Essential Oils

G. VERNIN and J. METZGER

1 A Short Survey of the Chemistry of Garlic

The medicinal value of Garlic (*Allium sativum L.)* was well known to the ancient Arab, Jewish, Greek, and Roman civilizations. Garlic also enjoys wide use as an important condiment and flavoring in the various cuisines of the world.

Garlic, believed to be native of the Central Asian Steppes, was probably brought to Europe by the Mongols, and became increasingly popular in the Mediterranean region.

During the Renaissance period, Paracelsus and Ambroise Paré recommended garlic as a preservative of the blood platelets and for protection against the cholera. Garlic is also known as an antiseptic, tonic, bactericide, vermifuge, expectorant, stomachic, and antihypertensive. The antibacterial activity of fresh garlic extracts was first recognized by Pasteur (1858).

Because of its noteworthy biological properties, numerous papers and reviews deal with the isolation, characterization, and identification of its antibacterial principle and the typical obnoxious odor components. The results are summarized in Table 1.

A century ago (1892), Semmler obtained 0.1 to 0.2% of volatile essential oil by steam distillation of fresh garlic cloves, with diallyl disulfide as the main component (80%).

Cavallito et al. (1944, 1945) discovered the unstable 2-propenyl 2-propenethiosulfinate known as allicin. This compound reacts with cysteine to give allylthiocysteine.

Between 1948 and 1950, Stoll and Seebeck identified allicin as the antibacterial principle of *Allium sativum* L. and the precursor of diallyl disulfide.

The presence of the corresponding methyl and propyl derivatives of alliin in garlic was demonstrated by Japanese workers (1953, 1954), who isolated the corresponding dialkyl thiosulfinates of allicin. The configuration of alliin was determined by X-ray diffraction by Hine and Rogers in 1956.

Using labeled sulfate and 35-S-labeled methionine, Suzuki et al. (1961) and Sugii et al (1963) respectively, were able to establish biosynthesis of the amino acids flavor precursors in garlic.

The thermal decomposition of allicin was studied by Brodnitz et al. (1971). The principal components were found to be diallyl mono-, di-, and trisulfides as well as two vinylic cyclic dithiins. At higher temperature, the percentage of these cyclic compounds increased. The structures of 3-vinyl-*6H*-dithiin (A) and 3-vinyl-*4H*-1,2-dithiin (B) were established by ^{1}H-NMR and GC-MS. However, according to Block et al (1986), the isomer A is 2-vinyl-1,3-dithiin (C), and the structural assignments of Brodnitz et al. must be revised.

A B C

The flavor components of garlic were previously studied by Oaks et al. (1964), and then by Zoghbi et al. (1984) and Vernin et al. (1986a).

The garlic essential oils are reported to consist primarily of diallyl, dimethyl, and allylmethyl mono-, di-, and trisulfides, a few minor components containing the n-propyl group, and cyclic mono- and polysulfides. Diallyl disulfide accounts for 30–50% of the total mixture.

Zhihui et al. (1988) investigated the chemical constituents of *Allium sativum* L. cultivated in Yunnan and Quijing (China). Among the 20 identified compounds, five of them have never been reported in garlic oil, namely 6-methyl-1-thia-2,4-cyclopentene, 4-methyl-1,2-dithiacyclopentene, 4-vinyl-1,2,3-trithia-5-cyclohexene, and allylmethyl pentasulfide. One of the main components seems to be 3-vinyl-1,2-(*6H*)-dithiin, but this assignment has to be reexamined.

Table 1. A short survey of the chemistry of garlic

	Authors
1. Antibacterial activity of fresh garlic extracts	Pasteur (1858)
2. First steam distillation of fresh garlic cloves (yield 0.1) to 0.2%) in which diallyl disulfide was the main component	Semmler (1892) cited by Guenther (1952)
3. Discovery of the unstable 2-propenyl-2-propene thiosulfinate (allicin)	Cavallito et al. (1945)
It reacts with cysteine to give allylthiocysteine $R\text{-}S\text{-}S\text{-}CH_2CH(NH_2)COOH$	
4. Antibiotic properties of allicin established	Small et al. (1947, 1949)
5. Isolation of the amino acid flavor precursor of the Allium species [(+)S-allyl-L-cysteine sulfoxide] or alliin	Stoll and Seebeck (1948)
$2\,R\text{-}S(\rightarrow O)\text{-}CH_2CH(NH_2)COOH + H_2O \xrightarrow[\text{pH 5 to 8}]{\text{Alliinase}} R\text{-}S(\rightarrow O)\text{-}SR + 2\,NH_3 + 2\,CH_3COOH$ Alliin → Allicin (with R : $CH_2{=}CH\text{-}CH_2$)	
6. Reaction between thiamine and ingredients of the plants of *Allium* genus. Detection of allylthiamine and its homologs	Matsukawa et al. (1953)
7. Determination of crystal and molecular structure (+)-S-methyl-L-cysteine-S-oxide	Hine and Rogers (1956)
8. The presence of the corresponding methyl and propyl derivatives of alliin in garlic demonstrated. The overall ratio of alkyl residues is ca. 85 : 13 : 2 (allyl, methyl, *n*-propyl)	Matsukawa et al. (1953); Yurugi (1954);
9. Thiosulfinates present in garlic (and onion) react with thiamine to give allylthiamine, which in the presence of cysteine regenerates thiamine and affords allylthiocysteine	Matsukawa et al. (1953)

Table 1. (*continued*)

	Author
10. Isolation of dialkyl thiosulfinates	Fujiwara et al. (1955)
11. Biosynthesis of amino acid flavor precursors in garlic from labeled sulfate and from 35-S-labeled methionine	Sugii et al. (1963); Suzuki et al. (1961)
12. Thermal decomposition of allicin at room temperature gives diallyl disulfide (66%), diallyl sulfide (11%), diallyl trisulfide (9%), and two vinylic cyclic disulfides (dithiines)	Brodnitz et al. (1971)
13. Methylallyl trisulfide as a platelet aggregation inhibitor	Ariga et al. (1981)
14. Ajoene (R-S-CH_2CH = CH-S-S-R with R = allyl) ↓ 0 (E/Z)-4,5,9-trithiadodeca-1,6,11-trien-9-oxide was found to be an inhibitor of platelet aggregation and to possess antithrombotic activity	Block et al. (1984)
15. Two thioacrolein dimers (dithiins) synthesized	Beslin (1983)
16. GC/MS analysis of several garlic essential oils	Vernin et al. (1986a); Zoghbi et al. (1984); Zhihui et al. (1988); Woon et al. (1988)
17. 2-Vinyl-*4H*-1,3-dithiin and 3-vinyl-1,2-dithiin found as potent antithrombotic agents	Block et al. (1986)
18. The two above compounds isolated from *Allium victoralis* L. (Japan) and prepared by decomposition of allicin in methanol at room temperature.	Nishimura et al. (1988)
19. Pyrolysis of diallyl disulfide affords diallyl polysulfides and numerous sulfur-containing-heterocyclic compounds	Block et al. (1988)
20. Quantitative determination of alliin by HPLC	
21. Reviews on garlic chemistry	Abraham et al. (1976); Van Straten et al. (1977); Shankaranarayana et al. (1982); Carson (1987); Block et al. (1988)

Woon et al. (1988) have recently studied the relationship between the sulfur fertilization and the quality and yield of Korean garlic. From the volatile oil obtained by steam distillation of garlic, 14 sulfur components were separated, identified, and quantitatively determined by GC and GC-MS.

Fertilizers SP-10 (10 kg/10 a) and SF-20 (20 kg/10 a) significantly increase the contents of organic sulfur components in the volatile oil and the yield of the cloves. Once again, a confusion between 3-vinyl-1,2(*6H*)-dithiin and 2-vinyl-1,3-dithiin must be noted.

Nishimura et al. (1988) reported the isolation and identification of two isomeric vinyl dithiins in Caucas extracts (*Allium victorialis* L. from Japan). Allicin, which was prepared by oxidation of diallyl disulfide with *m*-chloroperbenzoic acid, underwent decomposition in methanol at room temperature for 7 days to give 3-vinyl-(*4H*)-1,2-dithiin (B) and 2-vinyl-1,3-dithiin (C) as well as diallyl disulfide and diallyl trisulfide. The two isomers were isolated by column chromatography on silica gel and by preparative HPLC. Identification was achieved by IR, NMR (2D ^{1}H, ^{1}H COSY), and mass spectrometry.

By far the most important contributions devoted to garlic essential oils and their biological activity were made by Block et al. (1984–1988). In a recent paper (1988) they describe the pyrolysis of diallyl disulfide at 150 °C. They identified 32 compounds after pyrolysis, most of them being new ones. The same compounds were present in heated samples of commercial garlic oils. In order to explain the formation of these sulfur-containing heterocyclic compounds the authors suggest the following steps:

1. C-S homolysis of diallyl disulfide;
2. addition of the allyl dithio radical in the Markovnikov fashion to diallyl disulfide;
3. intramolecular atom abstraction-fragmentation of the radical adduct to produce thioacrolein; and
4. Diels-Alder addition to diallyl (n) sulfides (with n = 1,2,3).

The fact that thioacrolein works quite nicely as 4 $p\pi$ reactant in the Diels-Alder reaction opens the way to new applications of this reagent in heterocyclic synthesis. The respective mechanisms will be shown in the corresponding sections.

First of all, the goal of this chapter is to recall our previous work dealing with the GC-MS analyses (electron impact and chemical ionisation techniques, PCI and NCI) and mass fragmentometry techniques applied to the various garlic essential oils. We wish also to report the results of a more detailed GC-MS analysis of two essential oils originating from France (Provence) and Mexico, in order to detect minor compounds not previously reported.

2 Analytical Methods Used for Identification of Sulfur Compounds in Garlic Essential Oil

In the first section, analytical methods used for the identification of sulfur-containing compounds will be described. The second part will be devoted to the application of these techniques to the analyses of various garlic essential oils.

2.1 GC-MS (EI, PCI, NCI) of Garlic Essential Oils

The total ion current chromatograms of a Mexican garlic essential oil, obtained both by GC-MS (EI and PCI) are shown in Fig. 1.

The four main products are: diallyl disulfide (32%), diallyl trisulfide (31%), allylmethyl disulfide (7%), and diallyl sulfide (2%). Degradation products apparent as rounded peaks are more abundant upon EI than in PCI. As an example, methylallyl trisulfide present at the level of 14% in the EI mass spectrum is observed only in trace amounts in PCI spectrum, which implies that this compound could result from a reaction of allylmethyl disulfide upon EI. In the case of NCI

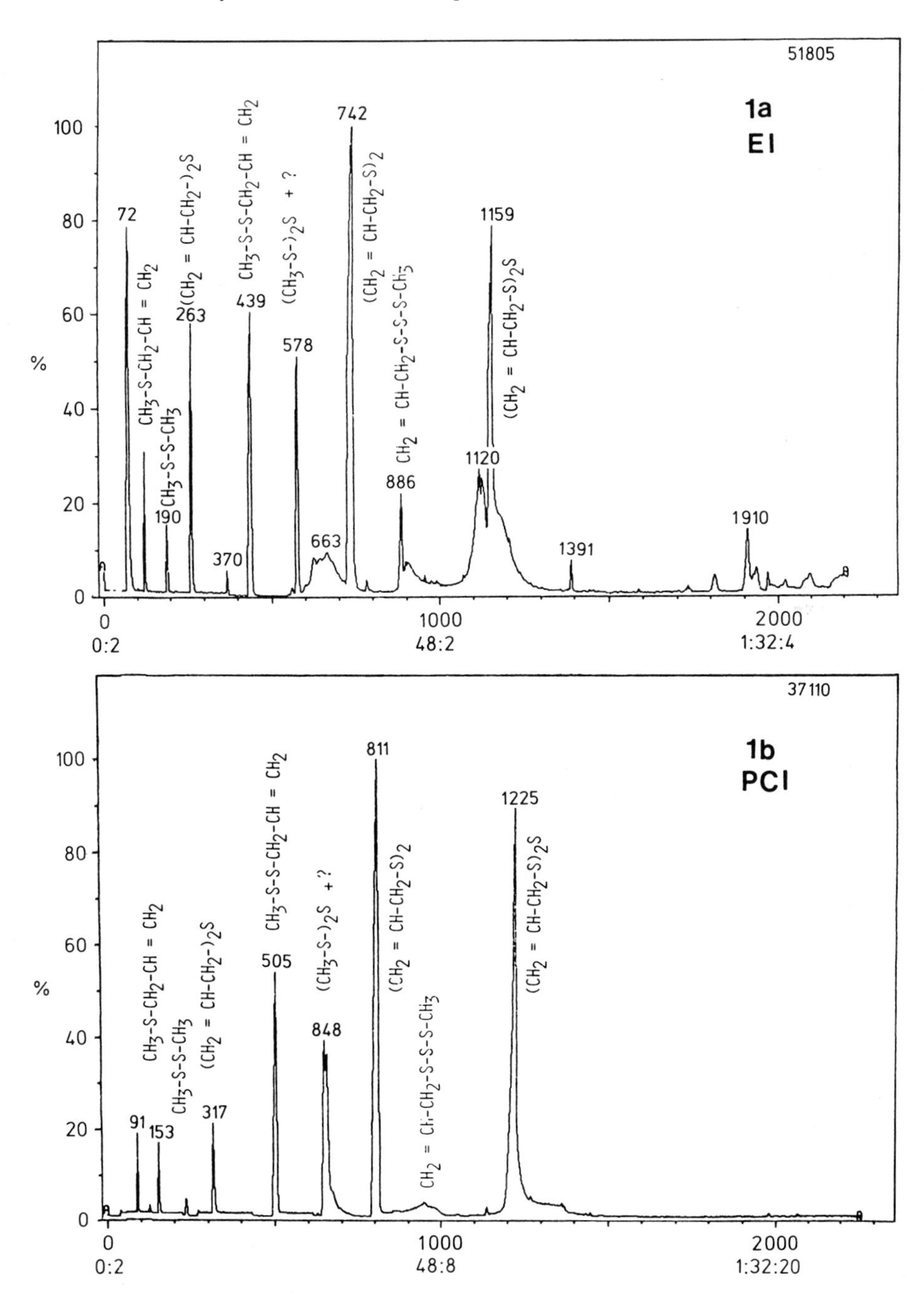

Fig. 1. Total ion current chromatograms of a Mexican garlic essential oil obtained by GC-MS (EI and PCI-isobutane) (FFAP column; VG 70–70F spectrometer)

(OH^-) not reported here, degradation products are even more apparent and a greater amount of diallyl trisulfide, arising from diallyl disulfide, is observed.

Owing to the high specificity and sensitivity of the Selected Ion Monitoring (SIM) mass fragmentometry technique, it was applied to the separation of the major volatile sulfur compounds of garlic essential oils (see Fig. 2).

Thus, in PCI mode, the selected ion at m/z : 115 allowed us to separate the diallyl mono-, di-, and trisulfide which possess this fragment in common. By monitoring the ions at m/z : 121, a selective separation of allylmethyl disulfide can be achieved (M + 1 molecular ion). The same situation arises with monitoring the ions at m/z 147 and 179 for diallyl di- and trisulfide, as shown on the mass fragmentograms.

The use of combined GC-MS with PCI (isobutane) and NCI (OH^-) yields interesting information on the quasi-molecular ions (M ± 1). These techniques have been applied in the analysis of flavors by Bruins (1987). Nevertheless, no information is available on sulfur-containing compounds.

We therefore applied these techniques to some aliphatic and cyclic sulfides of garlic essential oils. In the PCI mode, the quasi-molecular ion (M + 1) is the base peak for all compounds. The adduct cation $(M + i\text{-}C_4H_9)^+$ is also observed in most cases (see Fig. 3).

Upon NCI with hydroxyl ions as reactants, a different behavior was observed.

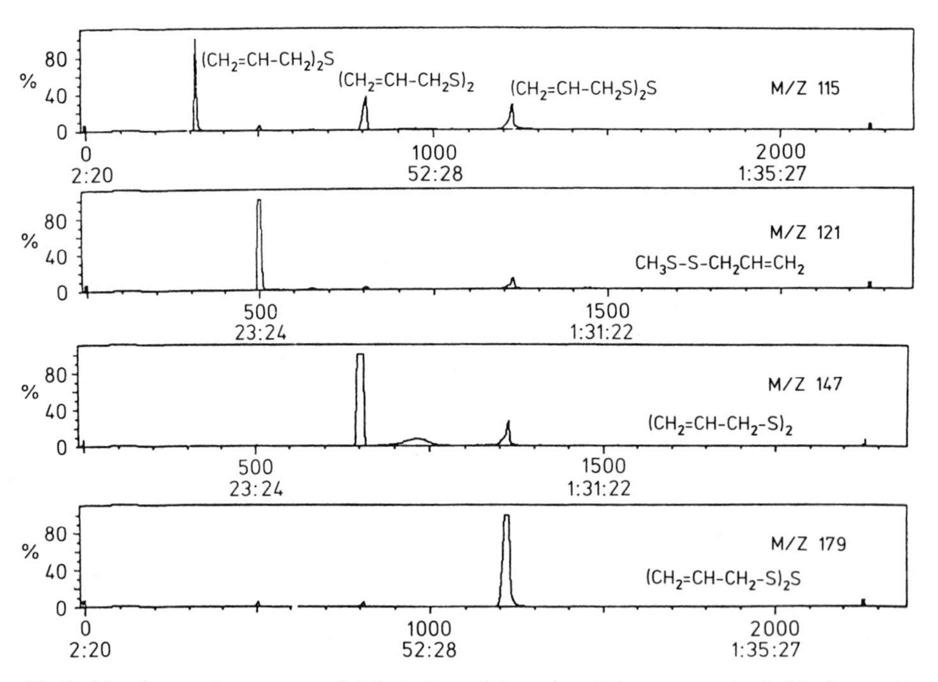

Fig. 2. Mass fragmentograms upon PCI/isobutane of the main sulfide components of a Mexican garlic essential oil

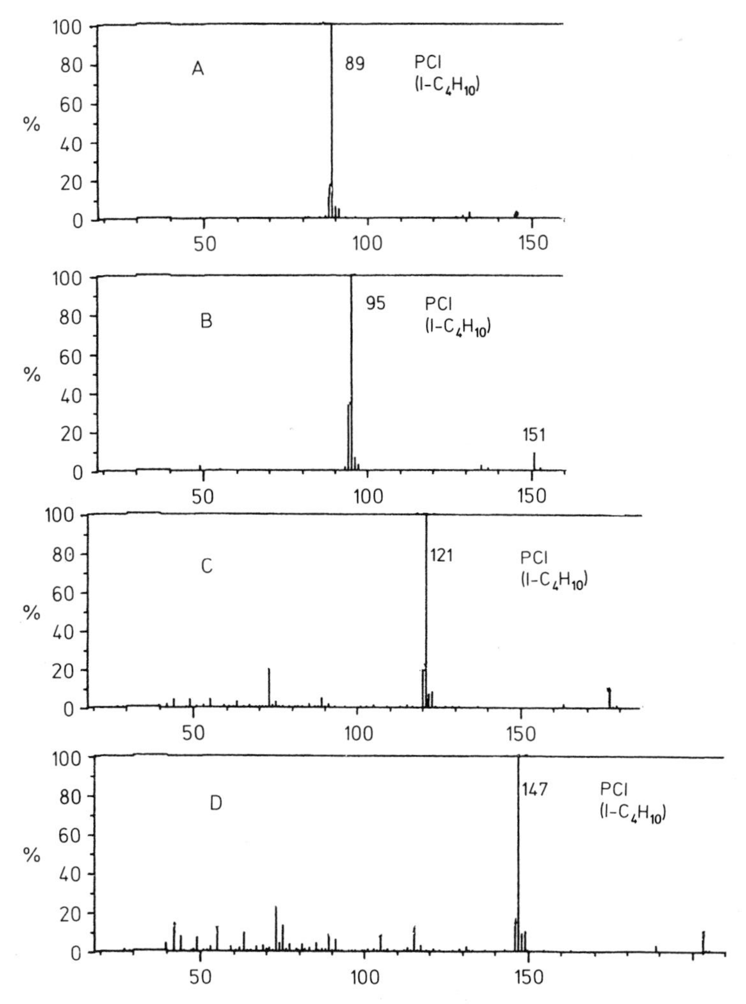

Fig. 3. Mass spectra upon PCI/isobutane of allylmethyl sulfide (**A**), dimethyl disulfide (**B**), allylmethyl disulfide (**C**), and diallyl disulfide (**D**)

With allyldimethyl sulfide (see Fig. 4) a deprotonated ion at m/z 87, as well as a predominant ion at m/z : 47 (CH_3S^-) were observed.

Dimethyl sulfide yields (NCI) weak ions at m/z : 93 and 78 corresponding to the quasi-molecular anion $(M - 1)^-$ and to the loss of a methyl group from the former, respectively.

Two predominant ions at m/z : 46 and 47 arise from the S - S homolytic breakdown of the $(M - 1)^-$ molecular anion (Fig. 5).

For diallyl disulfide, the most abundant constituent of garlic essential oil, the quasi-molecular ion $(M - 1)^-$ anion is observed (as well as for allylmethyl disulfide and diallyl trisulfide). Upon NCI, the more intense peak at m/z : 105 corresponds

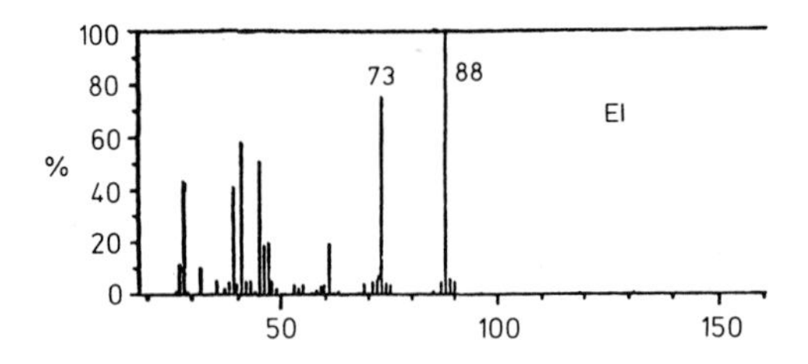

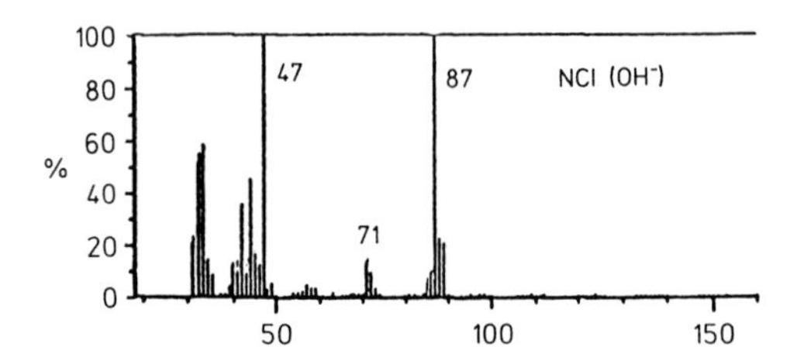

$CH_3\text{-}S\text{-}CH\text{-}CH{=}CH_2 \longrightarrow CH_3\text{-}S\text{-}CH{=}CH\text{-}CH_2\!:^-$

$^-HO\,H$ $(M-1) = \boxed{87}$

$CH_2{=}CH\text{-}CH_2\text{-}S\text{-}CH_3 \longrightarrow CH_2{=}CH\text{-}CH_2\text{-}OH + CH_3\text{-}S^- \ \boxed{(47)}$

HO^-

Fig. 4. Mass spectra (EI, NCI) of allylmethyl sulfide and the formation of ions at m/z = 87, 47, upon NCI

to the loss of allyl alcohol from the anionic molecule or of allyl carbene from the quasi-molecular ion. The base peak at m/z : 72 arises from the homolytic cleavage of the S - S bond in the quasi-molecular ion leading to an allylthio radical and a radical anion which cyclizes to give the more stable mesomeric cyclic form (see Fig. 6).

Mass spectra (EI, NCI) of allylmethyl disulfide and diallyl trisulfide are shown in Fig. 7a,b.

The last example deals with the isomeric vinyl dithiins which will be treated in more detail in the next section. They are present in trace amounts in garlic essential oils.

In Fig. 8a, mass spectra (both PCI and NCI) of 3-vinyl-1,2-(*4H*) dithiin are reported and in Fig. 8b those of 2-vinyl-1,3-dithiin are shown.

2.2 Kováts Indices as Filters and Their Properties

For many years, the Kováts indices (1958) have been used for identification purposes. They possess several interesting properties which are important in the analysis:

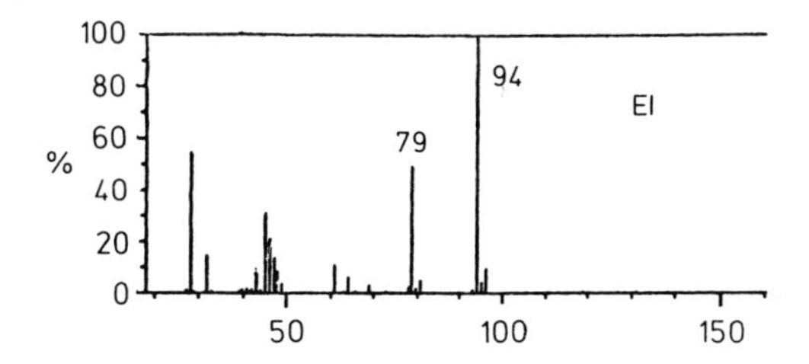

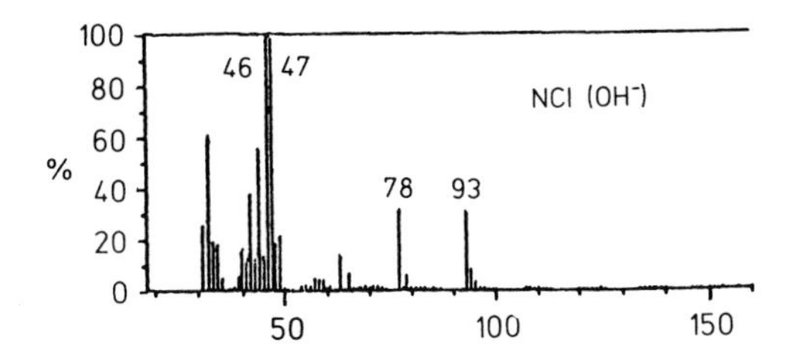

$CH_3\text{-}S\text{-}S\text{-}CH_2\text{-}H + {}^-OH \longrightarrow CH_3\text{-}S\text{-}S\text{-}CH_2{:}^- (M-1)^- = \boxed{93}$

$H_3C\text{-}S\text{-}S\text{-}CH_2{:}^- \longrightarrow {}^{\bullet}S\text{-}S\text{-}CH_2{:}^- \ \boxed{(78)}$

$CH_3\text{-}S\text{-}S\text{-}CH_2^- \longrightarrow CH_3\text{-}S{:}^- = \boxed{47}$ $\quad CH_3\text{-}S\text{-}S\text{-}CH_2{:}^- \longrightarrow S{=}CH_2^{\bullet -} = \boxed{46}$

Fig. 5. Mass spectra (EI, NCI) of dimethyl disulfide and the formation of ions at m/z = 93, 78, 47, and 46, upon NCI

— For a given column, they are reproducible, in contrast to other retention data (especially retention times), which depend on temperature.

— They possess additivity properties.

On the basis of the Kováts index of a parent molecule, it is possible to calculate the indices of its derivatives, by using the substituent increments:

$$KI[(PR(n)] = KI(PH) + \sum \Delta KI(R_i),$$

where $[PR(n)]$ indicates a parent molecule (PH) substituted by different or similar R groups; i.e., for the homologous series (ethers, sulfides, acetals, ketones, alcohols, esters, γ- and δ-lactones, etc.) which differ only by a methylene group, the addition of 100 units to the KI of the lower homologs led us to calculate the KI of the next compound.

— They also allow us to differentiate between compounds possessing similar mass spectra but with different retention times (homologous series, stereoisomers, position isomers).

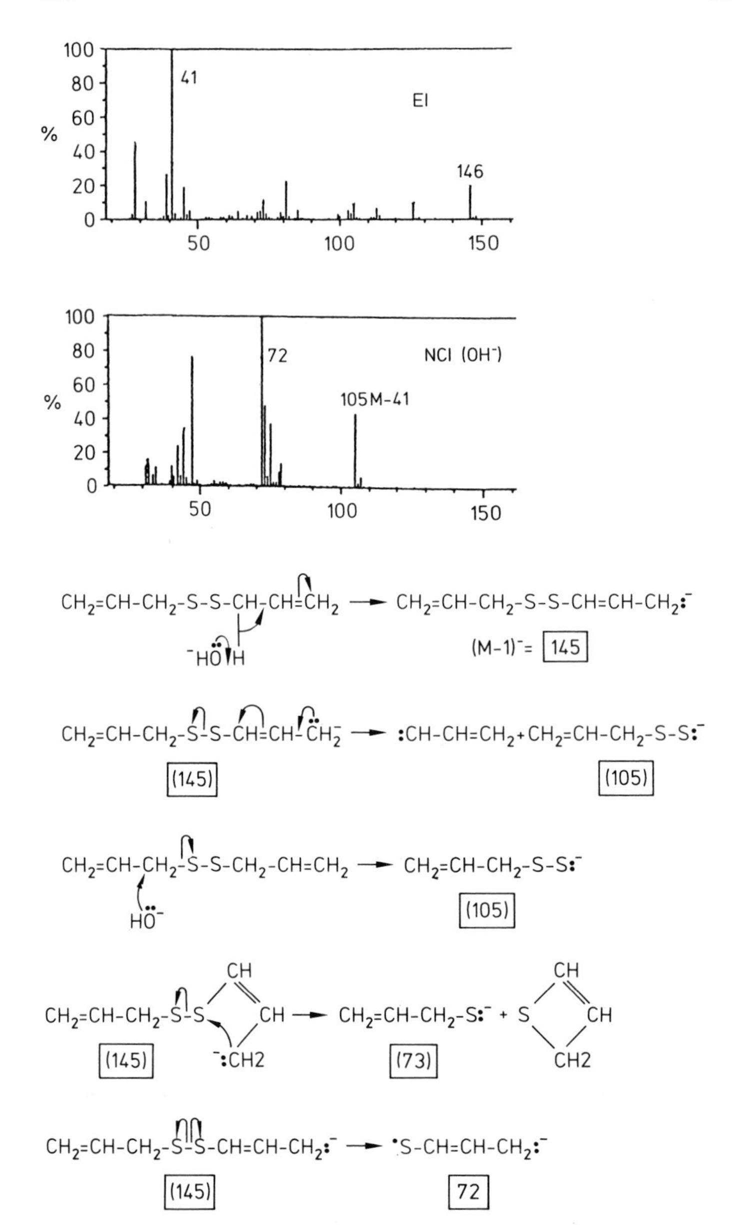

Fig. 6. Mass spectra of diallyl disulfide and the formation of ions at m/z = 145, 105, 73, and 72, upon NCI(OH⁻)

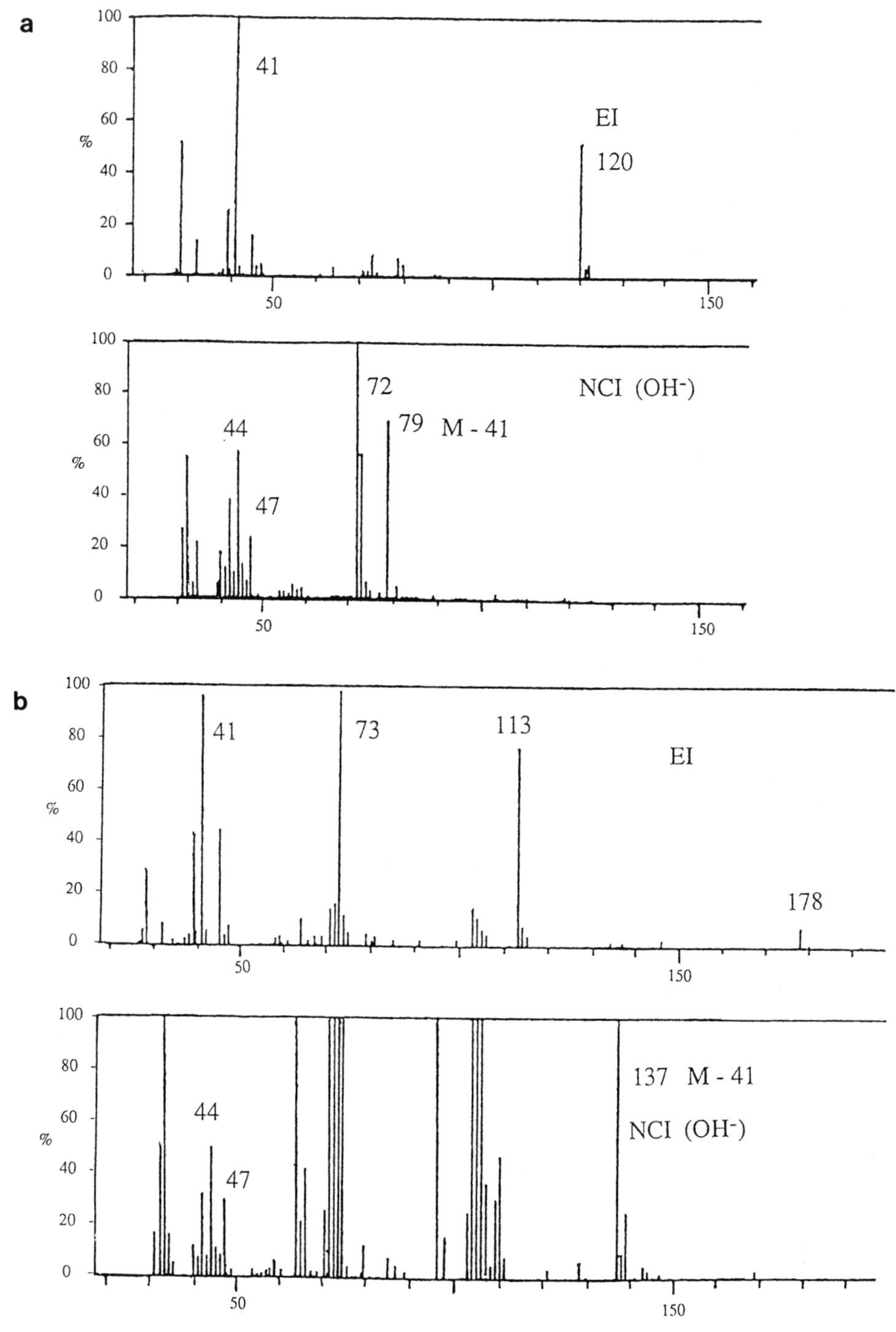

Fig. 7. **a** Mass spectra (EI, NCI) of allylmethyl disulfide. **b** Mass spectra (EI, NCI) of diallyl trisulfide

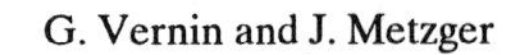

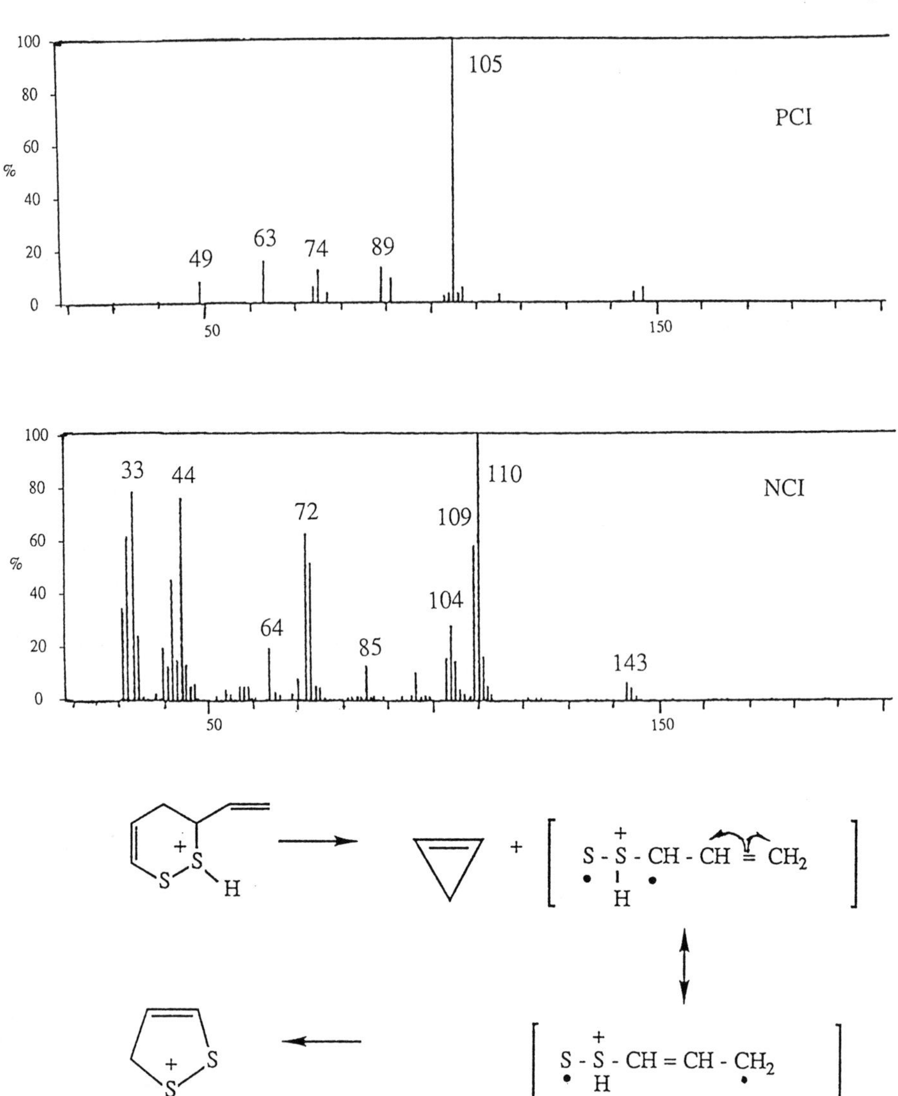

m/z 105 (100%)

Fig. 8a. Mass spectra (PCI, NCI) of 3-vinyl-1,2(4H)-dithiin. Suggested formation of the predominant ion at m/z = 105 upon PCI

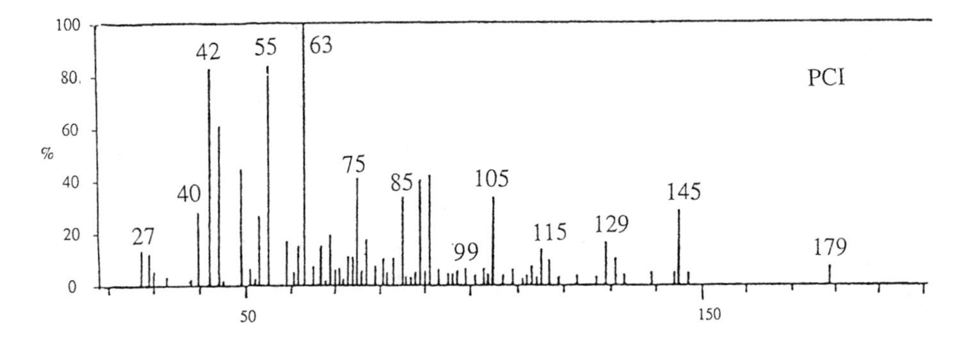

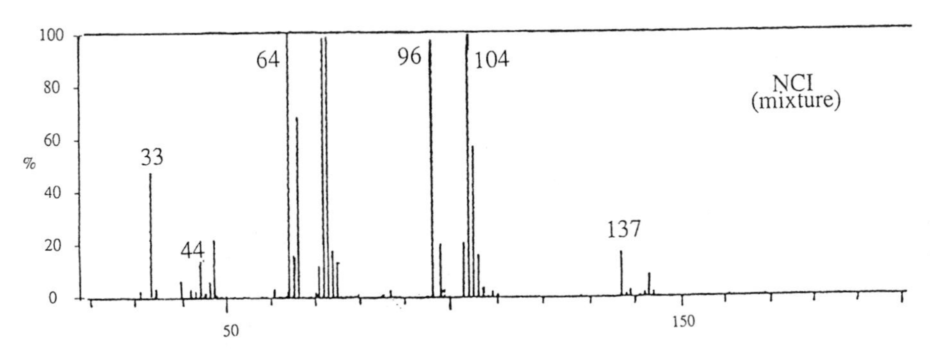

Fig. 8b. Mass spectra (PCI, NCI) of 2-vinyl-1,3-dithiin

— These indices are characteristic of a particular molecule. Furthermore, the Kováts index difference (KID) between polar (Carbowax 20M, FFAP) and apolar (OV-1) columns is characteristic of a given category of products.

Examples:

Compounds	DIK = KIP - KIA	KIP[a]	KIA[a]	DIK
Dialkyl sulfides	210 ± 30	760	520	240
Dialkyl disulfides	260 ± 50	1060	750	310
Dialkyl trisulfides	420 ± 30	1400	950	450
Alkylthiophenes	300 ± 20	1160	860	300

[a] For dimethyl derivatives

Kováts indices are calculated using the retention times of either the linear alkane series or linear ethyl esters having and even number of carbon atoms. In the programmed temperature mode, the formula proposed by Van Den Dool and Kratz (1963) can be used:

$$KI = 100\left[n + \frac{t'_{R(x)} - t'_{R(n)}}{t'_{R(n+1)} - t'_{R(n)}}\right],$$

where n is the number of carbon atoms in the alkane or in the acid chain of the ester which is eluted immediately before the compounds X and $t'_{R(x)}$, $t'_{R(n)}$ and $t'_{R(n+1)}$ are the reduced retention times of the compound X and the alkanes (or esters) with n and (n+1) carbon atoms, respectively. This formula is an approximation of a logarithmic formula.

This method requires addition of a reference to the mixture under study (alkanes or ethyl esters). In order to avoid this drawback in GC-MS, the scan numbers, which are proportional to the retention times, have been used. However, in the listing of the mass spectra, it is necessary to identify at least some compounds for which the Kováts indices are known (Jennings and Shibamoto 1980). A quasi-linear relationship exists between the scan numbers and the KI on the column used. With this in mind, a computer program called MBASIC.SCAN1 written in BASIC was prepared in our laboratory in order to derive this equation in a rapid fashion (Boniface et al. 1987). The main features of this program are shown in Table 2.

The menu contains three options:

1. linear equation search,
2. linear equation loading,
3. equation files printing

Table 2. The program MBASIC.SCAN1 and its Various Options

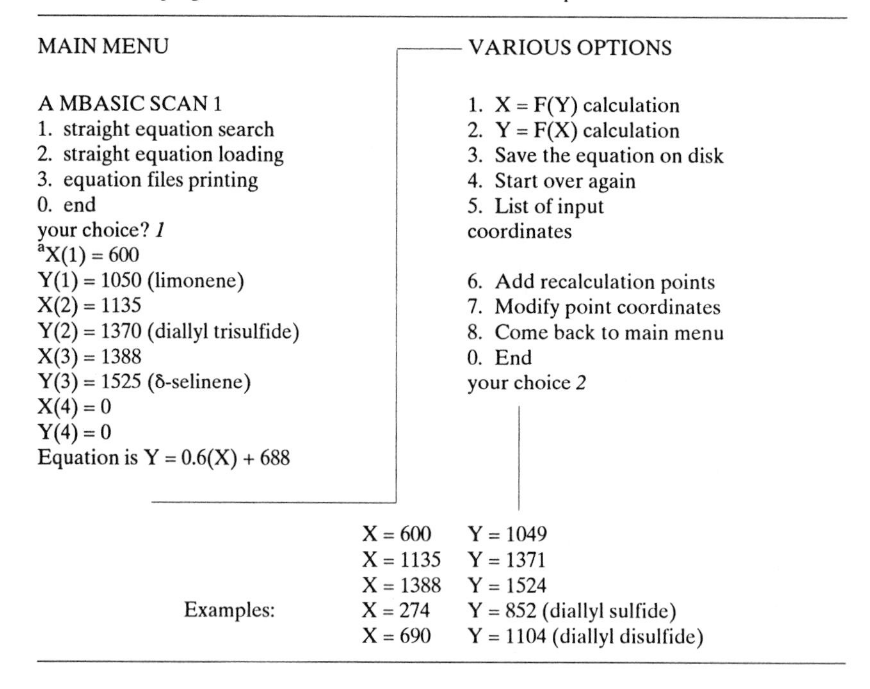

MAIN MENU	VARIOUS OPTIONS
A MBASIC SCAN 1	1. X = F(Y) calculation
1. straight equation search	2. Y = F(X) calculation
2. straight equation loading	3. Save the equation on disk
3. equation files printing	4. Start over again
0. end	5. List of input coordinates
your choice? *1*	
X(1) = 600 [a]	
Y(1) = 1050 (limonene)	6. Add recalculation points
X(2) = 1135	7. Modify point coordinates
Y(2) = 1370 (diallyl trisulfide)	8. Come back to main menu
X(3) = 1388	0. End
Y(3) = 1525 (δ-selinene)	your choice *2*
X(4) = 0	
Y(4) = 0	
Equation is Y = 0.6(X) + 688	

Examples:	X	Y
	X = 600	Y = 1049
	X = 1135	Y = 1371
	X = 1388	Y = 1524
	X = 274	Y = 852 (diallyl sulfide)
	X = 690	Y = 1104 (diallyl disulfide)

[a] X = Scans; Y = KIA (DB-1 column)

Let us choose the option (1) and provide the coordinates (scans and Kováts indices on the DB1 apolar column) for the three selected compounds : limonene (an artifact), diallyl trisulfide, and δ-selinene (an artifact). The linear equation is immediately given and the program prints the other various options. Option (2) allows us to calculate all other Kováts indices from the corresponding scans. The calculated values for the three compounds are identical to the experimental ones. As an example, the results for diallyl sulfide and disulfide are given.

The additivity of the Kováts indices previously mentioned was applied to the calculation of the KI values for volatile sulfur-containing compounds of garlic and onion for the most common groups of substances. In the case of heterocyclic sulfur-containing compounds, the problem is more complicated. Table 3 contains some of these values for polar and apolar columns.

As an example, the calculated KIA and KIP values for diallyl trisulfide, (Z)-allyl-1-propenyl sulfide, and allylpropyl trisulfide are in good agreement with experimental results. Furthermore, the method is of special interest for compounds belonging to the homologous series.

Table 3. Increments groups used for Kovts indices calculation for aliphatic (n) sulfide derivatives

Groups	KIA (Apolar)	KIP (Polar)	ΔKI(X)
CH_3	100	100	0
C_2H_5	200	200	0
C_3H_7	300	300	0
CH_2=CH-CH_2	280	330	50
(Z)-CH_3-CH=CH-	295	370	75
(E)-CH_3-CH=CH	310	410	75
-S-	290	500	210
-$(S)_2$-	560	840	280
-$(S)_3$-	775	1130	335
-$(S)_4$-	940	1375	435
-$(S)_5$-	1100	1620	520

Example : Comparison between experimental and calculated KI values for diallyl trisulfide and allyl-1-propenyl sulfide

Compounds	KIA		KIP	
	Exp.	Calc.	Exp.	Calc.
$(CH_2$=CH-$CH_2S)_2S$	1356	1325	1730	1790
(Z)-CH_2=CH-CH_2-S-CH=CH-CH_3	865	850	1195	1200
CH_2=CH-CH_2-$(S)_3$-$CH_2CH_2CH_3$	1363	1355	—	—

2.3 The SPECMA Bank

At present, MS data banks available commercially with mass spectrometers provide very little information about sulfur-containing aroma compounds. While 45 mass spectra were recorded with a Mexican garlic essential oil, only three of them have been identified by the EPA/NIH data bank : dimethyl di- and trisulfide and diallyl trisulfide; limonene and dibutyl phthalate are artifacts, and methylthiirane is actually allyl mercaptan (the listing is not reported here).

In order to overcome this difficulty, starting from 1982 we have developed our own data bank specialized for the identification of volatile aroma compounds in flavors and fragrances. The overall guiding principles for the bank design and the various stages which led to its realization have been published in several books and reviews (Petitjean et al. 1982, 1983; Vernin et al. 1983, 1986b, 1987). For more details, the reader can refer to these reviews.

The principle of the program originally written in PL1/CPM and then in TURBO PASCAL is shown in Fig. 9.

After the reading of data including the formula, molecular weight, limit damping mass, mass spectra, and Kováts indices, the program selects the area to be read. If a compound is not rejected by a filter, it is submitted to other filters and if it is accepted by all filters, it calculates the dissimilarity index (DI) between the unknown spectrum and that present in the bank. Above a threshold value mathematically fixed at 25–50, the program prints : "Unknown in the file". In an opposite case, it prints the identified compounds, in a sequence determined by increasing DI, and draws the corresponding spectra. In order to increase the likelihood of finding the compound, it is necessary to store several different mass spectra of a same compound in the memory of the computer.

Example: Search for an unknown product

The program has three main options (Table 4):

1. Search of unknown spectra (S);
2. Modification of the bank content (M);
3. Utility programs (U).

In the present case, we choose the S option. The program then requests the molecular weight range. We put in : 150–200 (or 178–178 if the molecular weight is known). After "INPUT MS PEAKS", peaks of the experimental mass spectrum (upon EI) are entered with their intensities relative to the base peak taken as 100, in any order. It is possible to correct an error at any time by re-entering the peak requiring a correction.

The unknown spectrum must be entered as completely as possible in order to decrease the number of possible answers. However, in order to avoid the ground noise and impurities, fragments with intensities below 10% are not taken into account.

If Kováts indices are known, they are also entered, The index difference (KID), being characteristic of a given group of compounds, may also be incorporated.

The computer then lists all the input data and, after a few seconds, it gives the list of possible compounds corresponding both to the experimental mass spectrum

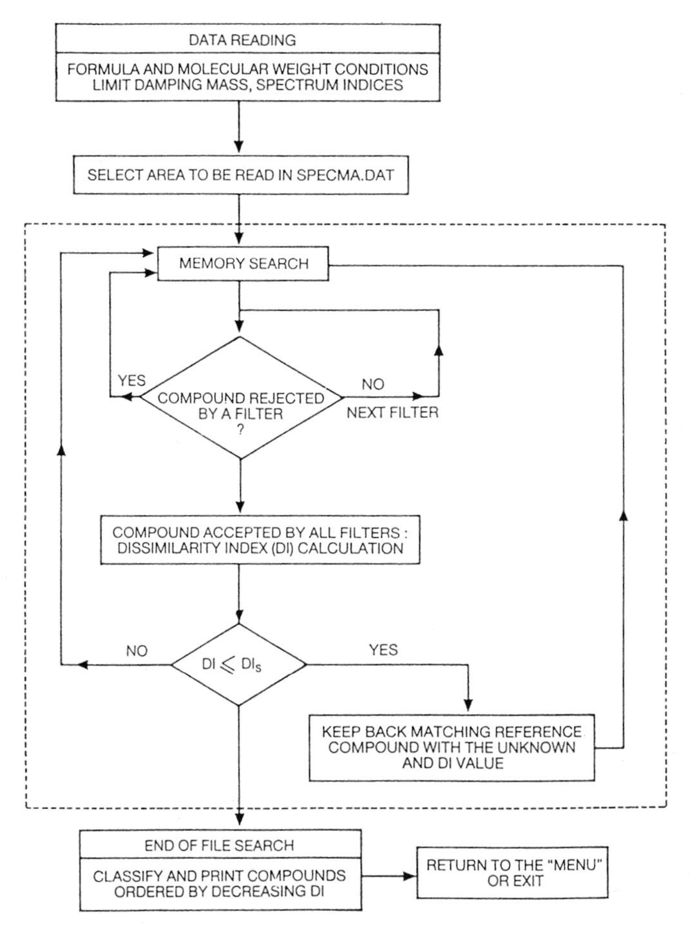

Fig. 9. Search procedure flowchart of the SPECMA bank (MS and KI). (Vernin et al. 1986b)

and Kováts indices in an order of increasing dissimilarity index. In the present case, diallyl trisulfide has been found with an excellent DI : 7–18. Using a revised version of the SPECMA bank, it is now possible to visually compare the sketch of an unknown spectrum with those in the memory of the computer (Fig. 10).

The various possibilities of the Utility commands are illustrated in Table 5.

Among all the options listed we choose the option 8 : the isotopic abundance calculation is of special interest for sulfur-containing compounds.

In Table 6 we compare the experimental and calculated values of the ratio (P + 2)/P for various sulfur-containing compounds of garlic.

The agreement between these values can act as another filter.

Table 4. Search example of an unknown compound by the SPECMA bank

S : Search of unknown spectra		
M : Modification of bank content		
U : Utility program		
E : End of the program		
Choose : (S, M, U, E) ?........................		S
Molar Formula Conditions : Y)es or N)o.........		
Input Molar Mass Range (Min Max):		150–200
Input Damping Limit Mass of Scanning?..........		40
Input MS peaks:		
Peak 1 : Mass? Intensity?		73–100
Peak 2 : Mass? Intensity?		41–95
Peak 3 : Mass? Intensity?		113–80
Peak 4 : Mass? Intensity?		45–40
Peak 5 : Mass? Intensity?		78–10
Input Kovats Indices:		
KI apolar (SE 30)..............................		1335
KI polar (Carbowax 20 M)		1710
Proposed Spectrum	M/E	I%
	178.0	10
	113.0	80
	73.0	100
	45.0	40
	41.0	95
Proposed KI	KIA	1335
	KIP	1710

Possible Compounds : 1			
Compound Number : 364			
Diallyltrisulfide		FMO	CEO
$C_6H_{10}S_3$ = 178			
KIA = 1300			
KIP = 1700			
Damping Mass = 40; m/e 28 and 32 absent;			
73 (100)			
113 (78)	103(15)	45(43)	41(95)
D.I. = 7/18			

3 GC-MS Analyses of Two Garlic Essential Oils Originating from France (Provence) and Mexico

3.1 Analyses and Composition

GC-MS analyses of garlic essential oils originating from France (Provence), Egypt, Mexico, Turkey, and China have been previously reported on a polar column (FFAP) but several minor compounds were unidentified (Vernin et al. 1986,1986a) In order to remedy to this deficiency, we reexamined in more detail

Nom : DIALLYL TRISULFIDE

Formule brute : C6 H10 S3 **(P.M. = 178) RN = 2050-87-5**

1ère PROPOSITION

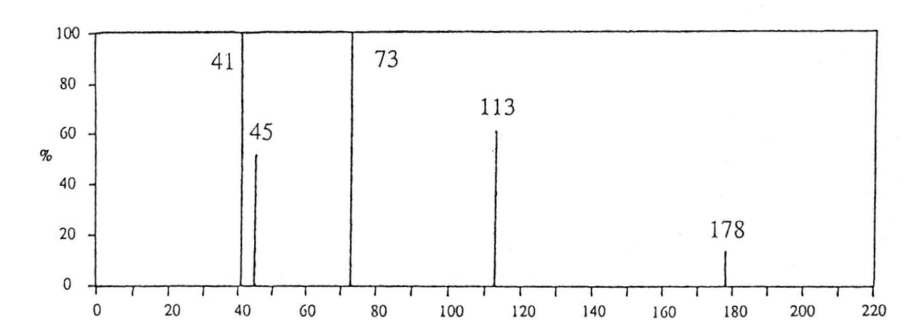

***** **DI = 918** *****

***** **DI -IKa = - 2.69%** *****

***** **DI -IKp = -0.58%** *****

Enregistrement N° 832

INCONNU

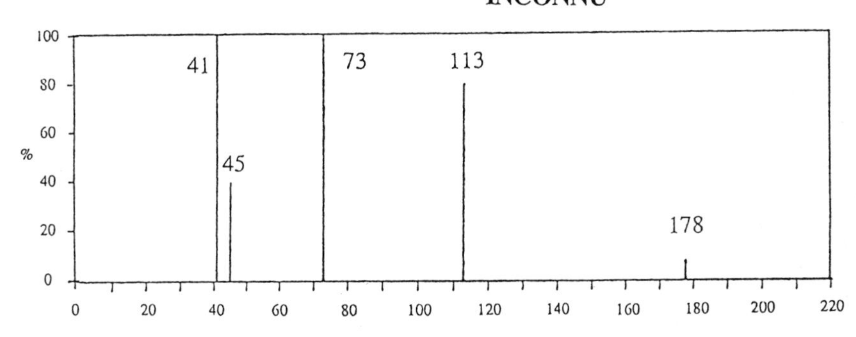

Fig. 10. Visual comparison between the unknown mass spectrum and that in the file

two of them, the first from a Mexican commercial sample and the second from France, using capillary DB1 and CPSIL 5CB apolar columns, respectively.

French garlic essential oil was obtained by steam distillation of crushed fresh garlic (1.2 kg) purchased at a local market in Marseilles.

The reconstructed chromatogram of the Mexican garlic essential oil is shown in Fig. 11.

If we compare it with that obtained on polar column (see Fig. 1a), we do not observe any rounded peaks after diallyl trisulfide. On the other hand, the compositon appears to be quite different on the two columns : allylmethyl di-, tri-, and tetrasulfides and diallyl tetrasulfide appear to be more abundant on the apolar column.

Table 5. Utility programs of the SPECMA bank (U option)

S : Search of unknown spectra
M : Modification of bank content
* U : Utility programs
E : End of the program

Choose : (S, M, U, E) ? *U*

0 : Catalog of compounds having the same: molar mass
1 : Catalog of compounds having the same: molar formula
2 : Catalog of compounds having the same: "registry number"
3 : Catalog of compounds having the same: reference
4 : Catalog of compounds having the same: Kovats indices
5 : Catalog of compounds having the same: name
6 : Spectrum drawing
7 : Reading of a recording
8 : Isotopic abundance calculation
9 : Exit of utility programs

Choose: (0, 1, 2, 3, 4, 5, 6, 7, 8, 9) ?

Table 6. Isotopic abundance calculation for (n) sulfide components of garlic essential oils

Compounds	E.F.	KIA	Percentage	
			(P + 2)/P Exp.	At low resolution Calc.
Allyl mercaptan	C_3H_6S	645	4.3	4.5
Allylmethyl sulfide	C_4H_8S	680	4.7	4.56
Dimethyl sulfide	$C_2H_6S_2$	710	11	8.94
Allylthio mercaptan	$C_3H_6S_2$	810	—	8.99
Diallyl sulfide	$C_6H_{10}S$	830	5.3	4.7
Allylmethyl disulfide	$C_4H_8S_2$	880	9.35	9.05
Dithiacyclopentene	$C_3H_3S_2$	910	10	8.98
Dimethyl trisulfide	$C_2H_6S_3$	925	12	13.4
Diallyl disulfide	$C_6H_{10}S_2$	1120	9	9.2
Methylpropyl trisulfide	$C_4H_{10}S_3$	1150	9.45	13.5
Allylmethyl trisulfide	$C_4H_8S_3$	1170	13.54	14.9
Trithiacyclohexene	$C_3H_4S_3$	1195	12.6	13.47
3-Vinyl-1,2-dithi-5-ine	$C_6H_8S_2$	1205	8.35	9.2
Dimethyl tetrasulfide	$C_2H_6S_4$	1230	—	17.9
Diallyl trisulfide	$C_6H_{10}S_3$	1355	—	13.71
Allylhydro tetrasulfide	$C_3H_6S_4$	1405	18	17.96
5-Vinyl-1,2,3,4-tetrathiacyclopentane	$C_3H_4S_4$	1455	15.7	17.96
Allylmethyl tetrasulfide	$C_4H_8S_4$	1460	18.2	18.04
Diallyl tetrasulfide	$C_6H_{10}S_4$	1640	18.95	18.23
Diallyl pentasulfide	$C_6H_{10}S_5$	—	—	22.75
Ajoene	$C_9H_{14}S_3$	—	—	14.28

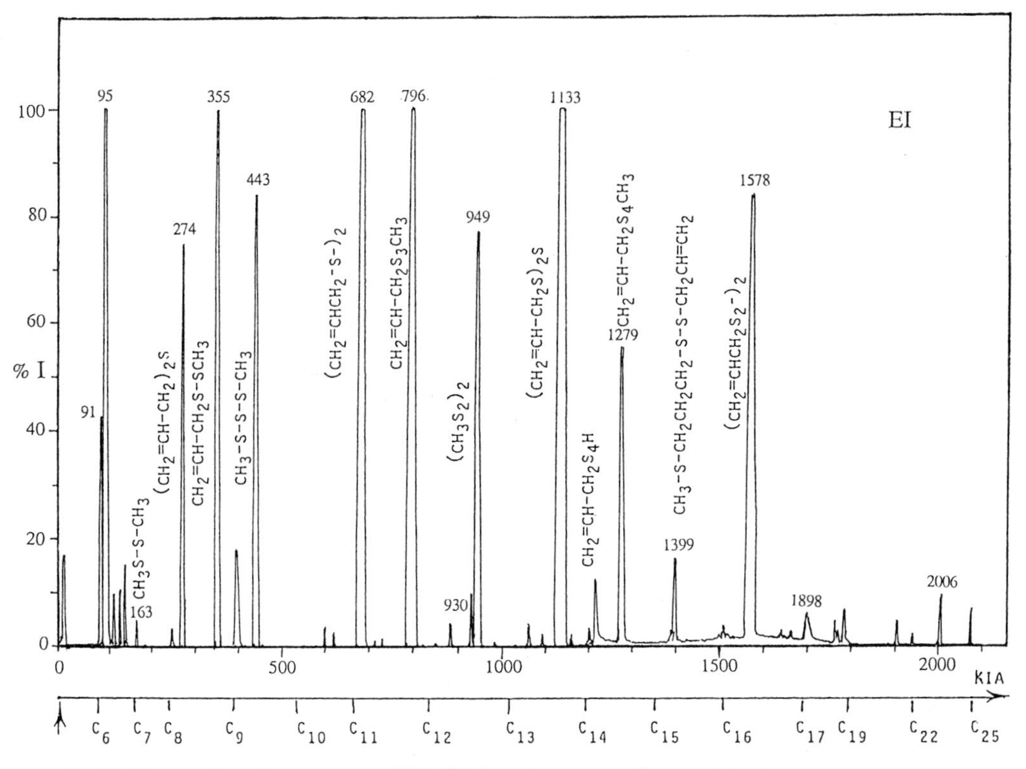

Fig. 11. The total ion chromatogram (TIC, EI) for a Mexican garlic essential oil

The examination of the two listings (France and Mexico) according to analytical methods described before, allows us to draw up the list of identified compounds.

They can be divided into two categories :

The first category includes sulfur-containing aliphatic compounds (Table 7) summarized below:

- RSH with R = CH_2=CH-CH_2; CH_3S-; CH_2=CH-CH_2S-
- $CH_3(S)_nCH_3$ with n = 2, 3, 4, 5
- CH_2=CH-$CH_2(S)_nCH_3$ with n = 1, 2, 3, 4
- CH_2=CH-$CH_2(S)_nCH_2$-CH=CH_2 with n = 1, 2, 3, 4, 5
- CH_2=CH-$CH_2(S)_nCH_2CH_2CH_3$ with n = 1, 2, 3
- CH_2=CH-$CH_2(S)_n$CH=CH-CH_3 with n = 2, 3 (two isomers)
- CH_3CH=CH$(S)_n$CH=CH-CH_3 with n = 3 (two isomers)
- $CH_3SCH_2CH_2CH_2SR$ with R = CH_3, CH_2-CH=CH_2

Most of them are well-known components of garlic essential oils, except mercaptans. A pentyl hydrosulfide (MW : 136, m/z = 71, 103, 55) has been reported as a constituent of chive (*Allium schoenoprasum* L.) by Hashimoto et al. (1983). Other

Table 7. GC-MS Analyses on apolar and polar columns of two garlic essential oils originating from France (Provence) (F) and Mexico (M), respectively. Sulfur-containing aliphatic compounds

Compounds	(1)	KIA[a]	KIP[b]	MW[c]	Main fragments[d] m/z	Occurrence[e] F	M
Allyl mercaptan	620	645	850	74	74,41,39,45	+	+
Methylthio mercaptan	650	670	1130	80	80,65,64	+	+
Allylmethyl sulfide	720	700	960	88	88,73,45,41,39	+	+
Dimethyl disulfide	750	710	1050	94	94,45,39,64,106	+	+
Allylthio mercaptan (dithia cyclopentane)	830	810	—	106	41,39,64,106	+	+
Diallyl sulfide (+ Z-Allyl-1-propenyl sulfide)	836	830	1135	114	45,41,39,73,72,114,99	+	+
(E)-Allyl-1-propenyl sulfide	850	—	—	114	45,41,39,114,99	+	—
Allylmethyl disulfide	870	880	1240	120	41,39,120,45	+	+
(E)-Methyl-1-propenyl disulfide	—	885	—	120	120	—	+
Dimethyl trisulfide	915	940	1330	126	126,79,45,64,111	+	+
Methylthio propenal ?	930	—	—	128	45,43,39,41,99	+	—
Allylpropyl disulfide	—	1065	1440	148	41,43,39,106,148	—	+
Diallyl disulfide	1125	1110	1470	146	39,41,45,81,146,105	+	+
(Z)-Allyl-1-propenyl disulfide	1130	—	—	146	41,39,45,146	+	—
(E)-Allyl-1-propenyl disulfide	1140	—	—	146	41,39,45,146	+	—
Methylpropyl trisulfide	1160	1140	1510	154	43,154,112	+	+
Allylmethyl trisulfide	1165	1180	1565	152	73,87,41,45,39	+	+
(Z)-Methyl-1-propenyl trisulfide	1180	—	1630	152	45,88,73,152	+	—
(E)-Methyl-1-propenyl trisulfide	1185	—	1640	152	45,88,73,152	+	—
Dimethyl tetrasulfide	1230	1250	—	158	158,79,45,47,64	+	+
(Z)-1-Propenylpropyl trisulfide	—	1315	—	180	41,73,43,115,64,180	—	+
Methylthiopropyl methyl disulfide	—	1345	—	(168)	89,41,45,75,61,73	—	+
Diallyl trisulfide	1360	1370	1760	178	41,73,113,45	+	+
Allylpropyl trisulfide	1365	1390	—	180	43,73,115	+	+
(Z)-Allyl-1-propenyl trisulfide	1372	—	—	178	41,45,39,144,73	+	—
(E)-Allyl-1-propenyl trisulfide	1380	—	—	178	41,45,39,114,73	+	—
1-Dipropenyl trisulfide (I) ?	1400	—	—	178	45,41,39,114,134	+	—
1-Dipropenyl trisulfide (II) ?	1410	—	—	178	45,41,39,114,134	+	—
Allylmethyl tetrasulfide	1440	1460	—	184	41,45,73,120	+	+
Methyl-1-propenyl tetrasulfide	1460	—	—	184	45,41,39,47,64,80,120	+	—
Methylthiopropyl allyl disulfide	—	1530	—	—	89,41	—	+
Diallyl tetrasulfide (+ BHT)	1640	1640	2000	210	41,39,73,146,64,210	+	+
6(or 7)-Methyl-4,5,6-trithia-undeca-1,10-diene (t_R = 62 mn, 20 s)	—	1680		220	41,73,115,64,39,45	—	+
Allyl-1-propenyl tetrasulfide	1660	—	—		41,39,45	+	—
Allylmethyl pentasulfide	1730	—	—	—	41,39,45,64,120,80	+	—
Diallyl pentasulfide	1848	—	—	—	41,39,45,64,146,210	+	—
7-Methyl-4,5,8,9-tetrathia dodeca-1,10-diene (t_R = 75 mn, 10 s)	—	1880	—	(252)	73,41,147,39,45,64	—	+

Several products (in trace amounts) on the apolar column remain unidentified. Their analytical data are summarized on p. 121

Table 7. (*continued*)

KIA	KIP	MW	Main fragments (m/z)
1045	—	130	41(100),39(43),45(36),74(50),130(6)
1060	—	130	41(100),39(39),74(47),45(32),130(0.2)
1240	—	128	39(100),45(96),41(70),43(50),71(30)128(30),103(10)

KIA	KIP	MW	Main fragments (m/z)
1270	—	160	45(100),41(65),43(29),95(29),131(29),98(35),59(35),47(24),99(24),71(24),72(24),44(18),58(18)
1300/1312 (two isomers)	—	162	41(100),39(38),45(29),97(10),162(5),64(5),131(4)
1460	—	?	45(100),41(93),39(69),47(52),64(22),80(13),120(3)
1480	—	?	41(100),45(91),39(88),72(35),71(29),74(22),132(4),165(7)
1700	—		41(100),69(84),45(78),29(81),59(27), +
1930	—	—	41(100),45(67),39(58),103(39),73(38),145(19)

Footnotes [a—e] see Table 8.
Note: Several other compounds have been identified: ethanol, n-pentanol, ethyl acetate, allylmethyl ketone, 2-hydroxy ethanal (?), (Z) and (E)-2-methyl-3-penten-1-als, ethylvinyl ketone, heptanal, an alcenal (KIP : 1830), octyl acetate, ethyl hexadecanoate, 3,5-dimethyl-2,5-dihydrothiophen-2-one. Aromatic hydrocarbons (benzene, toluene, p-cymene), terpenes and derivatives (limonene, δ-selinene), diterpenes (MW : 272 and 306), (E)-isoeugenol, 2-methylpyridine are thought to be artifacts arising from solvents and previous GC-MS analyses.

mercaptans (RS_3H, with R = butyl, hexenyl, and hexyl) have also been reported among a total of 45 volatile flavor compounds from onion by Boelens et al. (1971). Methylthiopropyl alkyl disulfides have been identified among volatile shallot oil components by Wu Chen and Chang May (1983). We also found them in onion essential oils (unpubl. results).

Alkyl-1-propenyl(n) sulfide derivatives can arise from (+S)-1-propenyl-L-cysteine sulfoxide precursor perhaps present in trace amounts in some garlic species, for example, in the French one.

The second category includes sulfur-containing heterocyclic compounds (Table 8).

With the exception of the two vinyl dithiin isomers, they have not been reported in garlic essential oils. As shown by Block et al. (1988), who identified 23 new compounds, they are mainly formed from thermal degradation of diallyl disulfide [or from other allyl (n) sulfide derivatives] or from allicin. Their identification, based only on their mass spectra, is not always an easy task, since they are quite different according to experimental conditions (apparatus, amount, scanning, molecule fragility) and authors. Thus, some of them are questionable or uncertain.

The composition of these two oils is given in Table 9. Percentages have been calculated from the two apolar columns.

The two oils have approximately the same percentage in diallyl (n) sulfide derivatives. The main feature is the low percentage in diallyl disulfide (usually

Table 8. GC-MS analysis on apolar and polar columns of two garlic essential oils originating from France (Provence)(F) and Mexico (M), respectively. Sulfur-containing heterocyclic compounds

Compounds (empirical formula)	KIA[a]	KIP[b]	MW[c]	Main fragments[d] m/z	Occurence[e] F	M
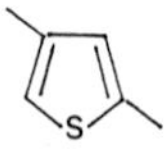 C_6H_8S	865	–	112	111,112,97,45	+	–
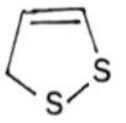 $C_3H_4S_2$	907	1500	104	103,104,45,71,59	+	+
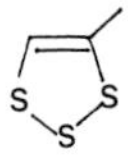 $C_3H_4S_3$	1195	–	136	71,45,72,136	+	–
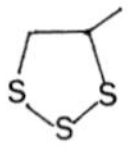 $C_3H_6S_3$	1210	–	138	138,73,45,61,64	–	+
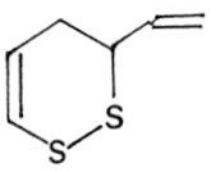 $C_6H_8S_2$	1215	1690	144	111,144,45,71,97	+	+
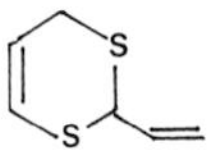 $C_6H_8S_2$	1230	1780	144	72,71,144,111,45	+	+
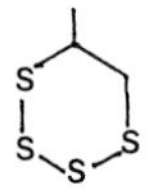 $C_3H_6S_4$	1400	–	170	41,170,106,128,64	+	+
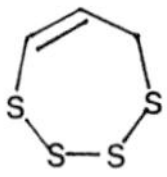 $C_3H_4S_4$?	1445	–	168	103,45,104,72,71,64	+	–
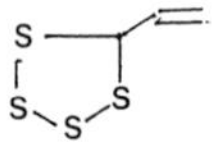 $C_3H_4S_4$?	1505	–	–	41,45,39,64,103,104	+	–

Table 8. (*continued*)

Compounds (empirical formula)	KIA[a]	KIP[b]	MW[c]	Main fragments[d] m/z	Occurrence[e] F	M
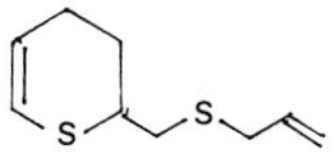 $C_9H_{14}S_2$	1550	–	186	41,45,39,117,99,81	+	–
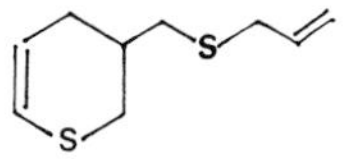 $C_9H_{14}S_2$	1560	–	186	117,45,41,39,99,81	+	–
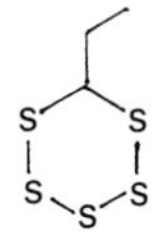 $C_3H_8S_5$	1710	–	202	41,138,73,74,45,64	+	+
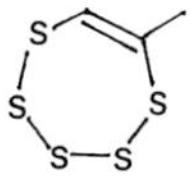 $C_3H_4S_5$	1730	–	200	45,71,72,136,64	+	–
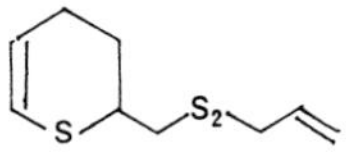 $C_9H_{14}S_3$	1780	–	-	41,113,39,79,71,85	+	–
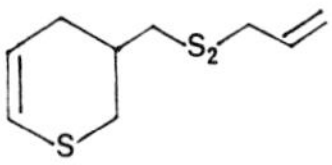 $C_9H_{14}S_3$	1830	–	218	41,113,45,79,85	+	–
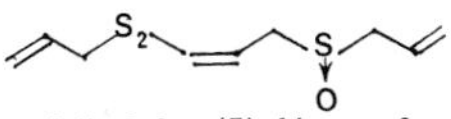 $C_9H_{14}S_3O$ (Z) Ajoene ?	1835	–	234	41,69,45,129	+	–
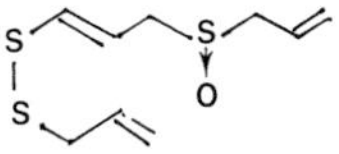 $C_9H_{14}S_3O$ (E) Ajoene ?	1845	–	234	41,69,129,45	+	–
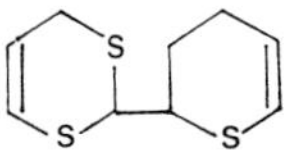 $C_9H_{12}S_3$	1925	–	216	45,39,41,99,97,111,85,216	+	–

Table 8. (*continued*)

Compounds (empirical formula)	KIA[a]	KIP[b]	MW[c]	Main fragments[d] m/z	Occurrence[e] F	M
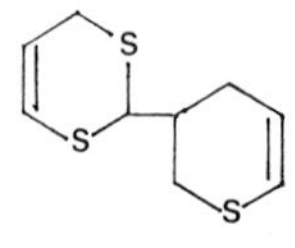 $C_9H_{12}S_3$	1935	–	216	45,111,119,97,216	+	–
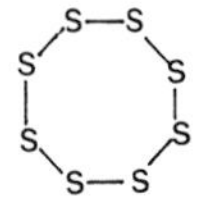 S_8	1952	–	256	64,128,160,192,96	+	–
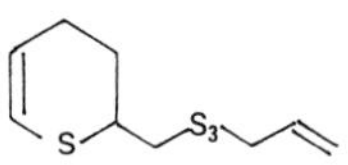 $C_9H_{14}S_4$	1955	–	(250)	113,79,41,45,39	–	+
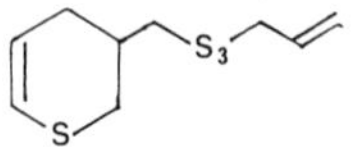 $C_9H_{14}S_4$	1970	–	(250)	41,39,45,145,99	+	+
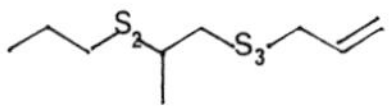 $C_9H_{16}S_5$	1980	–	284	41,45,73	+	–

[a]GC-MS analyses on apolar columns (KIA) were carried out: (1) by means of a Delsi 700 apparatus in conjunction with a Ribermag R-10–10 mass spectrometer, on a WCOT CPSIL 5 CB column (50 m × 0.32 mm i.d.) kept 5 mn at 70 °C and initially programmed from 70 to 150 °C at the rate of 2 °C/min and then programmed from 150 to 300 °C at the rate of 6°C/min (French garlic essential oil). (2) A Sigma Perkin Elmer apparatus was used in conjunction with a VG 70–70 F mass spectrometer, on a WCOT DB-1 capillary column (50 m × 0.22 mm i.d.) at programmed temperature from 60 to 250 °C, at the rate of 2 °C/mn (Mexican garlic essential oil).
[b]GC-MS analyses on polar column (KIP) were carried out as described above (2) on a FFAP column (50 m × 0.22 mm i.d.) at programmed temperature from 60 °C to 210 °C at the rate of 2 °C/mn. Kováts indices (KIA and KIP) were calculated as described in the Section 2.2.
[c]When molecular weights are not seen on the mass spectrum, they are indicated between parentheses.
[d]The base peak and the main fragments are given in decreasing order of intensity.
[e]The identification of compounds was carried out thanks to our SPECMA bank and files, and from literature data. Questionable identifications resulting from the absence of a reference spectrum in the file or from too great a dissimilarity index are indicated by a question mark.

Table 9. Percentage composition of garlic essential oils originating from France (Trets, near Marseilles) and Mexico

Compounds	Percentage	
	France (F)	Mexico (M)
Diallyl sulfide	1.2	2.4
Allylmethyl sulfide	4.2	4.1
Dithiacyclopentene (in mixture with (E)-propenylmethyl disulfide (F)	3.0	1.4
Dimethyl trisulfide	0.5	2.7
Diallyl disulfide	21.8	17.2
(Z)-Propenyl allyl disulfide	2.1	Traces
(E)-Propenyl allyl disulfide	6.0	Traces
Allylmethyl trisulfide	9.0	19
3-Vinyl-1,2-dithiin (+X)	5.45	0.24
2-Vinyl-1,3-dithiin	2.4	0.5
Dimethyl tetrasulfide	–	3.8
Diallyl trisulfide	24.2	26.4
(Z)-Allyl propenyl trisulfide	0.4	Traces
(E)-Allyl propenyl trisulfide	0.6	Traces
5-Methyl-1,2,3,4-tetrathiacyclohexane	1	1
Allylmethyl tetrasulfide	1.2	3
Diallyl tetrasulfide	4.85	8.2
(E)-Allyl propenyl tetrasulfide	0.3	–
Miscellaneous	11.8	10.06

about ≅ 50%), indicating some thermal degradation reactions of this product. They differ from the percentage in dimethyl trisulfide, allylmethyl tri- and tetrasulfides, and allylpropyl (n) sulfides, higher in the one originating from Mexico. A notable difference between the two oils rests on the presence of (Z) and (E) allyl-1-propenyl (n) sulfides in the French oil.

3.2 Mechanisms of Formation of Sulfide Derivatives

The formation of these new compounds is explained by Block et al. (1988) by a sequence involving:

1. a C-S homolysis of diallyl disulfide followed by a reversible terminal and internal addition of the allyldithio radical to diallyl disulfide;
2. an intramolecular hydrogen atom abstraction fragmentation of the intermediate formed by internal (Markovnikov) addition of the allyl-dithio radical, giving thioacrolein and the 1-(allyldithio)-2 propyl radical;
3. Diels-Alder self-condensation of thioacrolein acting as an heterodiene and its condensation to allyl mono-, di-, and trisulfides as shown in Scheme 1.

The self-condensation of thioacrolein affords the two vinyldithiins isomers 3 and 4, and the condensation with a third molecule of thioacrolein the trimers 5 and 6.

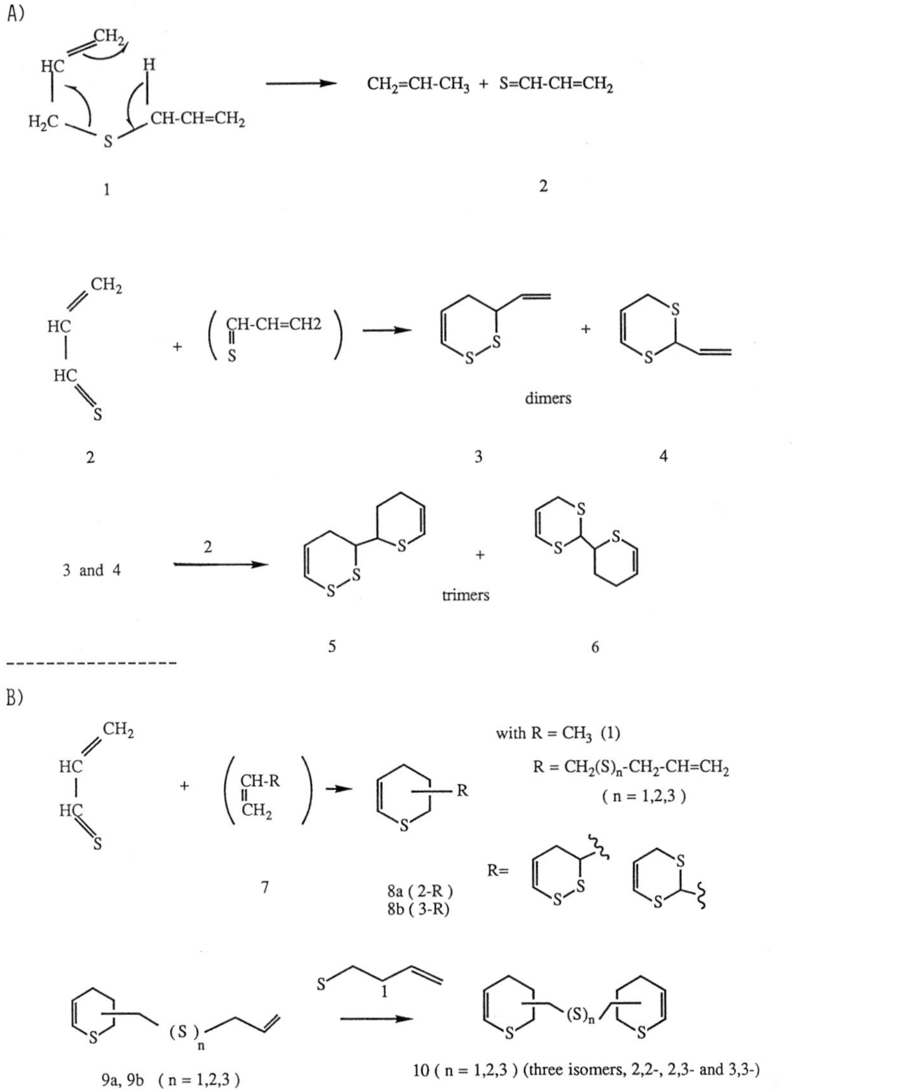

A)

CH_2=CH-CH_3 + S=CH-CH=CH_2

dimers

3 and 4

trimers

B)

with R = CH_3 (1)

R = $CH_2(S)_n$-CH_2-CH=CH_2 (n = 1,2,3)

8a (2-R)
8b (3-R)

9a, 9b (n = 1,2,3)

10 (n = 1,2,3) (three isomers, 2,2-, 2,3- and 3,3-)

Scheme 1. Diels-Alder self-condensation of thioacrolein (**A**) and condensation of thioacrolein with alkenes (**B**). (Block et al. 1988)

The synthesis of vinyl dithiins from thioacrolein was first reported by Beslin (1983). The isomer 1, 3 is a very thermally unstable compound at room temperature, and upon GC-MS it gives rise to several degradation products, mainly a compound of molecular weight 216 with main fragments at m/z 45, 39, 103, 97, 85, 59, 71 probably a trimer of thioacrolein, but the mass spectrum is quite different from that reported by Block et al. (1988) for that compound. It should be noted that these two isomers have been isolated in relatively large amounts by Nishimura et

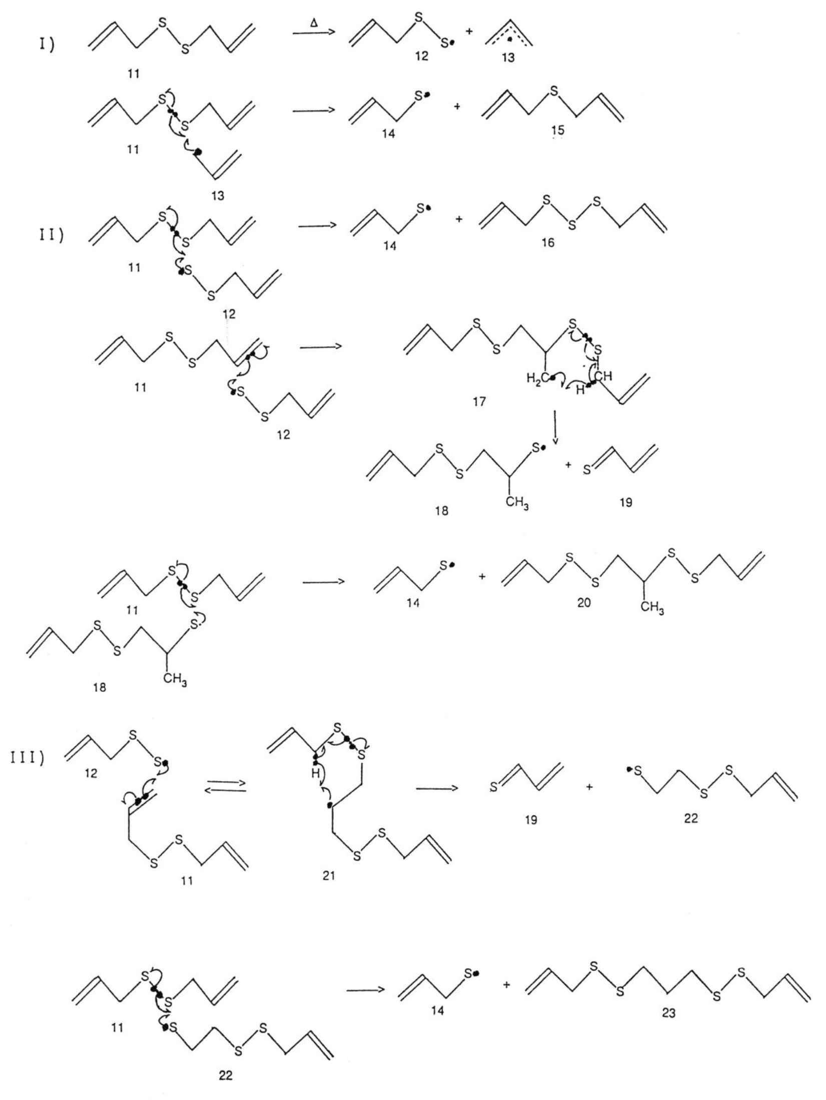

Scheme 2. Thermal degradation of diallyl disulfide and reversible terminal and internal addition of allyldithio radicals to diallyl disulfide. (Block et al. 1988)

al. (1988) by extraction with methylene chloride from leaves and stalks cut of young Caucasus garlic crushed into small pieces.

In another process, called headspace gas distillation extraction, these compounds were not found. However, we found these products in relatively large amounts (7 to 20%) in the dynamic headspace of fresh garlic cut in small pieces, while diallyl trisulfide was present only in trace amounts.

Probably, allicin trapped at room temperature on Tenax GC was thermally decomposed at 260 °C and changed into vinyl dithiins via thioacrolein:

$CH_2=CH-CH_2-S(O)S-CH_2-CH=CH_2 \longrightarrow CH_2=CH-CH_2SOH$
$+ CH_2=CH-CHS.$

Decomposition of allicin in methanol at room temperature for 7 days affords also vinyl dithiins, diallyl disulfide, and diallyl trisulfide (Nishimura et al. 1988).Since vinyl dithiins were present in trace amounts from thermal degradation of diallyl disulfide, it must be suggested that they are formed from allicin via thiyl radicals giving thioacrolein and then its dimers and trimers. By steam distillation of garlic, the formation of diallyl trisulfide is favored at the expense of vinyl dithiins.

Another interesting result is the formation (see Scheme 2) of compounds 20 and 23 in garlic essential oils as well as by thermal degradation of diallyl disulfide. According to Block et al. (1988), the thermal degradation of diallyl disulfide 11 (step I) affords allyldithio 12 and allyl radicals 13, the allylic C-S bond being weaker than the S-S bond by some 16 kcal/mol (in the case of diallyl trisulfide, the C-S and S-S bonds dissociation energies are very similar, so that in both homolytic cleavage should occur).

Addition of allyl and allyldithio radicals to the S-S bond of diallyl disulfide affords diallyl sulfide 15 and diallyl trisulfide 16, respectively.

Allyldithio radicals 12 can also react upon diallyl disulfide according to a Markovnikov addition *a process opposite to that encountered in intermolecular radical addition to simple olefins* (Step II). After rearrangement, the intermediate radical 17 gives thioacrolein 19 and a new radical 18. Addition of this latter to diallyl disulfide 11 generates the 6-methyl-4,5,8,9-tetrathiadodeca-1,11-diene 20. This process will be favored over the formation of the 4,5,9,10-tetrathiatrideca-1,12-diene 23 shown in step III. Addition of allyldithio radical to the terminal methylene of the allyl group of diallyl disulfide leads to an intermediate radical 21 and after rearrangement to radical 22. Addition of 22 to S-S bond of diallyl disulfide gives 23 and an allylthio radical.

4 Conclusion

Compounds present in garlic essential oils and all processed samples do not reflect the true composition of garlic cloves. It is also the case for medicinal preparations (tablets, capsules) and for flavourings (semolina or pulp) which contain in their headspace, after water dilution, a large amount of diallyl trisulfide. However, the antioxidant and lipoxygenase inhibitory activity of garlic essential oils is enhanced upon heating and concentration owing to the formation of higher boiling cyclic and acyclic organo-sulfur compounds as shown by Block et al. (1988).

Furthermore, this study allows us to show the usefulness of some analytical methods to identify sulfur-containing compounds based upon GC-MS upon EI and chemical ionization techniques, a specialized data bank, calculation of isotopic abundance and Kováts indices, respectively.

Acknowledgements. The authors wish to thank Pr. Beslin for providing samples of vinyldithiins, Mrs. G.M.F. Vernin for her assistance in gas chromatography analyses, Mrs. R.M. Zamkotsian for her collaboration in bibliography research and the mass spectra Centres of the Central Analytical Service of the CNRS (Lyon) and Marseilles for recording mass spectra. The authors are indebted to Pr. M. Chanon (Marseilles) and Pr. C. Párkányi (Boca Raton) for their kind assistance with the English manuscript corrections. Some figures have been reproduced with permission of Georg Thieme Verlag (Planta Medica, Vernin et al. (1986a).

References

Abraham KO, Schankaranarayana MI, Raghavan R, Natarajan CP (1976) *Allium* varieties : chemistry and analysis. Lebensm Wiss Technol 9:193

Ariga T, Oshiba S, Tamada T (1981) Platelet aggregation inhibitor in garlic. Lancet 1:150

Beslin P (1983) A facile synthesis of two thioacrolein dimers; a new entry to a flavor component in *Asparagus*. J Heterocyclic Chem 20:1753–1754

Block E, Ahmad S, Jain MK, Crecely RW, Apitz-Castro R, Cruz MR (1984) The Chemistry of alkyl thiosulfate esters. 8.(E, Z)-Ajoene, a potent antithrombotic agent from garlic. J Am Chem Soc 106:8295–8296

Block E, Iyer R, Grisoni S, Saha C, Belman S, Lossing FP (1988) Lipoxygenase inhibitors from the essential oil of garlic. Markovnikov addition of the allyl-dithio radical to olefins. J Am Chem Soc 110:7813–7827

Boelens H, de Valois PJ, Wobben HJ, Van der Gen A (1971) Volatile flavour compounds from onion. J Agric Food Chem 19:984–991

Boniface C, Vernin G, Metzger J (1987) Identification of informatisée de composés par analyse combiné spectres de masse-indices de Kovats : le programme SIMPA. Analysis 15:564–568

Brodnitz MH, Pascale JV, Van Derslice L (1971) Flavor components of garlic extracts, J Agric Food Chem 19:273–275

Bruins AP (1987) Gas chromatography-mass spectrometry of essential oils. In : Sandra P, Bicchi C (eds) Capillary gas chromatography in essential oil analysis : chromatographic methods. Hüthig, Heidelberg

Carson JF (1987) Chemistry and biological properties of onions and garlic. Food Rev Int 3:71–103

Cavallito CJ, Bailey JH (1944) Allicin the antibacterial principle of *Allium sativum*. I. Isolation physical properties and antibacterial action. J Am Chem Soc 66:1950–1951

Cavallito CJ, Buck JS, Suter CM (1944) Allicin the antibacterial principle of *Allium sativum*. II. Determination of chemical structures. J Am Chem Soc 66:1952–1954

Cavallito CJ, Bailey JH, Buck JS (1945) Antibacterial principle of *Allium sativum* III. Its precursor and "essential oil of garlic". J Am Chem Soc 67:1032–1033

Freeman GG (1975) Distribution of flavour components in onion (*Allium cepa* L.), leek (*Allium porum* L.) and garlic (*Allium sativum* L.) J. Sci Food Agric 26:471–481

Fujiwara M, Yoshimura M, Tsund S (1955) *Allium* thiamine a newly found derivative of vitamin B_1. III. *Allium* homologs in *Allium* plants. J. Biochem 42:591–601

Guenther E (1952) The essential oils, vol. 6. Van Nostrand, New York, pp 67

Hashimoto S, Migazawa M, Kameoka H (1983) Volatile flavor components of chive (*Allium schoenoprasum* L.) J Food Sci 48:1858–1873

Hine R, Rogers D (1956) Crystal and molecular structure of (+)-S-methyl-L-cysteine S oxide; a standard of absolute configuration for asymetric sulfur. Chem Ind 1428–1430; 51:5496 f

Jennings W, Shibamoto T (1980) Qualitative analysis of flavor and fragrance volatiles by glass capillary gas chromatography. Academic Press, New York

Kováts E (1958) Gas chromatographic characterization of organic compounds. I. Retention indexes of aliphatic halides, alcohols, aldehydes and ketones. Helv Chim Acta 41:1915–1932

Matsukawa T, Yurugi S, Matsuoka T (1953) Reaction between thiamine and ingredients of the plants of *Allium* genus. Detection of allyl thiamine and its homologs. Science (Jpn) 118:325–327

Nishimura H, Wijaya CH, Mizutani J (1988) Volatile flavor components and anti-thrombotic agents: vinyldithiins from *Allium victorialis* L. J Agric Food Chem 36:563–566

Oaks DM, Hartman H, Dimick KP (1964) Analysis of S compounds with electron capture/H flame dual channel gas chromatography. Anal Chem 36:1560–1565

Petitjean M (1982) Mass spectra and Kovats indices bank of heterocyclic flavouring compounds. Thesis Fac Sci ST-Jérôme, Marseilles

Petitjean M, Vernin G, Metzger J (1983) Mass spectra bank of volatile compounds In: Charalambous G, Inglett G (eds) Instrumental analysis of foods, vol. 1, Academic Press, New York pp 97–124

Semmler FW (1892) The essential oil of garlic. Arch Pharm 230:434–443

Shankaranarayana MS, Raghavan B, Abraham KO, Natarajan CP (1982) Sulfur compounds in flavours, Ch. III, In : Morton ID, MacLeod AJ (eds) Food flavours. Part A : Introduction. Elsevier. Amsterdam, pp 169–172

Small LD, Bailey JH, Cavallito CJ (1947) Alkyl thiosulfinates. J Am Chem Soc 49:1710–1713

Small LD, Bailey JH, Cavallito CJ (1949) Comparison of some properties of thiol sulfonates and thiol sulfinates. J Am Chem Soc 71:3565–3566

Stoll A, Seebeck E (1948) Allium compounds. I. Alliine, the true mother compound of garlic oil. Helv Chim Acta 31:189–210

Sugii M, Nagasawa S, Suzuki T (1963) Biosynthesis of S-methyl-L-cysteine and S-methyl-L-cysteine sulfoxide from methionine in garlic. Chem Pharm Bull (Tokyo) 11:134–136

Suzuki T, Sugii M, Kakimoto T (1961) New γ-glutamyl peptides in garlic. Chem Pharm Bull (Tokyo) 9:77–78

Van den Dool H, Kratz PD (1963) A generalization of the retention index system including linear temperature programmed gas liquid partition chromatography. J Chromatogr 11:643

Van Straten S, De Vrijer LL, De Beauveser JC (eds) (1977) Volatile compounds in Food. Central Institute for Nutrition and Food Research TNO Zeist, The Netherlands

Vernin G, Petitjean M, Metzger J (1983) Identification des composés volatils des flaveurs et fragrances par CPV-SM banque SPECMA. Conception et réalisation de la banque. Parf Cosm Arôm 51:43–52

Vernin G, Metzger J, Fraisse D, Scharff C (1986a) GC-MS(EI, PCI, NCI) computer analysis of volatile sulfur compounds in garlic essential oils. Application of the SIM technique. Planta Med 2:96–101

Vernin G, Petitjean M, Poite JC, Metzger J, Fraisse D, Suon KN (1986b) Mass spectra and Kovats indices data bank of volatile aroma compounds. Ch. VII, pp. 294–333. In: Vernin G, Chanon M (eds) Computer aids to chemistry. Ellis Horwood, Chichester, England

Vernin G, Petitjean M, Metzger J (1987) Gas Chromatography-Mass Spectrometry of essential oils. Part I. Computer matching techniques 10:287–328. In : Sandra P, Bicchi C (eds) Capillary gas chromatography in essential oils. Hüthig, Heidelberg

Woon CK, Young HJ, Shik WI (1988) A study of sulfur nutrition on the flavor components of garlic (*Allium sativum* L.) Han'guk toyang Piryo hakkoechi 21:183–193

Wu Chen JLP, Chang-May Wu (1983) Effect of heating of shallot essential oils (*Allium cepa* L.) *Aggregatum* G. IXth Int Congr Essential Oils, 13–17 March 1983. Essential Oil Technical Paper, Book 3, Singapore, pp 30–35

Yurugi S (1954) Vitamin B_1 and related compounds. Reaction between thiamine and ingredients of *Allium* genus plants. J Pharm Soc Jpn 74:519–524

Zhihui D, Jingkai D, Chongren Y, Yuichiro S (1988) Chemical constituents of garlic oil in Yunnan. Yun Nan Zhi Wu Yan Jui 10:223–226

Zoghbi MGB, Scott Ramos L, Maia JGS, da Silva ML (1984) Volatile sulfides of the Amazonian garlic busch. J Agric Food Chem 32:1009–1010

Analysis of Juniper and Other Forest Tree Oil

R.P. Adams

1 Introduction

Concurrent with the development of commercial gas chromatography in the late 1950's and early 1960's, botanists and chemists began to realize the value of a "new" suite of characters for the analysis and classification of plants — the terpenoids. Not only are the terpenoids under strong genetic control (Irving and Adams 1973), but the use of electronic digital integrators give quantitative characters amenable to multi-variate and geographic analyses (Adams 1970a; Adams 1972a). Given this "new" suite of quantitative chemical characters, the field of terpenoid chemosystematics rapidly expanded during the next two decades. The purpose of this chapter is to orient the novice to some of the basic procedures of analyses of terpenoids (particularly steam volatile components) and to show the utility of these compounds for the analyses of patterns of variation within and among forest tree taxa.

2 Sample Collection

Because the collection of wood samples has been examined in the chapter on cedarwood oil (Adams. this Vol.), I shall refer the reader interested in wood oils to that section. One should note that the wood and leaf oils are very different in *Juniperus* (and other Cupressaceae genera), whereas this is not the case in the pines, for example.

2.1 Sampling

Several general principles have been discovered over the past two decades in sampling for *Juniperus*, other conifers, and many (most?) forest trees. A preliminary study is generally needed if no work has been done on the genus or species of interest. One should take, for example, 3 (–5) leaf samples (100–200 g each) from each ordinal direction, from old and new foliage types (if present). Analyses of these 24 (–40) samples (3 reps, 4 sides, old/new leaves) will give an estimate of the within-tree variation. One can then usually decide if exposure, foliage type, etc. is a major factor (of course more samples may be needed depending on variability). In any case, it is desirable to get into the habit of always sampling in the same manner. For example, I collect eight to ten branchlets (15–20 cm long) for each juniper sample for steam distillation (plus material for herbarium voucher), from about chest high on the south-facing side of the tree. I collect these eight to ten branchlets from at least three major limbs. Fruiting structures (cones in the case of conifers) should be separated from the foliage before analyses, as they often differ in their volatile oil from the foliage (Hernandez et al. 1987; Vernin et al. 1988).

The investigator is faced with two options at this point: keep the foliage fresh until distillation or dry the foliage and extract it in an air-dried condition. The vast majority of researchers have chosen to keep the foliage fresh because some portion of the more volatile monoterpenes would be lost upon drying. Whether you can obtain reproducible results from dried foliage must be determined by doing a test on the species of interest.

With the junipers, we have found no difficulties in keeping the foliage at room temperature for a few days. Normally, we freeze the foliage as soon as possible and thaw it just before (or during) distillation. This may not be possible for broad-leafed deciduous trees, as the thin leaves tend to freeze together and stick together in a ball during distillation, resulting in poor extraction efficiency (see discussion on extraction below).

2.2 Sample Sizes

Sample sizes depend on the purpose of the study and on the variation among samples. For example, if the study is at the species level or higher, one generally finds qualitative differences, and numerous samples are not necessary. On the other hand, the examination of geographic variation or changes with seasons usually involves quantitative changes in composition, and larger sample sizes are needed. It is interesting to note that sampling 10 and then 20 *Juniperus virginiana* trees per population on successive years yielded the same geographic trends (Flake et al. 1969; 1973). Furthermore, the same geographic trend was found in *J. ashei* when using five samples as when using 15 samples per population (Adams and Turner 1970; Adams 1975a). It should be noted that both *J. ashei* and *J. virginiana* are dioecious (male and female trees), wind pollinated species. Different results might be expected with monecious and/or self-compatible species.

2.3 Diurnal, Seasonal, and Ontogenetic Variation

An important consideration in sampling is to collect materials that are comparable. Because one cannot make all the collections on one day at the same time, diurnal, seasonal, and ontogenetic variations must be considered.

2.3.1 Diurnal Variation

Hopfinger et al. (1979) found significant diurnal variation in leaf oil of Valencia orange trees. This variation is influenced by photosynthesis during the day (Fretz 1976); Lincoln and Langenheim 1977). In a study of diurnal variation in *J.*

scopulorum, analysis of 37 terpenoids revealed that 36 differed between trees, 11 varied between days, and 13 showed significant diurnal variation (Adams and Hagerman 1977). Four of these compounds are shown in Fig. 1. Notice that sabinene and methyl citronellate show patterns of gradual decline during the day into early evening (Fig. 1). In contrast, linalool, 4-terpineol, the eudesmols, elemol, etc., show increases starting at about 80% of their maximal values at 09.00 h and reaching their maximum values during the early evening, thence declining during the remainder of the night (Fig. 1). In order to judge the effects of these variations on genotype analyses, all of the 56 samples (four trees, seven sampling periods, 2 days) were subjected to principal coordinate analysis and the samples were found to cluster by genotypes (Fig 2). Thus, for *J. scopulorum*, it appears that the terpenoids are useful for the analysis of diurnal metabolic variation, but these variations do not mask the tree-to-tree variations. This study of diurnal variation was continued during the winter, using the same four trees of *J. scopulorum* (Adams 1979), and only one significant difference was found in the winter compared to 13 significant differences in diurnal variation in the summer. It was concluded that sampling during the dormant season (winter in that case) would greatly minimize or eliminate diurnal variation in *Juniperus* (and other conifers).

2.3.2 Seasonal Variation

Although wide seasonal variation of terpenoids is well known in herbaceous annuals such as the mints (Burbott and Loomis 1969) and sage (Fluck 1963), seasonal variation in the conifers is apparently much less. Von Rudloff (1972) examined the changes in volatile oils in leaves, buds, and twigs of white spruce

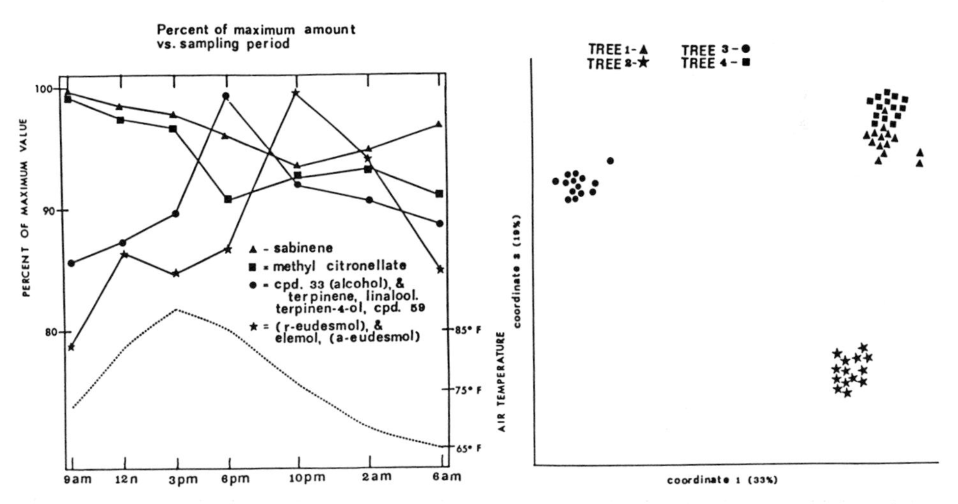

Fig. 1. *Left* diurnal variation in the percent maximum for four terpenoids representing the major trends in *J. scopulorum*. (Adams and Hagerman 1977)

Fig. 2. *Right* principal coordinate analysis for four *J. scopulorum* trees sampled 13 times over 2 days. Notice that the samples cluster by genotype, not by time sampled. (Adams and Hagerman 1977)

(*Picea glauca*). Figure 3 shows the seasonal variation in limonene and myrcene. The variation in camphor and bornyl acetate is depicted in Fig. 4. Notice that in all cases, the old shoots (2- to 4-year-old needles) are very constant throughout the year, with only a minor fluctuation during the onset of the growing season (May-June, Figs. 3,4). Sampling during the slow or no-growth time of the year (July-winter, in this case) effectively minimizes variation in white spruce. Seasonal variation in *Juniperus pinchotii* (Adams 1970b) indicated significant differences between summer and winter samples and that the variance among the summer samples was greater than the variance during the winter sampling. Adams (1970b) concluded that winter sampling of *J. pinchotii* was preferred. Analysis of seasonal variation in the terpenoids of *J. scopulorum* (Powell and Adams 1973) revealed significant seasonal variation and that seasonal variation was greater for components calculated on a weight basis (mg/g) than when calculated on a relative percent of the total oil basis. The use of relative percentage data was thus

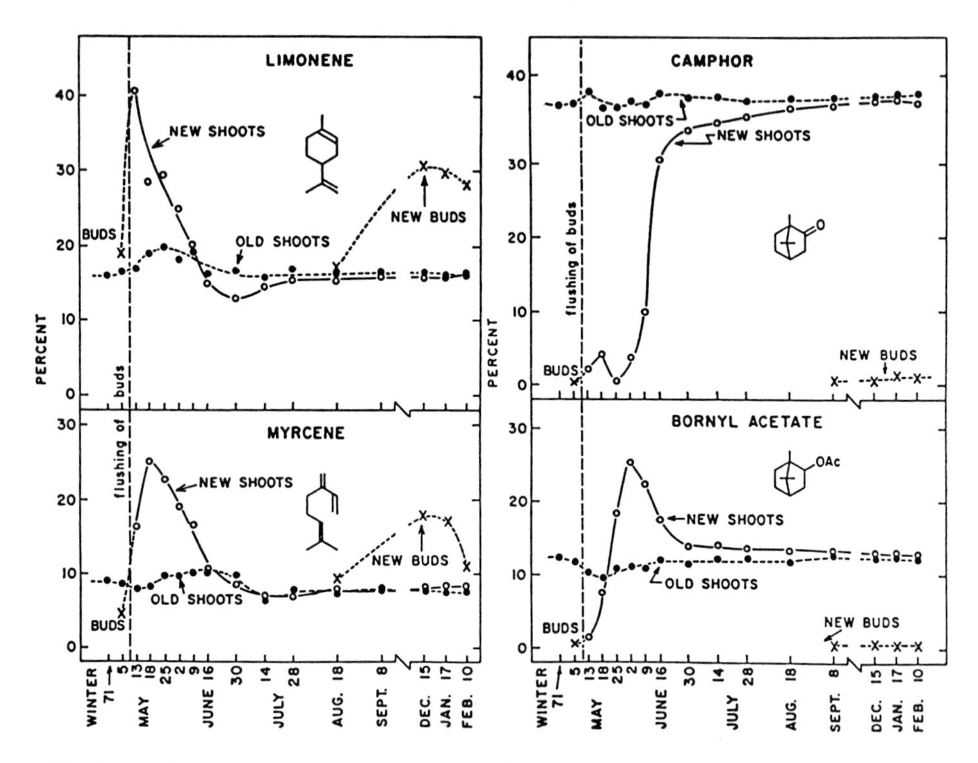

Fig. 3. *Left* changes in the relative percentage of limonene and myrcene in young and older white spruce (*Picea glauca*) shoot (needles) and buds. (von Rudloff 1972)

Fig. 4. *Right* comparisons of seasonal and ontogenetic variations in camphor and bornyl acetate from old and new shoots (needles) and buds in white spruce. (von Rudloff 1972)

encouraged for chemosystematic studies, although the weight basis would be preferred for biosynthesis studies.

Two recent papers deserve mention because they deal with seasonal variation in the terpenes of tropical rain forest trees. Whiffin and Hyland (1989) examined seasonal variation in *Litsea leefeana* and found that the three trees sampled showed no defined seasonal pattern (Fig. 5), with each tree showing its own pattern. Multi-variate clustering of the samples revealed that the terpenoid samples cluster by tree (Fig. 6). Whiffin and Hyland (1989) concluded that seasonal variation was not a problem in sampling Australian rain forest trees for volatile oil composition studies. A similar study on the Australian tree, *Angophora costata* (Leach and Whiffin 1989) concluded that quantitative changes in volatile oil composition over seasonal and diurnal periods were not significant in terms of chemosystematics studies, if immature leaves were excluded from the samples.

Nevertheless, it is prudent to minimize seasonal variation by sampling during the dormant (or least-growth) season, whether this be due to temperature or rainfall (i.e., samples taken during the middle or end of the driest season are least likely to be affected by metabolic changes).

2.3.3 Ontogenetic Variation

As previously mentioned, von Rudloff (1972) showed, rather dramatic changes in the volatile oils of new shoots (needles) in white spruce (Figs. 3, 4). In *Juniperus scopulorum* the young leaves (current new growth) were found to differ quantitatively from the mature leaves (Adams and Hagerman 1976) for 19 of 36 terpenoids

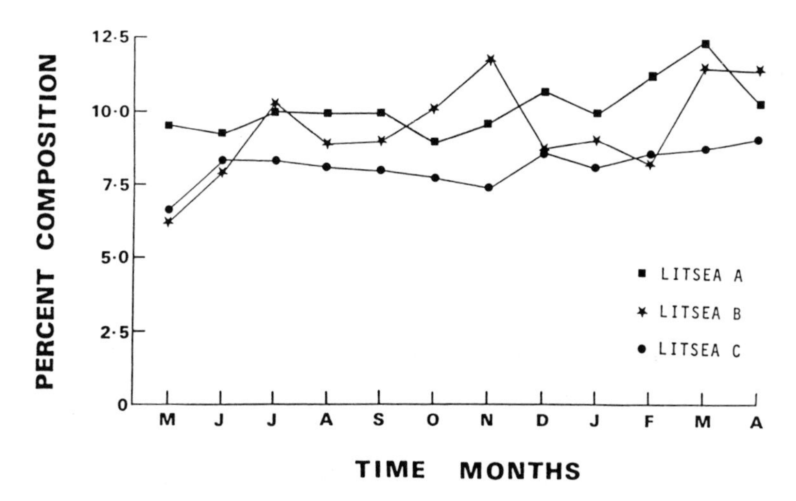

Fig. 5. Annual variation in a single terpenoid in three trees of *Litsea leefeana* in Australia. Note the rather constant, parallel levels from June to October, the Australian winter, and the larger variation during the rest of the year when growth is occurring. (Whiffin and Hyland 1989)

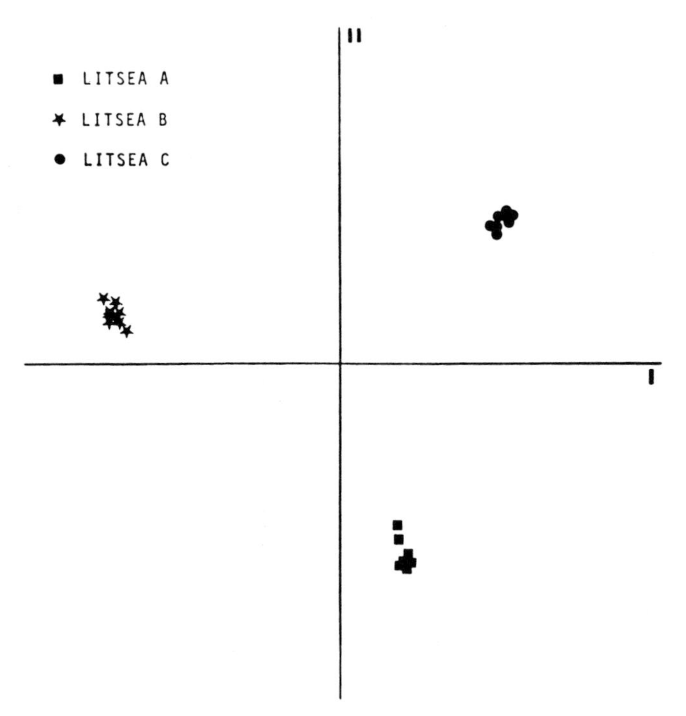

Fig. 6. Principal coordinate analysis of all the monthly samples of the three *Litsea leefeana* trees (graphed in Fig. 5). All of the monthly samples cluster by one of the three genotypes (*A*,*B*,*C*), not by month of collection. (Whiffin and Hyland 1989)

(Fig. 7). The elimination of juvenile or new growth from juniper leaf samples was recommended. However, analysis of the oil from juvenile and mature leaves of *J. horizontalis* (Adams et al. 1981c) revealed no significant differences in 39 terpenoids. Canonical variate analysis of the terpenoids of *J. scopulorum* and *J. virginiana* along with the mature foliage of *J. horizontalis* and coplotting an individual using the terpenoids from its juvenile leaves did not blur taxonomic distinctions (Fig. 8). Notice that individual 5A (adult leaves) and 5J (juvenile leaves) ordinate very much in the same position (Fig. 8).

Because almost all junipers are dioecious (male and female), an obvious question is "do males and females differ in their volatile leaf oils?". Examination of the oils from male and female *J. scopulorum* plants indicated (Adams and Powell 1976) that sexual differences are apparent during spring (rapid growth) but essentially no differences were found during the winter. Part of the differences in the spring may be due to the difficulty of finding and removing all of the small female cones, which may contribute a different oil profile. If one is working on a unisexual plant, this factor should be considered, but seems of no consequence in the junipers.

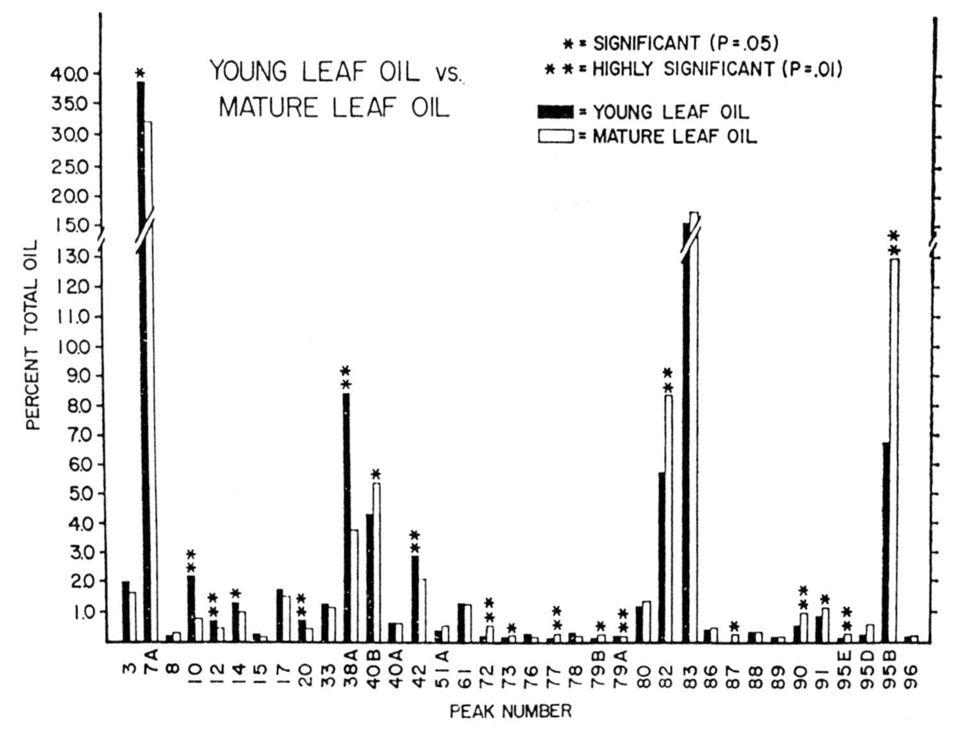

Fig. 7. Comparison of the major compounds in the leaf oils of young (juvenile) and mature (2–5 years old) leaves of *J. scopulorum* cv. *platinum* (Adams and Hagerman 1976)

For tropical evergreen trees, such as *Angophora costata*, Leach and Whiffin (1989) found that immature leaves must be eliminated from samples. That would appear to be good general advice.

One environmentally induced growth form of juniper has been analyzed. Near a natural burning coal vein in North Dakota, USA, the junipers are all very columnar in shape. These junipers were recognized (Fassett 1945) as a variety (*J. scopulorum* var. *columnaris* Fassett). Examination (Adams 1982a) of the volatile leaf oils from columnar and nearby, pyramidal (normal shaped) *J. scopulorum* trees revealed only one significantly different terpenoid. Canonical variate analysis showed that both the columnar and pyramidal *J. scopulorum* trees cluster together (Fig. 9). However, few transplant studies have been done where the terpenoids of conifer trees have been subsequently analyzed, so some caution must be advised in regards to the effects of edaphic factors on terpenes in trees.

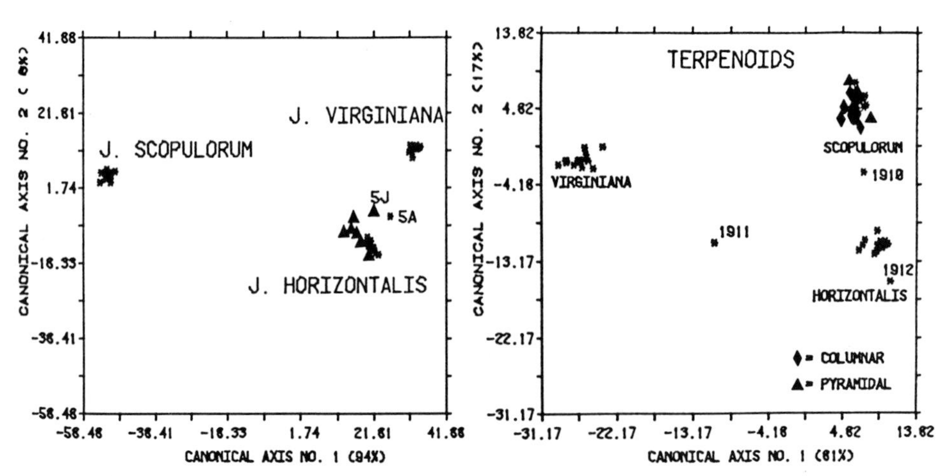

Fig. 8. *Left* canonical variate analysis of the volatile leaf oil from mature (*stars*) and juvenile foliage samples (*triangles*) of *J. horizontalis* co-analyzed with the oils from mature leaves of *J. scopulorum* and *J. virginiana.* Plant 5 (*5A* adult leaves; *5J* juvenile leaves) shows differentiation towards *J. virginiana.* (Adams et al.1981c)

Fig. 9. *Right* canonical variate analysis of the leaf terpenoids from columnar (caused by fumes from burning coal seams) and pyramidal-shaped (normal form) *J. scopulorum.* (Adams 1982a)

3 Oil Extraction

If wood is to be extracted, the reader is referred to the chapter on cedarwood oil (Adams, this vol.). Otherwise, I shall consider extraction of foliage. Providing some precautions are taken (see Adams, this Vol., notes on pH, etc.), steam distillation is a very useful method to obtain a good sample of the volatile oil. One might note, however, that if biosynthetic and/or physiological problems are to be addressed, one may want to treat the aqueous phase with β-glucodidase to release glycosidic bound terpenes (van den Dries and Svendsen 1989) to obtain a more quantitative estimate of the terpenes in the plant. The volatile oil can also be extracted by super critical fluids or solvent methods to minimize degradation (Cu et al. 1989). Recently, Craveiro et al. (1989) reported on the use of a microwave for steam distillation in 5 min. Although the method appears to be qualitative at present, further development could lead to quantitatively consistent extractions.

Assuming that steam distillation is used, one may want to separate the leaves from the woody stems, although this is not necessary in junipers because the small stems (3-8 mm diam.) do not contribute any quantity of oil (however, the weight of the stem wood may be very significant in computing yield data!). Normally, for junipers, we take a handful and orient all the stems downward to aid in the distribution of the upward steam column (see extraction apparatus in Adams, this Vol.). With junipers there is no need to cut up the leaves, but this is useful in conifers with resin ducts such as pine needles. Generally cutting the needles into 2-cm sections is sufficient.

Steam distillation for 2 h removes about 35% of the volatile oil in junipers and 24 h removes 95% of the oil. We normally do both. We remove the ether layer (see apparatus, Adams, this Vol.) after 2 h (and add new ether). The 2-h fraction is used for chemosystematic studies. After 24 h (22 h of additional distillation) we remove the ether layer. The excess ether from both fractions is evaporated in a hood with a jet of nitrogen. The samples are then weighed and the combined weights (i.e., 24 h total) are used to calculate the yield (see section on sample handing in Adams, this Vol.). After steam distillation, the foliage is placed in a tared, paper sack and dried for 48 h at 100 °C to determine dry wt. Oil yield is calculated as: oil wt./(dry distilled leaf wt. + oil wt.). Leaves that are dry and fragile or that have been frozen and thawed before distillation (and thus tend to stick together in a ball) are put into a nylon mesh bag, which is then put into the distillation chamber (see Adams, this Vol.). This prevents loose material from falling into the boiling water and/or clogging the steam intake. The reader is referred to Adams (this Vol.) for sample storage.

4 Chemical Analysis

The primary method for analysis is by gas chromatography either coupled to an electronic digital integrator or a computer.

4.1 Gas Chromatography

Gas chromatography has become an integral part of any essential oil analysis today. Bonded phase fused silica (or quartz) capillary columns, externally coated with polyamide resins to reduce breakage became commercially available about 1980. Third party manufacturers produce conversion kits so 1/4" injectors can be converted for use with capillary columns. One is almost without excuse for not converting to bonded phase, fused silica capillary columns today. For the nonpolar and moderately polar phases, the durability is tremendous. Columns last for years and can be washed out with solvents to regenerate them. Unfortunately, we cannot say that for the most popular phase for essential oil analyses, Carbowax 20M (PEG 20M). Our experience with the bonded Carbowax columns has been that the phases can be damaged easily by very small oxygen leaks and that the columns tend to change retention characteristics with age (Jennings 1978; Jennings and Shibamoto 1980; Adams 1980a) We have converted all of our primary analyses to a J & W DB-5, 30 m, 0.26 mm i.d., 0.25 micron coating thickness. The DB-5 phase is bonded methyl silicone with slight polarity and is equivalent to: BP-5; CP-Sil-8CB; DC-200; DC-560; Dexsil 300; GB-5; OV-3; OV-73; SE-52; SE-54; SPB-5; Ultra 2 and 007-2. Because extremely reproducible retention times (+/–2 s) are essential for identification of most terpenoids using GC/MS data systems (Adams 1989a), one's libraries become very valuable with time. If your column is changing characteristics, your library will not be useful. We have found less lot-to-lot variation among new (i.e., replacement) commercial nonpolar columns than polar capillary columns. Another plus for the nonpolar bonded columns is that analysis can be performed up to 350 °C, so phyto-steroids, fatty acids, and waxes can be analyzed on the same column.

4.1.1 Carrier Gas

Much has been written about the theoretical efficiencies of carrier gases (see Jennings 1987). If all other factors are kept constant, then one sees increased resolution from nitrogen (poorest), to helium to hydrogen (greatest resolution). However, many capillary gas chromatography systems are not well enough tuned to high resolution chromatography to see any difference between using helium and hydrogen as carrier gases. Hydrogen presents a problem of being explosive. Helium (zero grade) is generally expensive. Nitrogen is usually very cheap, but may be contaminated with water and oxygen. We use helium as a compromise (but mostly because we are interfaced with an Ion Trap Mass Spectrometer which requires helium). The most important consideration is: can you obtain gas that is free of water and oxygen? In any case, the carrier gas should be routed through an indicating oxygen/water trap for final cleaning. It is a good idea to have a sample of the gas analyzed for water and oxygen before deciding on a gas vender. The flow rate is easily checked by injecting butane (from a lighter recharger) or methane from a natural gas source with the GC oven at 200 °C. For helium, a flow rate of 20 to 50 cm/s is within the maximum theoretical plates range (see Jennings 1978 for theoretical efficiency plots for helium, hydrogen, and nitrogen). If your column is connected directly to the source of an ion trap, then a flow of about 1 ml/min of helium is required. A 30 m × 0.26 mm i.d. column will deliver 1 ml/min of helium at about 30–32 cm/s (well within the maximum theoretical plates range, Adams 1989a).

4.1.2 Sample Injecting

Assuming you have a capillary injector, either splitting or splitless, one needs to pay careful attention to several things. Choose the best quality septa available. Heat the injector to about 250 °C with the GC oven at room temperature overnight and then make a blank (no injection) run the next morning. This will show you your septum bleed. Remove the septum and see if it is still pliable. Keep a written tally sheet of your injections for each septum. Change the septum after about 25 injections. The bleed of oxygen through an old septum can cause severe column damage.

Sample sizes depend on the diameter and phase coating thickness of your column. For a 0.26 mm i.d., 0.25 micron coating column, we dilute the oil to 10% with ether (i.e., 90% ether, 10% oil), and inject 0.1 µl, which is split 20:1. This results in about 500 ng on column. Components that are 25% or more of the oil will be overloaded (have a leading peak edge) under these conditions, but this allows one to detect trace components. We use 1-µl syringes with the plunger in the needle and clean the syringes by first working the plunger back and forth into acetone and then heating in a syringe cleaner (cf. Hamilton) for 45–60 s. Regardless of how one cleans the syringe, you should check the procedure occasionally by injecting the "clean" syringe into the GC and making a blank run to look for carryover.

4.1.3 Temperature Programming

The separation of individual components can be greatly aided by temperature programming. A linear temperature program of 3°/min, from 60 to 240 °C is routinely used in our laboratory. If there are almost no monoterpenoids (as in cedarwood oils), a higher initial temperature can save analysis time; however, the retention time data will then not be comparable with a general library (i.e., that includes monoterpenes). Nonlinear temperature programming can be very effective in helping resolve difficult mixtures but, again, retention data may not be comparable to your library data. In the final analysis, one must experiment with the GC system to determine the best temperature program.

4.1.4 Detection

Detection has traditionally been by use of flame ionization detection (FID). FID sensitivity does vary, roughly according to carbon number. So, if exact mole concentrations are desired, response factors must be determined for each compound in the mixture. However, because data are often converted to a range-normalized basis, one seldom sees response factors and correction factors used in chemosystematic studies.

The other classical method for detection in GC analysis is thermal conductivity detection (TCD). Unfortunately, the development of micro-thermal conductivity detectors has not seemed to keep pace with the reduction in sample sizes in recent years, and one rarely reads of TCD being used with capillary columns.

The third and perhaps most rapidly growing detector is the ion trap detector (ITD or ion trap mass spectrometer, ITMS). Direct coupling of the capillary column to the source of the ion trap provides an efficient transfer and identifications can be made in concert with GC separation (Adams 1989a). Although little has been published on quantitation using the total ion counts, our experience has been favorable (Adams and Edmunds 1989).

5 Component Identification

The oils of juniper and other forest trees can be very complex, containing hundreds of terpenoids, aromatic compounds, and occasionally important amounts of aliphatic alcohols and aldehydes and more rarely, alkanes. Although liquid phase infrared (IR) analysis was the method of choice for identification in the 1960's, the introduction of capillary columns and the attendant small sample sizes has greatly reduced the use of liquid phase IR. Vapor phase IR identification promises to be of great use as libraries improve. However, the principal method of identification of known compounds is generally combined GC/MS or GC/MS/computer searches.

5.1 GC/MC Computer Searches

A large library of mass spectra is readily available from sources such as the US NBS (National Bureau of Standards, formerly the EPA/NIH data base) with thousands of spectra. Unfortunately, searches from these large data bases, with the current technology (i.e., simple matching coefficients and no retention data) do not yield reliable identifications (see Adams et al. 1979 for discussion). Vernin et al. (1988) report on the use of SPECMA MS data bank for the identification of terpenes of *J. communis* berries and needles which apparently utilizes both MS data and Kovats indices.

Our library system [LIBR(TP), available form Finnigan Corp.] uses ion trap mass spectra (ITMS) and retention times on DB5. Although the ITMS spectra are generally quite similar to quadrapole mass spectra (Adams 1989a), there can be significant differences, so a reference library of ion trap spectra is essential. We use cedrol for ion trap tuning because it is very sensitive to space charging effects (overloading) and tuning (Adams 1989a).

Various juniper species leaf oil compositions are given in Table 1. Hopefully, the publication of these retention times will be useful for building libraries and for identification.

6 Applications of Terpenoid Data

There are scores of applications for terpenoid data but I would like to focus on three major areas: analyses of hybridization and introgression, geographic variation, and specific or evolutionary studies.

6.1 Analyses of Hybridization and Introgression

6.1.1 Juniperus

One of the earliest cases of the use of terpenoids was for the reexamination of the classical case of putative introgressive hybridization between *J. ashei* and *J. viginiana* (Hall 1952). Subsequent studies using terpenoid data showed clinal variation in *J. virginiana* but neither hybridization nor introgression with *J. ashei* (Adams 1977; Adams and Turner 1970; Flake et al. 1973; von Rudloff et al. 1967; von Rudloff 1975).

Terpenoids have been used to document hybridization between *J. scopulorum* and *J. horizontalis* (von Rudloff 1975; Adams 1983a; Adams 1982b), between *J.*

Table 1. Representative volatile leaf oil composition for junipers (*J. communis* var. *depressa* (USA); *J. foetidissmia* (Greece); *J. flaccida* var. *martinezii* (Mexico); *J. procera* (Kenya) and *J. virginiana* var. *virginiana* (USA). Compounds are listed in order of elution from a DB5 column. Compounds in parenthesis are tentatively identified. Data expressed as % total oil using total ion counts (TIC). T = less than 0.1% of the total oil.

RT Compound	*Communis* var. *depressa*	Flaccida var. *martinezii*	*Foetidissima*	*Procera*	*Virginiana*
1. 214 2-Hexenal	T	—	0.1	0.2	T
2. 301 Tricyclene	T	0.5	—	T	T
3. 307 α-Thujene	—	0.6	1.3	T	T
4. 319 α-Pinene	14.1	13.5	2.6	12.5	1.4
5. 337 α-Fenchene	—	—	T	0.1	T
6. 340 Camphene	0.2	0.6	T	0.1	T
7. 348 Thuja-2,4(–10)-diene	T	0.1	—	—	—
8. 363 [Bicyclo(3,2,1)oct-2-ene, 3-methyl-4-methylene]	—	1.8	—	—	—
9. 379 Sabinene	0.2	8.5	19.6	T	6.7
10. 383 1-Octen-3-ol	—	—	—	0.3	—
11. 386 β-Pinene	2.1	1.1	—	1.2	T
12. 408 Myrcene	4.4	4.0	2.7	1.2	0.9
13. 427 2-Carene	T	T	—	—	T
14. 435 α-Phellandrene	T	0.9	0.2	—	—
15. 444 3-Carene	0.2	T	T	6.1	T
16. 457 α-Terpinene	T	1.1	4.3	T	T
17. 465 o-Cymene	—	—	—	T	—
18. 471 p-Cymene	T	1.2	0.5	T	—
19. 474 Sylvestrene	—	—	—	0.1	—
20. 481 Limonene	1.1	1.6	0.9	0.2	18.9
21. 482 β-Phellandrene	1.1	5.0	0.6	0.8	T
22. 485 1,8-Cineole	—	—	0.2	T	—
23. 498 *cis*-Ocimene	—	T	—	—	—
24. 519 *trans*-Ocimene	—	0.4	—	—	T
25. 535 Pentyl isobutyrate <n->	T	—	0.1	—	—
26. 545 Γ-Terpinene	T	2.0	6.5	T	T
27. 560 *trans*-Sabinene hydrate	T	0.5	1.8	—	T
28. 574 *cis*-Linalool oxide	—	T	0.1	—	—
29. 604 (Eucarvone)	—	0.1	—	—	—
30. 605 Fenchone	T	—	—	—	—
31. 608 Terpinolene	1.4	0.9	1.9	1.1	0.5
32. 609 p-Cymenene	—	0.3	—	—	—
33. 626 α-Pinene oxide	—	1.9	—	—	—
34. 629 *cis*-Sabinene hydrate	—	0.5	1.9	—	T
35. 632 Linalool	2.0	3.0	1.0	0.5	4.4
36. 642 α-Thujone	—	—	18.6	—	—
37. 643 Nonanal <n->	T	—	—	—	—
38. 645 Isopentyl-isovalerate	0.4	—	—	—	—
39. 661 1,3,8-p-Menthatriene	—	—	—	T	—
40. 664 endo-Fenchol	0.2	—	—	—	—
41. 667 β-Thujone	—	—	3.5	—	—
42. 682 *cis*-p-Menth-2-en-1-ol	T	—	1.2	T	T
43. 692 α-Campholenal	2.3	0.4	—	T	—
44. 724 *trans*-Pinocarveol	1.2	0.7	—	0.1	—

Table 1. *(continued)*

RT Compound	*Communis* var. *depressa*	Flaccida var. *martinezii*	*Foetidissima*	*Procera*	*Virginiana*
45. 725 *trans*-p-Menth-2-en-1-ol	T	—	1.2	T	T
46. 727 *cis*-Verbenol	0.6	—	—	—	—
47. 734 Camphor	—	11.4	—	0.2	3.7
48. 735 *trans*-Verbenol	3.4	0.5	—	—	—
49. 746 Camphene hydrate	1.1	—	—	—	T
50. 758 Citronellal	0.5	—	—	—	—
51. 766 β-Pinene oxide	—	0.2	0.2	—	—
52. 775 *cis*-3-Pinanone	0.4	—	—	—	—
53. 781 Pinocarvone	0.4	T	—	—	—
54. 789 Borneol	1.8	0.9	—	0.2	0.8
55. 792 p-Mentha-1,5-dien-8-ol	0.4	—	—	—	—
56. 804 Nonanol	—	—	—	T	—
57. 820 4-Terpineol	2.5	8.2	17.6	0.1	1.5
58. 837 p-Cymen-8-ol	0.3	0.4	0.1	0.1	—
59. 852 α-Terpineol	3.9	0.7	0.7	0.5	T
60. 864 Myrtenal	2.2	0.2	—	—	T
61. 865 *cis*-Piperitol	—	—	0.3	—	—
62. 867 Myrtenol	—	0.2	—	—	—
63. 869 Estragole	—	—	—	—	T
64. 887 *trans*-3-Pinanone	0.6	0.3	—	—	—
65. 894 Verbenone	0.9	0.4	—	—	—
66. 896 *trans*-Piperitol	—	—	0.4	—	—
67. 923 *trans*-Carveol	0.6	0.1	—	—	—
68. 950 Citronellol	4.0	—	0.1	—	2.3
69. 967 Myrtenyl acetate	—	0.5	—	—	—
70. 968 Thymol methyl ether	0.1	—	—	—	—
71. 984 Carvone	0.3	0.1	—	—	T
72. 1009 *cis*-Myrtanol	0.1	0.4	—	—	—
73. 1011 Carvenone	0.3	—	—	—	—
74. 1011 Piperitone	—	0.4	—	—	T
75. 1018 Geraniol	0.4	—	—	—	—
76. 1023 Linalyl acetate	—	0.4	—	—	—
77. 1026 *trans*-Myrtanol	0.9	—	—	—	—
78. 1035 Methyl citronellate	1.3	—	0.1	—	—
79. 1099 Bornyl acetate	5.2	2.3	0.1	0.4	2.1
80. 1101 Safrole	—	—	—	—	10.9
81. 1113 Thymol	—	T	—	—	—
82. 1117 *cis*-Sabinyl acetate	—	—	0.9	—	—
83. 1119 *trans*-Verbenyl acetate	—	0.9	—	—	—
84. 1137 Carvacrol	—	0.1	—	—	—
85. 1229 *cis*-Isosafrole	—	—	—	—	6.7
86. 1240 4-Terpinenyl acetate	—	—	—	—	T
87. 1264 α-Terpinenyl acetate	—	0.5	—	—	—
88. 1275 Citronellyl acetate	0.5	—	—	—	—
89. 1303 Neryl acetate	0.2	—	—	—	—
90. 1334 α-Copaene	—	T	—	—	—
91. 1352 Geranyl acetate	1.9	—	—	—	—
92. 1371 β-Cubebene	—	T	—	—	—
93. 1375 β-Elemene	0.2	—	—	—	—
94. 1403 Methyl eugenol	—	—	—	—	2.9

Table 1. *(continued)*

RT Compound	*Communis* var. *depressa*	Flaccida var. *martinezii*	*Foetidissima*	*Procera*	*Virginiana*
95. 1442 Caryophyllene	0.2	0.2	0.1	0.5	T
96. 1467 Thujopsene	—	—	—	—	T
97. 1519 α-Cadinene	—	0.3	—	—	T
98. 1525 Geranyl acetone	0.2	—	—	—	—
99. 1527 α-Humulene	0.2	—	0.1	0.7	—
100. 1537 *cis*-β-Farnesene	0.3	—	—	—	—
101. 1577 β-Cadinene	—	0.3	—	—	T
102. 1594 Germacrene D	0.9	—	—	0.3	T
103. 1602 ar-Curcumene	0.2	—	—	—	—
104. 1634 α-Zingiberene	0.6	—	—	—	—
105. 1643 α-Muurolene	0.1	T	0.1	—	T
106. 1667 β-Bisabolene	0.8	—	—	—	—
107. 1676 Γ-Cadinene	0.2	1.0	0.1	—	T
108. 1695 Calamenene (1S,*cis*-)	—	T	—	—	—
109. 1700 δ-Cadinene	0.8	0.9	0.3	—	0.8
110. 1759 Elemol	—	0.7	0.1	4.3	8.2
111. 1772 Elemicin	—	T	—	—	T
112. 1777 Γ-Elemene	0.3	—	—	—	—
113. 1796 *trans*-Norelidol	3.8	—	—	—	—
114. 1821 Germacrene D-4-ol	1.2	—	0.1	—	T
115. 1825 Spathulenol	1.2	—	—	—	—
116. 1837 Caryophyllene oxide	0.2	0.2	T	0.5	—
117. 1876 Cedrol	—	—	3.2	—	—
118. 1898 β-Oplopenone	0.2	—	0.5	—	T
119. 1944 Cubenol	—	1.0	—	—	0.9
120. 1951 Γ-Eudesmol	—	0.2	—	1.4	2.8
121. 1973 τ-Cadinol	0.2	T	0.3	—	T
122. 1976 τ-Muurolol	0.2	0.2	0.3	—	2.4
123. 1984 Torreyol (=δ-cadinol)	0.1	T	0.1	—	—
124. 1993 β-Eudesmol	—	0.3	—	2.3	1.7
125. 2000 α-Eudesmol	—	0.3	—	3.8	3.1
126. 2003 α-Cadinol	0.8	—	1.0	—	—
127. 2034 (Elemol acetate)	—	—	—	1.3	—
128. 2079 epi-α-Bisabolol	3.2	—	—	—	—
129. 2141 *cis,cis*-Farnesol	0.4	—	—	—	—
130. 2159 *trans,trans*-Farnesol	1.6	—	—	—	—
131. 2201 *trans,cis*-Farnesol	0.4	—	—	—	—
132. 2306 Acetoxyelemol <8-α->	—	—	—	3.5	3.5
133. 2660 Manool <epi-13->	—	0.9	—	0.2	—
134. 2717 Manoyl oxide	—	0.9	0.2	0.5	—
135. 2841 Abietatriene	—	0.3	0.2	1.3	—
136. 2845 Manool	0.3	0.3	—	—	—
137. 2891 Abietadiene	—	—	—	15.4	—
138. 3253 (*cis*-) Totarol	—	—	—	0.6	—
139. 3275 (*cis*-) Abietal	—	—	—	1.7	—
140. 3300 (*trans*-) Totarol	—	—	0.4	21.4	—
141. 3330 Ferruginol	—	—	—	2.4	—

horizontalis and *J. virginiana* (Palma-Otal et al. 1983), between *J. scopulorum* and *J. virginiana* (Adams 1983a; Flake et al. 1978), and suggested between *J. virginiana* and *J. virginiana* var. *silicicola* (Adams 1986).

6.1.2 Other Forest Trees

The literature on the use of terpenoids is so vast that space does not permit any significant review. However, a few examples in various genera may be illustrative. In the spruce (*Picea*) one must mentioned the very early study on introgression of white and Engelmann spruce along the Bow River in Alberta (Ogilvie and von Rudloff 1968). Hybridization in *Pinus* has been subject of numerous studies including Zavarin et al. (1980); Snajberk et al. (1982); Bailey et al. (1982); and Neet-Sarqueda et al. (1988).

The analysis of hybridization between *Eucalyptus crenulata* and *E. ovata* provides a good example of the use of terpenoids for an angiosperm forest tree (Simmons and Parsons 1976). The intermediate chemical profiles of the hydrids are also shown in their analyses (Simmons and Parsons 1976).

6.2 Studies of Geographic Variation

6.2.1 Juniperus

Over the past 20 years several important studies have been made on geographic variation in *Juniperus*. Geographic variation studies of *J. phinchotii* (Adams 1972b, 1975b) revealed fairly uniform populations except where the taxon is sympatric with *J. erythrocarpa* in the trans-Pecos Texas area. Previous suggestions of hybridization with *J. ashei*, *J. depeana*, and *J. monosperma* were refuted.

Populational studies in *J. ashei* (Adams and Turner 1970; Adams 1975a; 1977) revealed ancestral populations of *J. ashei*. Fig. 10 shows that the composite of the terpenoid pattern, factored by principal coordinate analysis, accounted for 50% of the variation among populations. The divergent populations were shown to bear affinities to the sibling, ancestral species, *J. saltillensis* in Mexico. Furthermore, the observed terpenoid pattern (Adams 1977) supports the post-glacial migration into the Edwards plateau and northward (Fig. 11). Without the use of continuous, quantitative terpenoid data, analysis of the ancestral history of *J. ashei* would have been almost impossible.

Another major study on Pleistocene refugia and recolonization was performed on *J. scopulorum* (Adams 1983a). Canonical variate analysis and contour mapping of the population coordinate scores were used to describe infraspecific variation and correlate these patterns with the Wisconsin maximum ice advance and subsequent recolonization by *J. scopulorum* (Adams 1983a).

Comer et al. (1982) showed that even the oil from juvenile leaves could be utilized if very controlled conditions were used. They analyzed the leaf oils from seedlings (juvenile foliage) of *J. virginiana* and *J. scopulorum* grown in a common garden and determined patterns of geographic variation between *J. scopulorum*

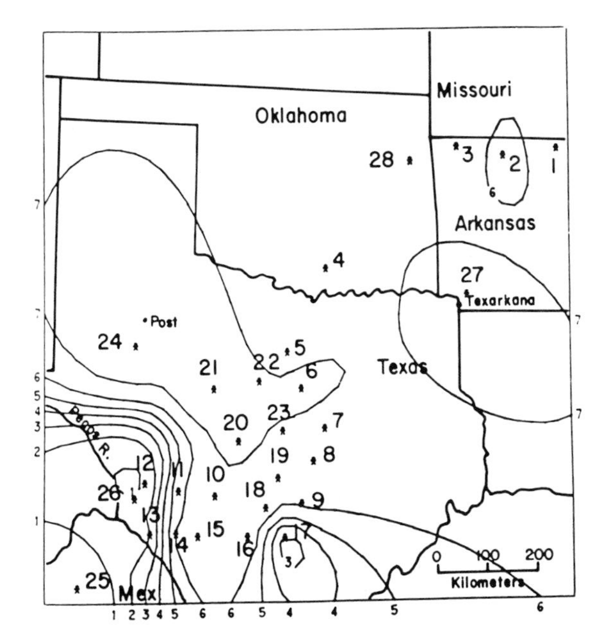

Fig. 10. Principal coordinate analysis of the leaf terpenoids of *J. ashei* showing that the major geographic trend (50% of the variation) is due to the ancestral (relictual) genotypes in populations 12, 13, 25, and 26 arising from the Mexican highlands. (Adams 1977)

and *J. viginiana* throughout the Great Plains of the United States (Comer et al. 1982).

Finally, I would like to mention a study on *J. silicicola* and *J. virginiana* in the southeastern United States that resulted in the substantiation of the recognition of *J. virginiana* var. *silicicola* (Adams 1986). The two taxa are scarcely distinct morphologically or chemically. As can be seen in Fig. 12, canonical variate analyses of the terpenoids clearly shows that the two taxa form a continuum. In fact one population (WT from Texas) that had previously been called *J. silicicola* (due to the flattened tree crowns), is clearly most similar to *J. virginiana* from nearby Bastrop, Texas (note WT and BT, Fig. 12). Due to the clinal gradation into *J. virginiana*, specific status for *J. silicicola* could not be supported but because *J. silicicola* occupies unique sites for junipers (sand dunes) and thus probably has some unique physiological genes for that adaptation, the varietal status (*J. virginiana* var. *silicicola*) was chosen for that taxon.

6.2.2 Other Forest Trees

Terpenoids have been used in numerous studies on geographic variation of other forest trees. Many of these studies have been on conifers such as *Abies grandis* (Zavarin et al. 1977), *Picea glauca* (Wilkinson et al. 1971), *Pseudotsuga menziesii*

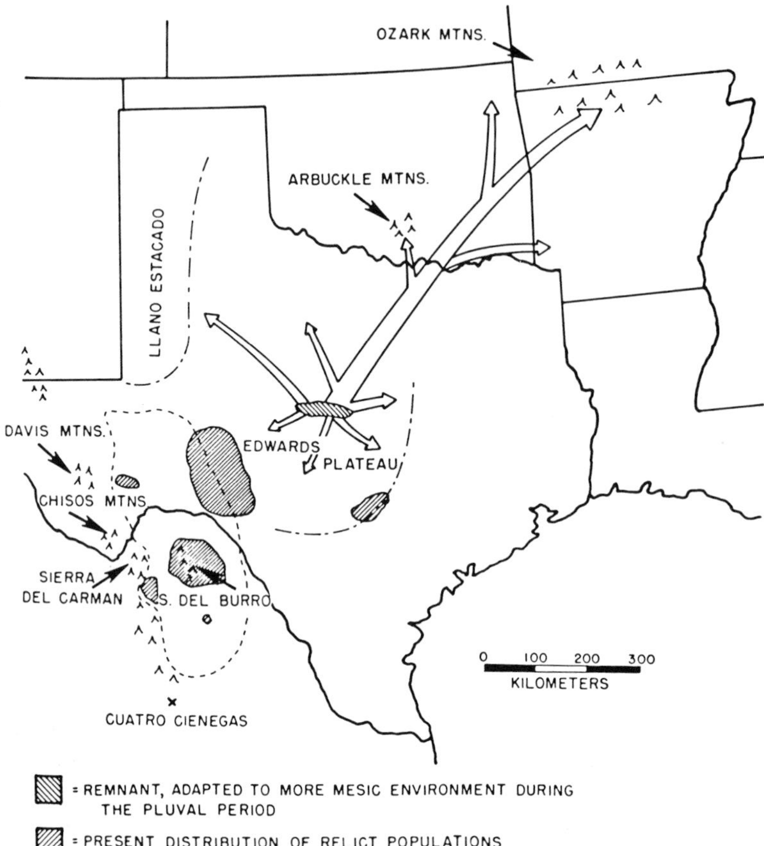

Fig. 11. The post-glacial migration of *J. ashei* into its present distribution, based on the terpenoids and morphology. The congruence of the terpenoids and morphology greatly strengthens the postulate. (Adams 1977)

(von Rudloff 1973, 1975; Zavarin and Snajberk 1975) and *Pinus* (Adams and Edmunds 1989; Hunt and von Rudloff 1977; Smith and Preisler (1988); Zavarin et al. (1989).

For example, von Rudloff (1975) found that a clinal pattern from the interior range of *Pseudotsuga menziesii* (Douglas fir) to the coastal locations (Fig. 13). Notice cpd. 7 (β-pinene) ranges from about 5% in the interior population to over 40% in the coastal population. Likewise, bornyl acetate (cpd. 33) is over 30% in the interior population and decreases to almost zero in the coastal populations (Fig. 13).

Analyses (Zavarin et al. 1989) of the wood monoterpenes of pinyon pine (*Pinus edulis*) from 18 locations revealed that the populations from Arizona and

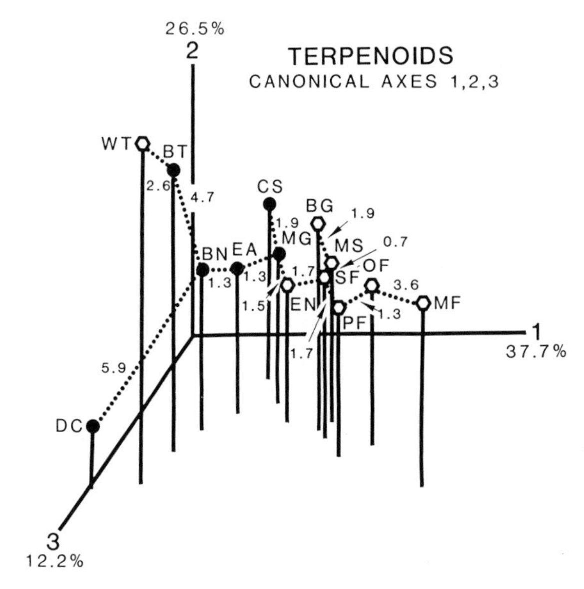

Fig. 12. Ordination of populations of *J. virginiana* (*solid circles*) and putative *J. virginiana* var. *silicicola* (*open hexagons*). Note the putative var. *silicicola* from West Columbia, Texas (WT), is clearly most similar to *J. virginiana* from Bastrop, Texas (BT) and is not *J. virginiana* var. *silicicola*. (Adams 1986)

New Mexico differed considerably from populations in other areas (Fig. 14). These populations were named the Apache chemical race. Interestingly, a presumed, recently established (ca. a few hundred years ago) population in northern Colorado (popn. 26, Fig. 14), clustered strongly with the Arizona and New Mexico, Apache race group. Because the native Americans used (and continue to use) pinyon nuts for food, it is thought that this outlying population was established during north-south migrations by the native Americans. This paper serves as a good example of the potential use of volatile oil characters to address anthropological questions as well as the broad applicability of terpenoid data.

6.3 Taxon Level Differences and Evolutionary Studies

6.3.1 Juniperus

One of the common applications of terpenoid data is for the determination of species limits and for identification of unknown plants. A recent example is that of the cultivated tree referred to as *J. excelsa* at the Royal Botanic Garden, Kew, London. Foliage was collected and compared with native trees in Greece and the tree of unknown origin at Kew was definitely established as being *J. excelsa* (Adams 1990).

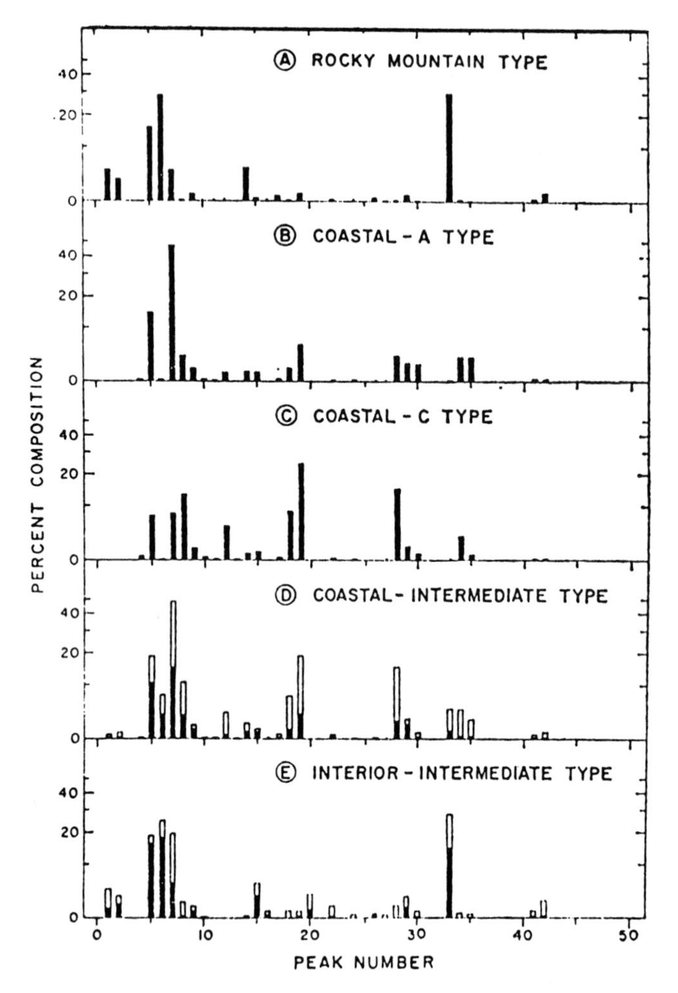

Fig. 13. Bar histograms showing mean leaf oil compositions of the rocky mountain and coastal varieties of Douglas fir and intermediate types. (von Rudloff 1975)

The leaf terpenoids have been used in North America to analyze forms of *J. monticola* (now *J. sabinoides*) (Adams et al. 1980a), varieties of *J. flaccida* (Adams et al. 1984a), varieties of *J. deppeana* (Adams et al. 1984b), and ancestral and derived species (*J. saltillensis* and *J. ashei*) (Adams et al. 1980b).

A number of comparisons have been made between species such as: *J. durangensis* and *J. jaliscana* (Adams et al. 1985a); *J. comitana*, *J. gamboana* and *J. standlevi* (Adams et al. 1985b); *J. lucayana* and *J. saxicola* (Adams et al 1987); *J. blancoi*, *J. horizontalis*, *J. virginiana*, and *J. scopulorum* (Adams et al., 1981a); *J. californica*, *J. monosperma*, *J. occidentalis* and *J. osteosperma* (Adams et al. 1983); *J. oxycedrus*, *J, thurifera* and *J. sabina* (Hernandez et al. 1987); *J. dahurica*, *J.*

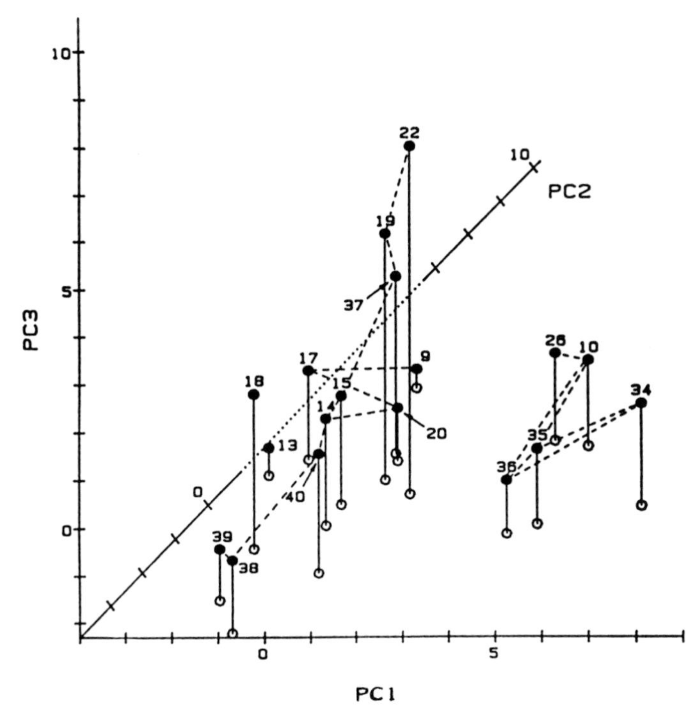

Fig. 14. Principal component analysis of biosynthetically-transformed monoterpenes for populations of pinyon pine (*Pinus edulis*) showing clustering by region. The cluster of 10, 26, 34, 35, and 36 was named the Apache chemical race. (Zavarin et al. 1989)

pseudosabina, *J. sabina* and *J. sibirica* (Satar 1984), and *J. erythrocarpa*, *J. monosperma* and *J. pinchotii* (Adams et al. 1981b).

Several studies have compared numerous species spanning a region: Caribbean junipers (Adams and Hogge 1983; Adams 1983b; Adams 1989b) and the junipers of Mexico and Guatemala (Zanoni and Adams 1976).

The Caribbean junipers are interesting in that they are a monophyletic group of relative recent origin, and they differ by only a few morphological characters. However, there has been considerable divergence in their leaf volatile oils. The leaf oils have proved very useful in analysis of the origin of the group (Adams, 1989b). Figure 15 shows the proposed routes of speciation for the group, based on principal coordinate analysis and a minimum spanning network using terpenoid data. The pathways and relationships in Hispaniola are so complex that additional characters and/or more sampling will be needed to resolve the situation.

The fact that terpenoids can be readily quantitated is a key factor in their use, as they are amenable to multivariate analysis. Underlying trends can be discovered and random variation reduced by multivariate analyses.

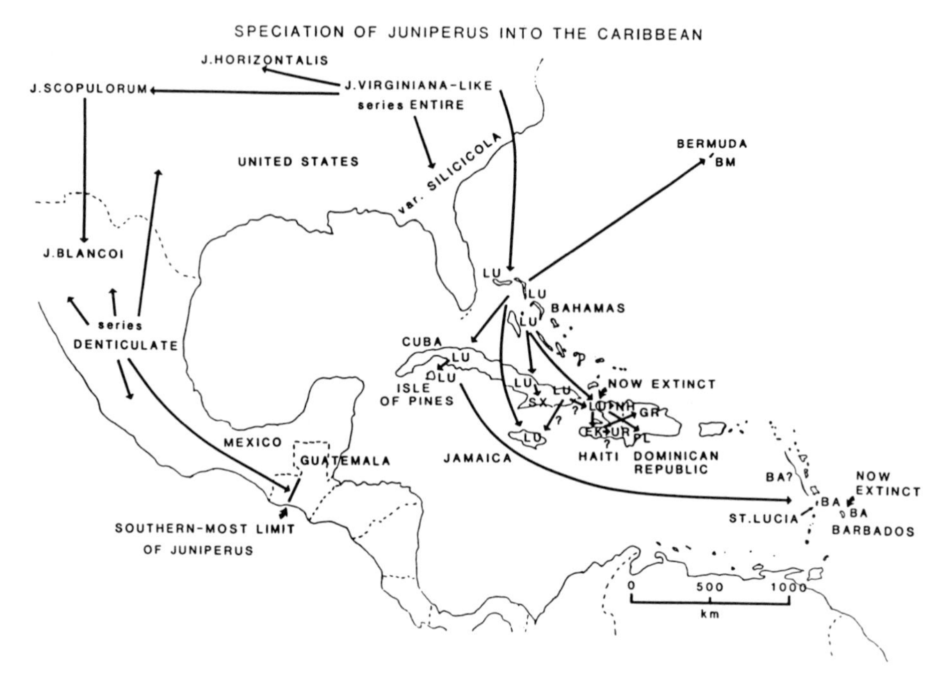

Fig. 15. Proposed speciation of *Juniperus* into the Caribbean from *J. virginiana* (or ancestral stock). The analysis was primarily based on terpenoid data because of the lack of morphological features separating the taxa. (Adams 1989b)

6.3.2 Other Forest Trees

Langenheim and her students (see Langenheim et al. 1982) have carefully researched both specific differences and co-evolution with herbivorous insects in *Hymenaea* (a tropical legume genus). They have found the leaf resin terpenoids to be of considerable taxonomic use, but also of use for the analyses of co-evolution with insects. Work on co-evolution has resulted in the formation of a subdiscipline, chemical ecology. The number of papers in chemical ecology is now so great that is beyond the scope of this review.

Carman and Sutherland (1979) analyzed the leaf diterpenes of several accessions of *Cupressus macrocarpa* and *C. arizonica* and individual samples from *C. tortulosa* and *C. sempervirens*. All of the 27 trees were non-native and had been placed into cultivation in Australia. The identities of two of the *C. macrocarpa* were uncertain. The diterpenes generally divided the individuals into two groups: high in isophyllocladene, phyllocladene, and isohibaene, or high in abietadiene and abietatriene. Although the study was limited in samples and particularly in the knowledge concerning the origin of the trees sampled, it does provide some indication that diterpenes could be useful for systematic studies.

Pauly et al. (1983) examined the volatile leaf oils from two closely related *Cupressus* species (*C. dupreziana*, *C. sempervirens*) and found their leaf oils to be almost identical. They concluded that *C. dupreziana* was a subspecies, derived from *C. sempervirens*. The leaves were ground (which I would not recommend) and steam distilled for 3 h (cf. 24 h by Carman and Sutherland 1979). Only a trace of the diterpenoid manool was found, so perhaps the extractions were too short to give a good yield of diterpenes.

Lastly, it should be noted that von Rudloff has published numerous papers on specific differences in the leaf oils of conifers (see von Rudloff 1975 for a review). Figure 16 shows specific differences in the volatile leaf oils of North

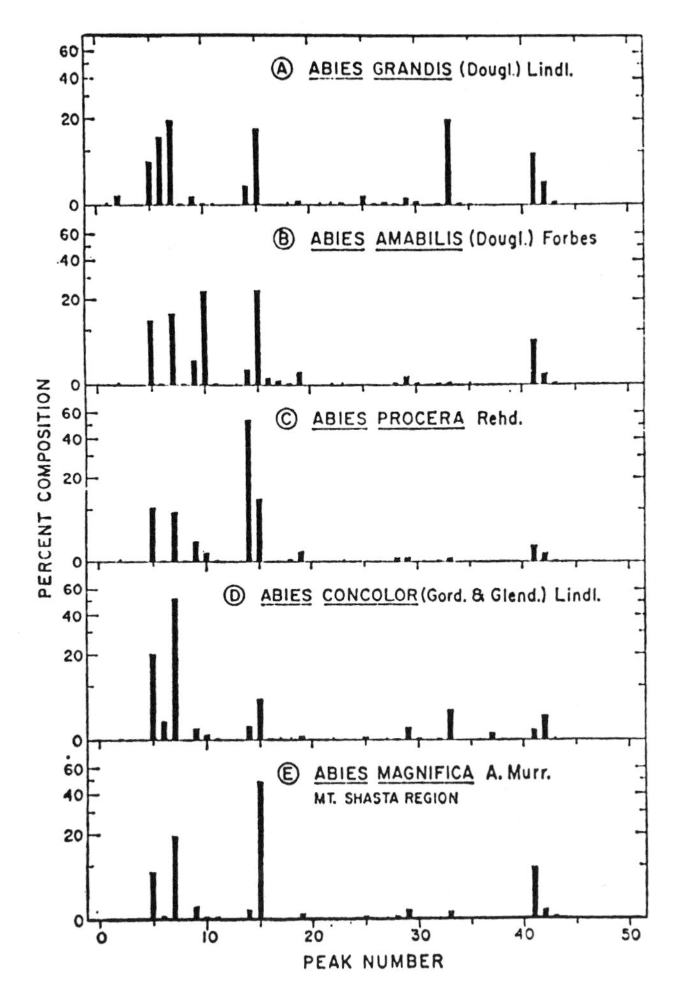

Fig. 16. Bar histograms showing average leaf oil terpene compositions for *Abies* species (firs) Each species has a unique profile of components which can be used for chemotaxonomic studies. (von Rudloff 1975)

American firs (*Abies*). This is but one example of many that he presents to show the utility of volatile leaf oils in analyzing evolutionary patterns among species (von Rudloff 1975). One should note, however, that his work was preceded by detailed baseline studies on ontogenetic, seasonal, and sampling studies. Although we can now be more assured of the impact of ontogenetic, seasonal, and sampling studies, one must still do some preliminary research before embarking on a terpenoid-based study.

References

Adams RP (1970a) Differential systematics and contour mapping of geographical variation. Syst Zool 19:385–390

Adams RP (1970b) Seasonal variation of terpenoid constituents in natural populations of *Juniperus pinchotii* Sudw. Phytochemistry 9:397–402

Adams RP (1972a) Numerical analyses of some common errors in chemosystematics. Brittonia 24:9–21

Adams RP (1972b) Chemosystematic and numerical studies of natural populations of *Juniperus pinchotii* Sudw. Taxon 21:407–427

Adams RP (1975a) Gene flow versus selection pressure and ancestral differentiation in the composition of species: analysis of populational variation in *Juniperus ashei* Buch. using terpenoid data. J Mol Evol 5:177–185

Adams RP (1975b) Numerical-chemosystematic studies of infraspecific variation in *Juniperus pinchotii* Sudw. Biochem Syst Ecol 3:71–74

Adams RP (1977) Chemosystematics — analysis of populational differentiation and variability of ancestral and modern *Juniperus ashei*. Ann Mo Bot Gard 64:184–209

Adams RP (1979) Diurnal variation in the terpenoids of *Juniperus scopulorum* (Cupressaceae)—summer versus winter. Am J Bot 66:986–988

Adams RP (1982a) The effects of gases from a burning coal seam on morphological and terpenoid characters in *Juniperus scopulorum* (Cupressaceae). Southwest Nat 27:279–286

Adams RP (1982b) A comparison of multivariate methods for the detection of hybridization. Taxon 31:646–661

Adams RP (1983a) Infraspecific terpenoid variation in *Juniperus scopulorum* evidence for Pleistocene refugia and recolonization in western North America. Taxon 32:30–46

Adams RP (1983b) The junipers (*Juniperus*: Cupressaceae) of Hispaniola: comparisons with other Caribbean species and among collections from Hispaniola. Moscosa 2:77–89

Adams RP (1986) Geographic variation in *Juniperus silicicola* and *Juniperus virginiana* of the southeastern United States: multivariate analyses of morphology and terpenoids. Taxon 35:61–75

Adams RP (1989a) Identification of essential oils by ion trap mass spectroscopy. Academic Press, New York

Adams RP (1989b) Biogeography and evolution of the junipers of the west Indies. In: Woods CA (ed) Biogeography of the West Indies. Sandhill Crane Press, Gainesville, FL, pp 167–190

Adams RP (1990) The chemical composition of the leaf oils of *Juniperus excelsa* M.-Bieb. J Ess Oil Res 2:45–48

Adams RP, Edmunds GF Jr (1989) A reexamination of the volatile leaf oils of *Pinus ponderosa* Dougl. ex. P. Lawson using ion trap mass spectroscopy. Flav Frag J 4:19–23

Adams RP, Hagerman A (1976) The volatile oils of young versus mature leaf oils of *Juniperus scopulorum:* chemosystematic significance. Biochem Syst Ecol 4:75–79

Adams RP, Hagerman A (1977) Diurnal variation in the volatile terpenoids of *Juniperus scopulorum* (Cupressaceae). Am J Bot 64:278–285

Adams RP, Hogge L (1983) Chemosystematic studies of the Caribbean junipers based on their volatile oils. Biochem Syst Ecol 11:85–89

Adams RP, Powell RA (1976) Seasonal variation of sexual differences in the volatile oil of *Juniperus scopulorum*. Phytochemistry 15:509–510

Adams RP, Turner BL (1970) Chemosystematic and numerical studies in natural populations of *Juniperus ashei* Buch. Taxon 19:728–751

Adams RP, Granat M, Hogge L, von Rudloff E (1979) Identification of lower terpenoids from gas chromatograph-mass spectral data by on-line computer method. J. Chromatogr Sci 17:75–81
Adams RP,von Rudloff E, Hogge L, Zanoni TA (1980a) The volatile terpenoids of *Juniperus monticola* f. *monticola*, f. *compacta*, and *f. orizabensis*. J Nat Prod 43:417–419
Adams RP, von Rudloff E, Zanoni TA, Hogge L (1980b) The terpenoids of an ancestral/advanced species pair of *Juniperus*. Biochem Syst Ecol 8:35–37
Adams RP, von Rudloff E, Hogge L, Zanoni TA (1981a) The volatile terpenoids of *Juniperus blancoi* and its affinities with other entire leaf margined junipers of North America. J Nat Prod 44:21–26
Adams RP, Zanoni TA, von Rudloff E, Hogge L (1981b) The southwestern USA and northern Mexico one-seeded junipers: their volatile oils and evolution. Biochem Syst Ecol 9:93–96
Adams RP, Palma MM, Moore WS (1981c) Examination of the volatile oils of mature and juvenile leaves of *Juniperus horizontalis*: Chemosystematic significance. Phytochemistry 20:2501–2502
Adams RP, von Rudloff E, Hogge L (1983) Chemosystematic studies of the western North American junipers based on their volatile oils. Biochem Syst Ecol 11:189–193
Adams RP, Zanoni TA, Hogge L (1984a) The volatile oils of *Juniperus flaccida* var. *flaccida* and var. *poblana*. J. Nat Prod 47:1064–1065
Adams RP, Zanoni TA, Hogge L (1984b) Analyses of the volatile leaf oils of *Juniperus deppeana* and its infraspecific taxa: chemosytematic implications. Biochem Syst Ecol 12:23–27
Adams RP, Zanoni TA, Hogge L (1985a) The volatile leaf oils of two rare junipers from western Mexico: *Juniperus durangensis* and *Juniperus jaliscana*. J Nat Prod 48:673–675
Adams RP, Zanoni TA, Hogge L (1985b) The volatile leaf oils of the junipers of Guatemala and Chiapas, Mexico *Juniperus comitana*, *Juniperus gamboana* and *Juniperus standlevi*. J Nat Prod 48:678–681
Adams RP, Almirall AL, Hogge L (1987) The volatile leaf oils of the junipers of Cuba: *Juniperus lucayana* Britton and *Juniperus saxicola* Britton and Wilson. Flav Frag J 2:33–36
Bailey, DK, Snajberk K, Zavarin E (1982) On the question of natural hybridization between *Pinus discolor* and *Pinus cembroides*. Biochem Syst Ecol 10:111–119
Burbott AJ, Loomis WD (1969) Evidence of metabolic turnover of monoterpenes in peppermint. Plant Physiol 44:173–179
Carman RM, Sutherland MD (1979) Cupressene and other diterpenes of *Cupressus* species. Aust J Chem 32:1131–1142
Comer CW, Adams RP, Van Haverbeke DR (1982) Intra- and interspecific variation of *Juniperus virginiana* L. and *J. scopulorum* Sarg. seedlings based on volatile oil composition. Biochem Syst Ecol 10:297–306
Craveiro AA, Matos FJA, Alencar JW, Plumel MM (1989) Microwave extraction of an essential oil. Flav Frag J 4:43–44
Cu J-Q, Perineau F, Delmas M, Gaset A (1989) Comparison of the chemical composition of carrot seed essential oil extracted by different solvent. Flav Frag J 4:255–231
Fassett NC (1945) *Juniperus virginiana*, *J. horizontalis* and *J. scopulorum* V. Taxonomic treatment. Bull Torrey Bot Club 72:480–482
Flake RH, von Rudloff E, Turner BL (1969) Quantitative study of clinal variation in *Juniperus virginiana* using terpenoid data. Proc Natl Acad Sci USA 62:487–494
Flake RH, von Rudloff E, Turner BL (1973) Confirmation of a clinal pattern of chemical differentiation in *Juniperus virginiana* from terpenoid data obtained in successive years. In: Runeckles VC, and Marbry TJ (eds) Terpenoids: Structure, biogenesis, and distribution. Academic Press, New York, pp 215–228 Recent advances in phytochemistry series, Vol 6
Flake RH, Urbatsch L, Turner BL (1978) Chemical documentation of allopatric introgression in *Juniperus*. Syst Bot 3:129–144
Fluck H (1963) Intrinsic and extrinsic factors affecting the production of natural products. In: Swain T (ed) Chemical plant taxonomy. Academic Press, London, pp 167–186
Fretz TA (1976) Effect of photoperiod and nitrogen on the composition of foliar monoterpenes of *Juniperus horizontalis* Moench. cv. Plumosa. J. Am Soc Hortic Sci 101:611–613
Hall MT (1952) Variation and hybridization in *Juniperus*. Ann Mo Bot Gard 39:1–64
Hernandez EG, Martinez MDCL, Villanova RG (1987) Determination by gas chromatography of terpenes in the berries of the species *Juniperus oxycedrus* L., *J. thurifera* L. and *J. sabina* L. J Chromatogr 396:416–420
Hopfinger JA, Kumamoto J, Scora RW (1979) Diurnal variation in the essential oils of Valencia orange leaves. J Bot 66:111–115

Hunt RS, von Rudloff E (1977) Leaf-oil-terpene variation in western white pine populations of the Pacific northwest. For Sci 23:507–516

Irving RS, Adams RP (1973) Genetic and biosynthetic relationships of monoterpenes. In: Runeckles VC, Mabry TJ (eds) In: Terpenoids: structure, biogenesis, and distribution. Academic Press, New York, Recent advances in phytochemistry series, vol 6.

Jennings W (1978) Gas chromatography with glass capillary columns. Academic Press, New York

Jennings W (1987) Analytical gas chromatography. Academic Press, New York

Jennings W, Shibamoto T (1980) Qualitative analysis of flavor and fragrance volatiles by glass capillary gas chromatography. Academic Press, New York

Langenheim JH, Lincoln DE, Stubblebine WH, Gabrielli AC (1982) Evolutionary implications of leaf resin pocket patterns in the tropical tree *Hymenaea* (Caesalpinioideae: Leguminosae). Am J Bot 69:595–607

Leach GJ, Whiffin T (1989) Ontogenetic, seasonal and diurnal variation in leaf oils and leaf phenolics of *Angophora costata*. Aust Syst Bot 2:99–111

Lincoln DE, Langenheim JH (1977) Effect of light and temperature on monoterpene yield and composition in *Satureja douglasii*. Biochem Syst Ecol 6:21–32

Neet-Sarqueda C, Plummettaz Clot A.-C, Becholey I (1988) Mise en évidence de l'hybridation introgressive entre *Pinus sylvestris* L. et *Pinus uncinata* DC. en Valais (Suisse) par deux methodes multi-variées. Bot Helv 98:161–169

Ogilvie RT, von Rudloff E (1968) Chemosystematic studies in the genus *Picea* (Pinaceae). IV. The introgression of white and Englemann spruce as found along the Bow River. Can J Bot 46:901–908

Palma-Otal M, Moore WS, Adams RP, Joswiak GR (1983) Morphological chemical and biogeographical analyses of a hybrid zone involving *Juniperus virginiana* and *J. horizontalis* in Wisconsin. Can J Bot 61:2733–2746

Pauly G, Yani A, Piovetti L, Bernard-Dagan C (1983) Volatile constituents of the leaves of *Cupressus dupreziana* and *Cupressus sempervirens*. Phytochemistry 22:957–959

Powell RA, Adams RP (1973) Seasonal variation in volatile terpenoids of *Juniperus scopulorum* (Cupressaceae). Am J Bot 60:1041–1051

Satar S (1984) Mono- and sesquiterpenes in the essential oil of mongolian *Juniperus* species. Pharmazie 39:66–67

Simmons D, Parsons RF (1976) Analysis of a hybrid swarm involving *Eucalyptus crenulata* and *E. ovata* using leaf oils and morphology. Biochem Syst Ecol 4:97–101

Smith RH, Preisler HK (1988) Xylem monoterpenes of *Pinus monophylla* in California and Nevada. Southwest Nat 33:205–214

Snajberk K, Zavarin E, Derby R (1982) Terpenoid and morphological variability of *Pinus quadrifolia* and the natural hybridization with *Pinus monophylla* in the San Jacinto Mountains of California. Biochem Syst Ecol 10:121–132

van den Dries, Svendsen AB (1989) A simple method of detection of glycosidic bound monoterpenes and other volatile compounds occurring in fresh plant material. Flav Frag J 4:59–61

Vernin G, Boniface C, Metzger J, Ghiglione C, Hammoud A, Suon K, Fraisse D, Parkanyi C (1988) GC-MS-SPECMA bank analysis of *Juniperus communis* needles and berries Phytochemistry 27:1061–1064

von Rudloff E (1972) Seasonal variation in the composition of the volatile oil of the leaves, buds, and twigs of white spruce (*Picea glauca*). Can J. Bot 50:1595–1603

von Rudloff E (1973) Geographical variation in the terpene leaf composition of the volatile oil of Douglas fir. Pure Appl Chem 34:401–410

von Rudloff E (1975) Volatile leaf oil analysis in chemosystematic studies of North American conifers. Biochem syst Ecol 2:131–167

von Rudloff E, Irving R, Turner BL (1967) Reevaluation of allopatric introgression between *Juniperus ashei* and *Juniperus virginiana* using gas chromatography. Am J Bot 54:600

Whiffin T, Hyland BPM (1989) The extent and systematic significance of seasonal variation of volatile oil composition in Australian rainforest trees. Taxon 38:167–177

Wilkinson RC, Hanover JW, Wright JW, Flake RH (1971) Genetic variation in the monoterpene composition of white spruce. For Sci 17:83–90

Zanoni TA, Adams RP (1976) The genus *Juniperus* in Mexico and Guatemala: Numerical and chemosystematic analysis. Biochem Syst Ecol 4:147–158

Zavarin E, Snajberk K, (1975) *Pseudotsuga menziesii* chemical races of California and Oregon. Biochem Syst Ecol 2:121–129
Zavarin E, Snajberk K, Critchfield WB (1977) Terpenoid chemosystematic studies of *Abies grandis*. Biochem Syst Ecol 5:81–93
Zavarin E, Snajberk K, Derby R (1980) Terpenoid and morphological variability of *Pinus quadrifolia* and its natural hybridization with *Pinus monophylla* in Northern Baja California and adjoining United States. Biochem Syst Ecol 8:225–235
Zavarin E, Snajberk K,Cool L (1989) Monoterpenoid differentiation in relation to the morphology of *Pinus edulis*. Biochem Syst Ecol 17:271–282

Cedar Wood Oil — Analyses and Properties

R.P. ADAMS

1 Introduction

Cedarwood oil is an important natural product for components used directly in fragrance compounding or as a source of raw components in the production of additional fragrance compounds. The oil is used to scent soaps, technical preparations, room sprays, disinfectants, and similar products, as a clearing agent for microscope sections, and with immersion lenses (Guenther 1952).

The price varies but has generally been about $4.50/lb. for Virginia cedarwood oil, $3.50/lb. for Texas cedarwood oil and $1.50-$1.75/lb. for Chinese cedarwood oil. The Chinese cedarwood oil, although almost identical in composition, is less valued because its fragrance is very different from the Texas and Virginia cedarwood oils. The commercial cedarwood oils are obtained from three genera of the Cupressaceae: *Juniperus* (Texas and Virginia oils); *Cupressus* (China) and *Cedrus* (Morocco, India) according to Bauer and Garbe 1985. The heartwood oils of the Cupressaceae are well known for having the same components across the family (i.e., evolutionally conserved), so the occurrence of similar oils in different genera should not be surprising.

The world production (1984) has been reviewed by Lawrence (1985), who reported the following (source and metric tons): Texas (*J. ashei* Buch.)—1400; Virginia (*J. virginiana* L., S.E. United States)—240; China (*Cupressus funebris* Endl.)—450; India (Himalaya, *Cedrus atlantica* Menetti)—20; Morocco (Atlas Mtns., *Cedrus deodora* Loud.)— 7; Kenya (*J. procera* Endl.)— no production at present.

Cedarwood oils have not been examined thoroughly or systematically. Many of the analyses of cedarwood oil were done by Runeberg (1960a-e, 1961) and associates (Pilo and Runeberg 1960; Pettersson and Runeberg 1961) (Table 1). For many years *J. ashei* was reported to contain only alpha-cedrene and cedrol (Guenther 1952; erroneously referred to as *J. mexicana* in Erdtman and Norin 1966 and Walker 1968, see Zanoni and Adams 1979 for nomenclatural discussion). However, more recently, Kitchens et al. (1971) reported beta-cedrene, thujopsene, widdrol, pseudocedrol, beta-chamigrene, prim cedrol, widdrene, isowiddrene, alpha-chamigrene, and cuparene (three isomers). Unfortunately, as is typical of most of the reports on identifications, not enough data were given by Kitchens et al. (1971) to allow confirmatory studies of their work.

Juniperus californica Carr.was reported (Pettersson and Runeberg 1961) to have cedrol as the major component (52%) of the heartwood volatile oils (Table 1) with a considerable amount of thujopsene (26%). A more recent study (Adams 1987) reported low yields (Table 1) of cedrol from two chemical races (A and B, Vasek and Scora 1967) of *J. californica.* It is presumed that Pettersson and Runeberg (1961) may have analyzed the wood of *J. occidentalis* Hooker instead of *J. californica.*

Juniperus communis L. was reported to have mostly thujopsene (37%); however, it is not clear how the author (Bredenberg 1961) arrived at these percentages

Table 1. Literature reports on the composition of the volatile wood oils of *Juniperus* species. Approximate percent concentration of key components was obtained when possible from the original literature cited

Species	[a]ACDR	BCDR	THJP	CPRN	CDRL	WDDL	Reference
J. ashei	+	+	+	+	+	+	Guenther 1952; Windemuth 1945
(=*J. mexicana* in part)							Kitchens et al. 1971
J. ashei	1.8	1.6	60.4	2.8	19.0	1.1	Adams 1987
J. californica	2.6		26.0	1.0	52.0	0.2	Pettersson and Runeberg 1961
J. californica'A'	4.9	2.7	19.7	6.4	8.0	8.0	Adams 1987
J. californica'B'	3.9	1.9	18.7	4.7	9.3	9.2	Adams 1987
J. cedrus			82.4	3.7	2.2	2.6	Runeberg 1960a
J. chinensis			11.6	4.3	72.9	6.0	Pilo and Runeberg 1960
J. communis			37.0	3.0	2.0	1.0	Bredenberg 1961
J. conferta			+	+	+		Doi and Shibuya 1972
J. deppeana	16.9	3.9	14.9	3.9	26.4	1.0	Adams 1987
J. erythrocarpa	1.9	1.6	67.9	3.0	8.5	0.5	Adams 1987
J. excelsa					+		Rutowski and Vinogradova 1927
J. foetidissma	58.3				8.3	5.0	Runeberg 1961
J. horizontalis	+		+	+	+	+	Narasimhachari and von Rudloff 1961
J. monosperma	2.7	1.8	61.0	3.8	4.1	1.7	Adams 1987
J. occidentalis	+				+		Kurth and Lackey 1948
J. occidentalis var.							
occidentalis	8.8	2.6	18.9	1.5	38.9	1.6	Adams 1987
australis	3.3	1.3	20.1	1.5	38.2	1.6	Adams 1987
J. osteosperma (= *J. utahensis*)	12.7		47.8	12.5		13.5	Runeberg 1960b
J. osteosperma	4.0	1.8	40.0	2.6	13.2	1.5	Adams 1987
J. phoenicea			79.3	2.9	7.2	0.1	Runeberg 1960c
J. pinchotii	2.8	1.2	4.8	0.1	4.4	—	Adams 1987
J. procera	41.8			2.5	41.8		Pettersson and Runeberg 1961
J. recurva	3.5	0.9	5.1	1.8	49.0	16.7	Oda et al. 1977
J. semiglobosa		+			+		Goryaev et al. 1962
J. thurifera	23.3		15.5	3.9	27.1		Runeberg 1960d
J. scopulorum	4.3	2.4	57.9	6.1	6.1	3.0	Adams 1987
J. virginiana	35.0		30.0	2.0	4.0	2.0	Runeberg 1960e
J. virginiana	27.2	7.7	27.6	6.3	15.8	1.0	Adams 1987

[a]ACDR = alpha-cedrene; BCDR = beta-cedrene; THJP = thujopsene; CPRN = cuparene; CDRL = cedrol; WDDL = widdrol.

Juniperus horizontalis Moench is a prostrate plant that forms mats. Due to its low wood biomass, its oil composition is primarily of academic interest. Narasimhachari and von Rudloff (1961) reported that *J. horizontalis* contained alpha-cedrene, thujopsene, cuparene, cedrol, and widdrol, but relative concentrations were not reported (Table 1). *Juniperus occidentalis* was examined by Kurth and Lackey (1948), who merely reported that the wood contained alpha-cedrene and cedrol. A more recent analysis of both varieties (Table 1; Adams 1987) showed the varieties to be high in cedrol and thujopsene.

Juniperus osteosperma (referred to as *J. utahensis* Lemm. by Runeberg 1960b) had 47.8% thujopsene, with about equal amounts of alpha-cedrene, cuparene, and widdrol (Table 1). Adams (1987) found that the taxon was high in thujopsene, but reported that cedrol was also a major component (Table 1).

Juniperus virginiana L. wood was not directly analyzed by Runeberg (1960e). Using a commercial sample of cedarwood oil said to be from *J. virginiana*, he found mostly alpha-cedrene and thujopsene with a very small amount of cedrol (4%) (Table 1). However, the commercial cedarwood oil may have been precipitated or fractionally distilled to remove cedrol because Adams (1987) stated that *J. virginiana* wood (collected from native trees in Texas) contained about 16% cedrol (Table 1). Wenninger et al. (1967) analyzed the sesquiterpene hydrocarbons of American cedarwood oil (*J. virginiana*?) and reported that the oil contained 55-65% sesquiterpene hydrocarbons, with alpha-cedrene and thujopsene as the major components. Runeberg (1960e) stated that the highest yield of oil, about 3.5% of the wood (dry wt.?), was obtained from sawmill waste from older tree (i.e., trees with a greater ratio of heartwood to sapwood). Guenther (1952) obtained only a 0.2% yield by distilling sapwood of *J. virginiana*; he noted that young trees (commonly called sap cedars) yielded less than 1% oil, compared with older trees (commonly called virgin cedars), which yielded 3.5%.

2 Sample Collection

It is assumed that the reader is not only interested in the analysis of commercial cedarwood oil, but also in investigating the cedarwood oil from trees. In general, at least five trees should be sampled (obviously ten is preferred). Unfortunately, little is known about seasonal variation of wood oils. Depending on the local customs and laws, one may be faced with only a few options in collecting wood samples. I have even resorted to visiting a local woodworking shop in Ethiopia to obtain wood sample of *J. procera*. The ideal situation is to visit a site where trees are being cut for firewood, posts, lumber, etc. and obtain wood blocks directly. Failing this, one may look for broken limbs and cut off the stump section near the stem. I have also used a drill with a large wood bit (2 cm) to obtain wood from the trunk. One must be very careful to keep track of both sap- and heartwood shavings if biomass and yields are to be determined.

If a wood block is available, a wood planer is useful to produce fresh wood shavings for steam distillation. A power drill can also be used to produce wood chips for distillation, but care must be taken so the bit does not get hot and cause a loss of the oil. Cedarwood oil appears to be very stable in intact wood blocks, as cedarwood cut in the 1930's is still being collected and distilled near Junction, Texas, with the oil being apparently acceptable.

3 Oil Extraction

Commercially, cedarwood oil is extracted in a variety of manners ranging from home-built stills to a newly designed process by Texarome Inc., Leakey, Texas (Figs. 1, 2). Distillation times vary from 8 h or more in traditional stills, but only 30 s in the Texarome process (K. E. Harwell, pers. commun). One should bear in mind that changing the distillation time, sizes of wood chips extracted or temperature and pressure of the steam can make vast differences in the cedarwood oil composition, so that comparisons of various commercial oils may vary as much by distillery as by region or species utilized.

For laboratory use, one should be most careful about placing the wood into water and boiling out the oil. Several studies (Fischer et al. 1987; Koedam and Looman 1980; Koedam et al. 1981; Schmaus and Kubeczca 1985) have shown that plants produce acidic conditions when boiled and this leads to terpene rearrange-

Fig. 1. Texarome's new cedarwood oil plant near Leakey, Texas. Note *Juniperus ashei* on the hillside in the background. (Photo courtesy Texarome, Inc.)

Fig. 2. Close-up of modern oil extraction equipment at Texarome, Inc. Residence time is 30 to 60 s. (Photo courtesy Texarome, Inc.)

ments and decompositions. This is shown for cedarwood oil in Fig. 3. The initial pH of the water in the boiling flask was 7.12. After 2 h boiling the wood chips directly in the flask the pH was 6.17 and the composition was quite changed (Fig. 3, upper). Steam distillation using an apparatus with the plant material suspended above the steam generator flask (Fig. 4) resulted in the chromatogram in Fig. 3 (lower). In this case, the pH of the water in the steam generator flask was 8.62 after steam distillation. The shift in the base line (Fig. 3, upper) is indicative of decomposition. Note particularly the low yield of α- and β-cedrenes (peaks 6, 7). There is a large increase in the oxygenated sesquiterpenoids (peaks 30 and upward).

Fischer et al. (1987) discuss the fact that the original (in situ) flavor components of marjoram may be quite different from those of the commercial oils. However, if one is to work within the legal and market framework that has already been established for cedarwood oil, it seems that practical work will be forced to use steam distillation extraction. Von Rudloff (1967) examined the use of direct distillation (plant material in boiling water), a Markham-type device, and a modified Clevenger-type circulatory apparatus. He preferred the modified Clevenger-type circulatory apparatus and that is essentially what I recommend (Fig. 4). I have added ball joints so the apparatus is easier to align and the ether trap can be adjusted. Notice that the plant material is placed in the cylindrical part so that only steam comes in contact with the plant material. An external heating jacket can be added to the cylindrical part to increase the distillation efficiency if

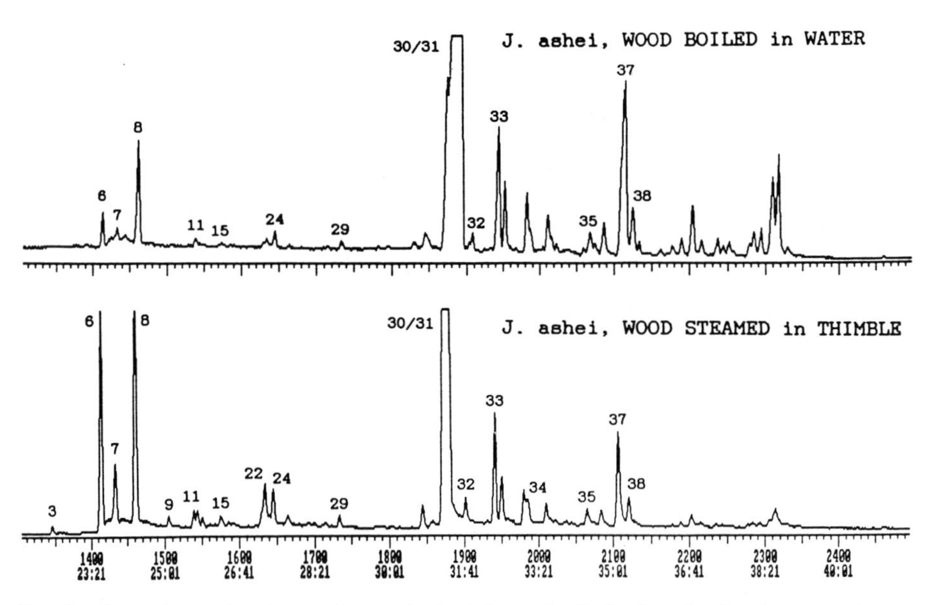

Fig. 3. Comparison of cedarwood oils obtained by hydrodistillation, boiling in water (*upper* chromatogram) and steamed in suspension (*lower* chromatogram). These and later chromatograms run on J & W DB5 silica capillary, 60–240 °C, 3°/min

desired. The condenser has also been modified so that the water jacket completely covers the ether trap area. This has resulted in much less loss of ether during distillation.

I prefer ether as the terpene trap because the ether can be evaporated by a stream of nitrogen in a hood and almost none of the terpenes are lost. Pentane could be substituted for ether. The use of hexane is discouraged because its higher boiling point results in the loss of volatile terpenes during concentration. The condenser (lower portion) should be filled with water until the water overflows into the distillation chamber. Then, the ether is placed on top of the water layer. As the distillate condenses, the oil is trapped in the ether (pentane) and the water condensate goes into the lower layer and thence back into the distillation chamber. The low density of ether allows one to trap oils that have a density greater than water. The apparatus can be run without attendance and any terpenes lost in the water are automatically volatilized as the condensate flows back into the distillation chamber.

When using the apparatus with finely ground or small wood chips, I have found it useful to place the ground wood into a sandwich of nylon screen (as used for window screens) and then place the elongated sandwich into the cylindrical chamber. If loose, finely ground material is placed directly into the cylindrical chamber, it will pack down and block the steam. Channels will form and the distillation will not proceed regularly. In addition, the distillate water, returning from the condenser, will accumulate on top of the plug and one faces the danger of

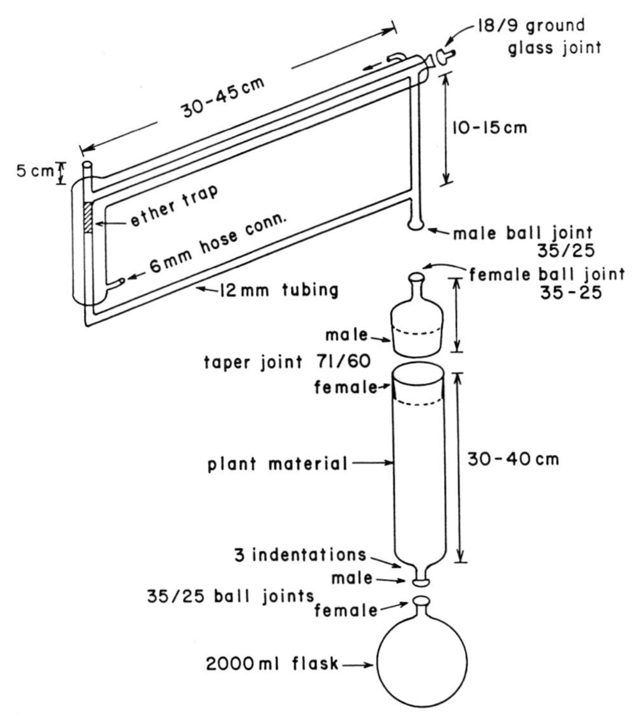

Fig. 4. Simplified Clevenger-type circulatory steam distillation apparatus recommended for cedarwood oil and general terpene extraction. Note the plant material is suspended during distillation and the oil is collected in an ether trap

running the steam generator flask to dryness. Care should be taken when handling the ether, and the entire apparatus should be placed in a well-ventilated hood when used.

The oil samples can be dried over anhydrous sodium sulfate to remove water in the ether if desired. We routinely preweigh our vials (with either compression or screw caps but in either case, using Teflon-coated caps), and evaporate the ether in a hood with nitrogen. A GC run is then used to determine the percent ether remaining in the sample and the final weight of the oil is then calculated. The samples should be stored at -20 °C or colder for long-term storage. Sealing the samples under nitrogen is also advisable for very long storage. Although decomposition of various oil samples has been mentioned to me by many colleagues, we have not experienced a problem over the past 25 years. Either our cedar and juniper oils are very stable or the aforementioned procedures mitigate decomposition. I expect that those who distill directly in water obtain oils that are quite acidic, and this may be the reason that oil decomposition is a problem. In any case, one can not assume that there will be no decomposition during long-term storage (months to years).

4 Chemical Analysis

Traditionally, cedarwood oils are defined (Walker 1968) on the basis of several physical properties: specific gravity at 15 °C (or 20 °C) 0.94-0.99; optical rotation –16 to –60°; refractive index at 20 °C 1.48-1.51; and solubility (at 20 °C) in 90 or 95% ethanol (varies with source). Although this treatment will focus on the individual chemical components, one should be aware of the practical use of the aforementioned physical properties.

4.1 Gas Chromatography

Gas chromatography has become an integral part of any essential oil analysis today. For a detailed discussion see Adams (Chap. 7, this Volume) for information on columns, carrier gases, sample injection, temperature programming and detection. All of our primary analyses are on a J & W fused silica capillary columns, DB-5, 30 m, 0.26 mm i.d., 0.25 micron coating thickness.

5 Identification

Early work on the identifications of terpenoids used component trapping from preparative GC, with subsequent liquid infrared (IR) spectral analysis for identification. The introduction of capillary columns have reduced the samples to the point that those techniques are no longer practical. The more recent development of vapor phase IR with on-the-fly analysis offers considerable promise as libraries are being compiled. However, the most practical method of identification is generally combined GC/MS or GC/MS/computer searches.

5.1 GC/MS

A large library of mass spectra is readily available from sources such as the US NBS (National Bureau of Standards, formerly the EPA/NIH data base) with thousands of spectra. Unfortunately, searches from these large data bases, with the current technology (i.e., simple matching coefficients and no retention data) do not yield reliable identifications (see Adams et al. 1979 for discussion). Although numerous papers have been written on analyses (see introduction), only the major components can be easily, unequivocally identified.

Analyses of the three major cedarwood oils are shown in Fig. 5 and a detailed list of components and retention times (on DB5) are given in Table 2. Notice that the three oils share the major components (α-cedrene, β-cedrene, thujopsene, cedrol, and widdrol). Although the minor components vary quantitatively among the oils, there is a remarkable uniformity. The off-flavor of the Chinese cedarwood oil (*C. funebris*) is apparently due to minor components.

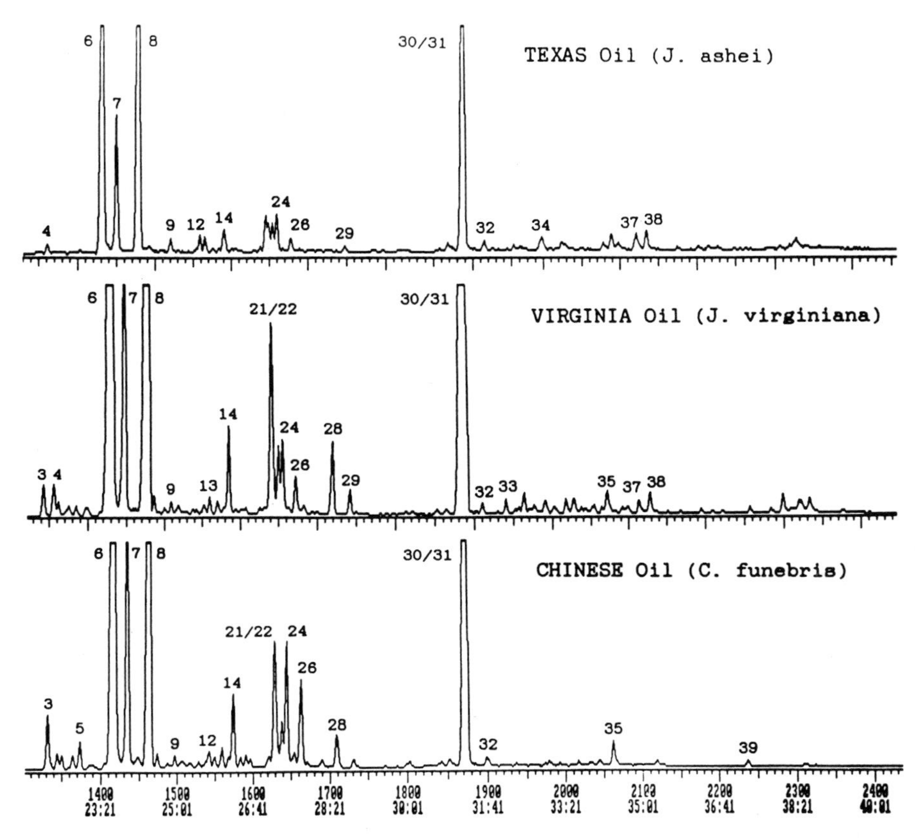

Fig. 5. Comparison of commercial cedarwood oils on a DB5 column. The *peak numbers* are identified in Table 2

Ion trap mass spectra (ITMS) for the major components are given in Figs. 6 and 7. Although the ITMS spectra are generally quite similar to quadrapole mass spectra (Adams 1989), there is a large reduction in ion 151 in both cedrol and widdrol on the ion trap. It might be noted that cedrol is very sensitive to space charging effects (overloading) and tuning on the ion trap. We use cedrol as a tuning standard on the ion trap due to its sensitivity (Adams 1989).

6 Properties

The general properties of cedarwood oils have been mentioned in the introduction. In this section, I would like to focus on several of the more unusual bioactivity properties.

Table 2. Cedarwood oil compositions from Texas (*Juniperus ashei*), Virginia (*J. virginiana*) and China (*Cupressus funebris*)

RT[a] Compound	Texas	Virginia	China
1. 734 Camphor	0.2	—	—
2. 990 Carvacrol, methyl ether	—	—	0.7
3. 1341 Sesquiterpene	—	0.7	1.7
4. 1354 Sesquiterpene	0.3	0.7	0.5
5. 1384 Sesquiterpene	—	0.2	0.8
6. 1421 α-Cedrene	30.7	21.1	26.4
7. 1441 β-Cedrene	5.5	8.2	9.2
8. 1467 Thujopsene	25.0	21.3	29.9
9. 1507 α-Himachalene	0.5	0.2	0.2
10. 1538 cis-β-Farnesene	—	0.1	0.1
11. 1547 Thujopsadiene	0.1	—	—
12. 1551 α-Acoradiene	0.7	0.2	0.6
13. 1558 β-Acoradiene	0.6	0.3	0.3
14. 1581 β-Chamigrene	1.1	1.8	2.2
15. 1585 Γ-Himachalene	0.1	—	—
16. 1594 Γ-Curcumene	0.1	0.1	0.2
17. 1602 ar-Curcumene	0.1	0.1	0.4
18. 1608 β-Selinene	—	0.1	0.2
19. 1624 Valencene	0.1	0.1	—
20. 1631 (β-Alaskene)	0.2	0.1	0.1
21. 1633 α-Selinene + ?	1.5	3.0	3.1
22. 1643 β-Himachalene	1.4	2.1	1.4
23. 1646 (α-Chamigrene)	1.2	1.6	1.4
24. 1652 Cuparene	1.7	1.6	3.4
25. 1667 β-Bisabolene	—	—	0.4
26. 1675 α-Alaskene (=Γ-acoradiene)	0.7	0.9	2.6
27. 1701 *trans*-β-Farnesene	—	—	0.1
28. 1719 Sesquiterpene	—	1.6	1.1
29. 1739 Sesquiterpene alcohol	0.3	0.6	0.3
30. 1876 Cedrol	19.1	22.2	9.6
31. 1878 Widdrol	1.6	2.3	9.5
32. 1907 6-Isocedrol	0.4	0.2	0.1
33. 1944 Cubenol	0.2	0.1	—
34. 1966 *trans*-3-Thujopsanone	0.8	—	—
35. 2072 α-Bisabolol	0.4	0.6	0.8
36. 2085 8-Cedren-13-ol	0.9	—	—
37. 2116 Sesquiterpene alcohol	0.9	0.3	—
38. 2128 Sesquiterpene alcohol	0.8	0.6	—
39. 2246 Cedryl acetate	—	—	0.1
40. 2597 Cembrene	—	—	0.1
41. 2891 Abietadiene	—	—	0.3

[a]Compounds are listed in order of their retention times (RT) on a J β W DB5 capillary column. Compounds in parenthesis are tentatively identified.

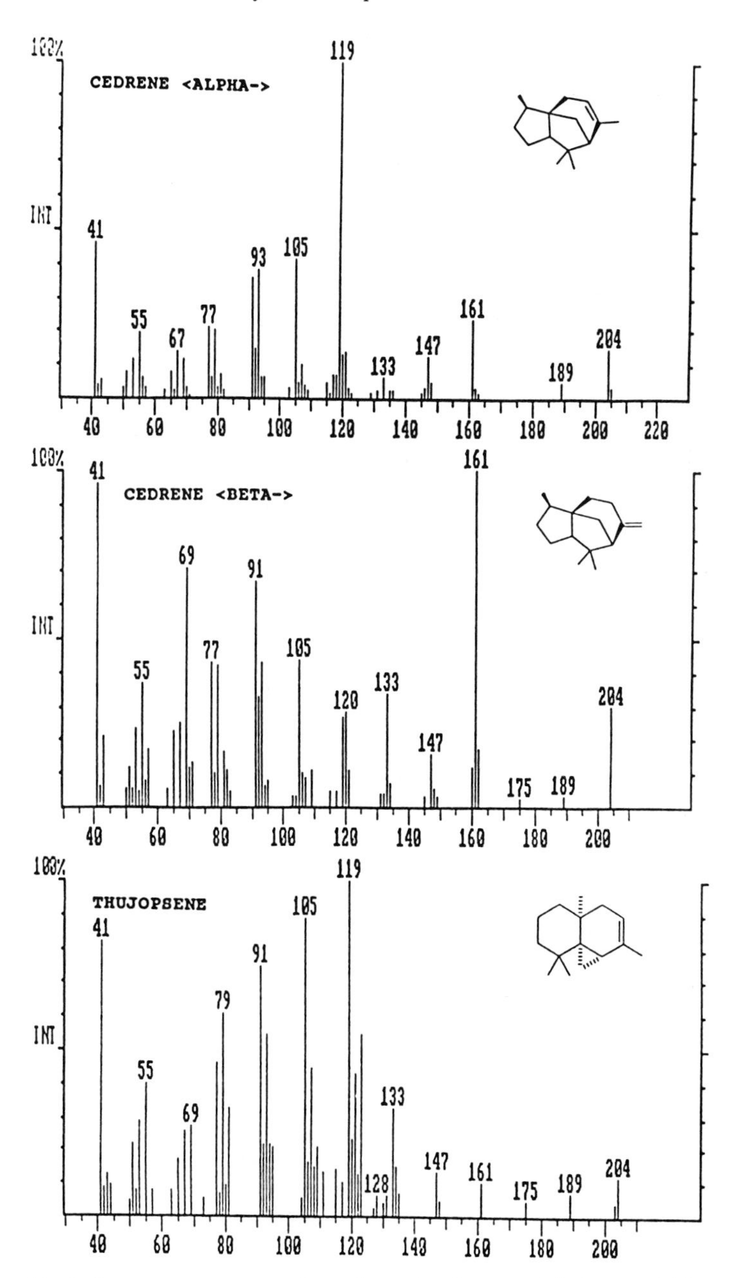

Fig. 6. Ion trap mass spectra of α-cedrene, β-cedrene, and thujopsene (*cis*)

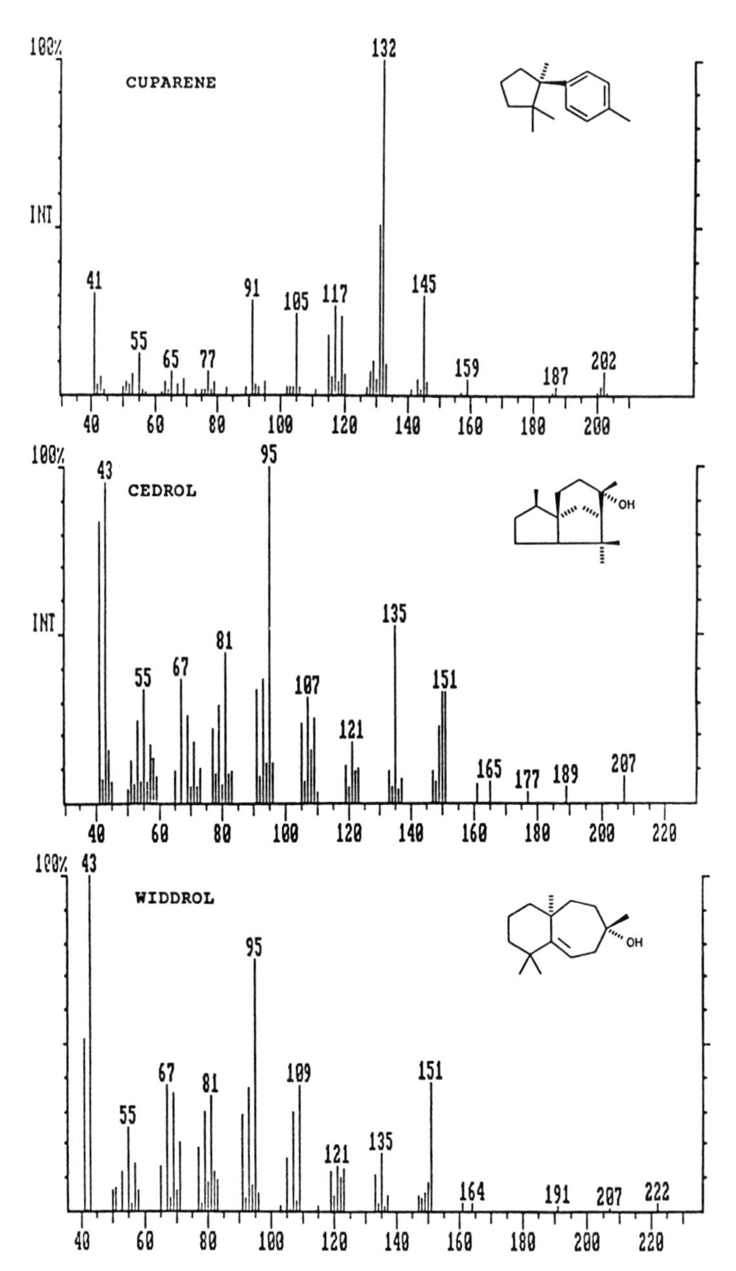

Fig. 7. Ion trap mass spectra of cuparene, cedrol, and widdrol

6.1 Antimicrobial Activities

Hexane and methanol extracts of heartwood, bark/sapwood, and leaves of 12 taxa of *Juniperus* from the United States were assayed for anti-fungal and anti-bacterial activities (Clark et al. 1990). The hexane extracts of the heartwood (which contains the cedarwood oil) of several junipers appear comparable in antibacterial activity to streptomycin. No anti-fungal activities comparable to amphotericin B were found in either the hexane or methanol extracts of heartwood. Additional research is needed to isolate and determine the anti-bacterial components.

6.2 Insecticidal Activities

Oda et al. (1977) examined the insecticidal activities of several extracts of the heartwood of *Juniperus recurva* from Nepal. The insecticidal activities were found in the steam volatile fraction(i.e., cedarwood oil). Detailed examination of individual components revealed the following LD_{50} µg/mosquito: α-cedrene— 33.5; β-cedrene— not active; thujopsene,— 4.5; acoradiene— not active; β-chamigrene— not active; curparene— not active; 8, 14-cedranoxide— 10.7; 8-cedren-13-al— not active; cedrol— 21.2; widdrol— not active; 8-cedren-13-ol acetate— not active; 8-cedren-13-ol- 6.6; 8S,13- and 14-cedrane-diols— not active. Clearly the most insecticidal components were thujopsene and 8-cedren-13-ol. Again, additional research is warranted on both *Juniperus* and *Cupressus* (and other Cupressaceae species) for natural insecticidal compounds to replace chlorinated pesticides of current usage.

6.3 Termiticidal Activities

The control of termites is a world wide problem. Current preservatives use arsenic, and chlorinated and copper-based products, all of which are toxic to humans and/or carcinogenic. Carter (1976) found that termites (*Reticulitermes flavipes*) could not survive on sawdust from *Juniperus virginiana* or on filter paper treated with a pentane extract (cedarwood oil) of *J. virginiana* sawdust.

Subsequently, Adams et al. (1988) found extremely high termiticidal activities in the heartwood sawdust from all 12 of the United States junipers examined. Hexane extracts of the heartwoods revealed that treated paper showed termiticidal activities for seven of the taxa. Additional research is continuing (McDaniel and Adams in prep.) to determine if the extracts are anti-feedants and/or toxic to termites. Because the junipers are used for posts in the United States, it is obvious that wood preservatives are in the wood. These same kinds of observations about wood rotting should be used to select promising species for additional termiticidal (and wood rotting) tests around the world (particularly in the Cupressaceae).

References

Adams RP (1987) Investigation of *Juniperus* species of the United States for new sources of cedarwood oil. Econ Bot 41:48–54

Adams RP (1989) Identification of essential oils by ion trap mass spectroscopy. Academic Press, New York

Adams RP, Granat M, Hogge LR, von Rudloff E (1979) Identification of lower terpenoids from gas-chromatography-mass spectral data by on-line computer method. J Chromatogr Sci 17:75–81

Adams RP, McDaniel CA, Carter FL (1988) Termiticidal activities in the heartwood, bark/sapwood and leaves of *Juniperus* species from the United States. Biochem Syst Ecol 16:453–456

Bauer K, Garbe D (1985) Common fragrance and flavor materials. VCH Verlagsgesellschaft, Weinheim, W. Germany

Bredenberg JB (1961) The chemistry of the order Cupressales. 36. The ethereal oil of the wood of *Juniperus communis* L. Acta Chem Scand 15:961–966

Carter FL (1976) Responses of subterranean termites to wood extractives. Mater Org Beih 3:357-364

Clark AM, McChesney JD, Adams RP (1990) Antimicrobial properties of the heartwood, bark/sapwood and leaves of *Juniperus* species. Phytother Res 4:15–19

Doi K, Shibuya T (1972) Sesquiterpenes of *Juniperus conferta*. Phytochemistry 11:1174

Erdtman H, Norin T (1966) The chemistry of the Cupressales. In: Zechmeister L (ed) Fortschritte der Chemie organischer Naturstoffe. Springer, Berlin Heidelberg, New York, pp 207–287

Fischer N, Nitz S, Drawert F (1987) Original flavour compounds and the essential oil composition of Marjoram (*Majorana hortensis* Moench). Flav Frag J 2:55–61

Goryaev MI, Tolstikov GA, Ignatova LA, Dembitshii AD (1962) Natural β-cedrene. Dokl Akad Nauk (USSR) 146:1331 (Abstr in Chem Abstr 58:9149 1963)

Guenther E (1952) The essential oils, Vol 6. Reprinted 1976 by Robert E. Kreiger, Huntington, New York

Kitchens GC, Dorsky J, Kaiser K (1971) Cedarwood oil and derivatives. Givaudanian 1:3–9

Koedam A, Looman A (1980) Effect of pH during distillation on the composition of the volatile oil from *Juniperus sabina*. Planta Med 1980 Suppl:22–28

Koedam A, Scheffer JJC, Svendsen AB (1981) Comparison of isolation procedures for essential oils. IV. Leyland cypress. Perf Flav 5:56–65

Kurth EF, Lackey HB (1948) The constituents of Sierra juniper wood (*Juniperus occidentalis* Hooker). J Am Chem Soc 70:2206–2209

Lawrence BM (1985) A review of the world production of essential oils (1984) Perf Flav 10:1–16

Narasimhachari N, von Rudloff E (1961) The chemical composition of the wood and bark extractives of *Juniperus horizontalis* Moench. Can J Chem 39:2572–2581

Oda J, Ando N, Nakajima Y, Inouye Y (1977) Studies on insecticidal constituents of *Juniperus recurva* Buch. Agric Biol Chem 41:201–204

Pettersson E, Runeberg J (1981) The chemistry of the order Cupressales. 34. Heartwood constituents of *Juniperus procera* Hochst. and *Juniperus californica* Carr. Acta Chem Scand 15:713–720

Pilo C, Runeberg J (1960) The chemistry of the order Cupressales 25. Heartwood constituents of *Juniperus chinensis* L. Acta Chem Scand 14:353–358

Runeberg J (1960a) The chemistry of the order Cupressales. 30. Heartwood constituents of *Juniperus cedrus* Webb. & Benth. Acta Chem Scand 14:1991–1994

Runeberg J (1960b) The chemistry of the order Cupressales. 27. Heartwood constituents of *Juniperus utahensis* Lemm. Acta Chem Scand 14:797–804

Runeberg J (1960c) The chemistry of the order Cupressales. 31. Heartwood constituents of *Juniperus phoenicea* L. Acta Chem Scand 14:1995–1998

Runeberg J (1960d) The chemistry of the order Cupressales. 29. Heartwood constituents of *Juniperus thurifera* L. Acta Chem Scand 14:1985–1990

Runeberg J (1960e) The chemistry of the order Cupressales. 28. Heartwood constituents of *Juniperus virginiana* L. Acta Chem Scand 14:1288–1294

Runeberg J (1961) The chemistry of the order Cupressales. 35. Heartwood constituents of *Juniperus foetidissima* Willd. Acta Chem Scand 15:721–726

Rutowski BN, Vinogradova IV (1927) Untersuchung der Zusammensetzung russischer ätherischer Öle. Trans Shi Chem-Pharm Inst (USSR) No. 16:142 (Abstr in Chem Zentralbl 1927, II:1311)

Schmaus G, Kubeczka K-H (1985) The influence of isolation conditions on the composition of essential oils containing linalool and linalyl acetate. In: Svendsen AB, Scheffer JJC (eds) Essential oils and aromatic plants. Nijhoff/Junk, Dordrecht

Vasek FC, Scora RW (1967) Analysis of the oils of western American junipers by gas-liquid chromatography. Am J Bot 54:781–789

von Rudloff E (1967) Chemosystematic studies in the genus *Picea* (Pinaceae). Can J Bot 45:891–901

Walker GT (1968) Cedarwood oil. Perf Ess Oil Res 59:346–350

Wenninger JA, Yates RL, Dolinsky M (1967) Sesquiterpene hydrocarbons of commercial copaiba balsam and American cedarwood oils. J Am Oil Chem Soc 50:1304–1313

Windemuth N (1945) The volatile oil of *Juniperus mexicana* Schiede. Pharm Arch 16:17

Zanoni TA, Adams RP (1979) The genus Juniperus (Cupressaceae) in Mexico and Guatemala: synonymy, key, and distributions of the taxa. Bull Soc Bot Mex 38:83–121

Analysis of Croton Oil by Reversed-Phase Overpressure Layer Chromatography

A.D. KINGHORN and C.A.J. ERDELMEIER

1 Introduction — Overpressure-Layer Chromatography as a Separatory Technique

Although the first paper on overpressure-layer chromatography (OPLC) (also known as overpressured thin layer chromatography) appeared over 10 years ago (Tyihák et al. 1979), many applications of this planar chromatographic technique have since been published (Witkiewicz and Bladek 1986; Tyihák 1987). OPLC involves many of the elements of thin layer chromatography (TLC), but, unlike the latter technique, the vapor phase is completely eliminated so as to leave the sorbent layer covered with an elastic membrane under external pressure. When forced by a pump, the mobile phase migrates through the sorbent layer due to the cushion system at overpressure (Tyihák et al. 1979; Witkiewicz and Bladek 1986; Tyihák 1987). Details of the apparatus required for OPLC, as well as of the theoretical basis of this technique, have appeared in the literature (Hauck and Jost 1983; Witkiewicz and Bladek 1986; Tyihák 1987). Higher efficiencies, increased solute loading capacity, and reduced solvent development times can be obtained by OPLC when compared to capillary controlled systems such as TLC. If large numbers of samples have to be analyzed, or if some of the components in a mixture are difficult to detect, OPLC also has advantages over high-performance liquid chromatography (HPLC) (Witkiewicz and Bladek 1986; Tyihák 1987). In addition to being used for the analytical separation of organic molecules, OPLC is very useful for their preparative purification (Nyiredy et al. 1986; Erdelmeier et al. 1987).

Several applications of OPLC for the separation of plant-derived natural products of various polarities have been described in the literature, including methods for the analysis of anthraquinones (Nyiredy et al. 1986), *Camptotheca* alkaloids (Erdelmeier et al. 1986), canthin-6-one alkaloids (Erdelmeier et al. 1987), cardiac glycosides (Erdelmeier et al. 1987), flavonoid glycosides (Dallenbach-Toelke et al. 1986), furocoumarins (Nyiredy et al. 1986), ginsenoside triterpene glycosides (Dallenbach-Toelke et al. 1986), iridoid and secoiridoid glycosides (Erdelmeier et al. 1987; Nyiredy et al. 1986), *ent*-kaurene glycosides (Fullas et al. 1989), lignans (Erdelmeier et al. 1986), a sesquiterpenoid (Fullas et al. 1989), simaroubolides (Erdelmeier et al. 1986), and steroidal saponins (Fullas et al. 1989).

2 Phorbol Ester Constituents of Croton Oil

The purgative, skin-irritant and tumor-promoting constituents of the seed oil of *Croton tiglium* L. (Euphorbiaceae) (croton oil) are a complex mixture of esters of the tetracyclic diterpene, phorbol (Fig. 1, **1**) (Hecker and Schmidt 1974). Although

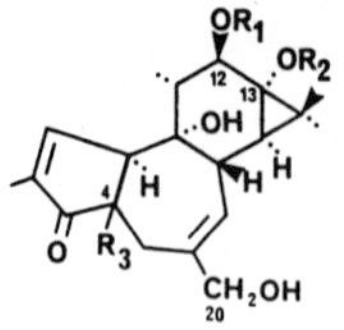

Compound	R1	R2	R3
1	H	H	β-OH
2	H	H	α-H
3	Tetradecanoate	Acetate	β-OH
4	Decanoate	Acetate	β-OH
5	Dodecanoate	Acetate	β-OH
6	Hexadecanoate	Acetate	β-OH
7	2-Methylbutyrate	Dodecanoate	β-OH
8	2-Methylbutyrate	Decanoate	β-OH
9	Tiglate	Decanoate	β-OH
10	Acetate	Dodecanoate	β-OH
11	2-Methylbutyrate	Octanoate	β-OH
12	Tiglate	Octanoate	β-OH
13	Acetate	Decanoate	β-OH
14	Tiglate	Butyrate	β-OH
15	Tiglate	Dodecanoate	β-OH
16	Butyrate	Dodecanoate	β-OH
17	Tiglate	Isobutyrate	β-OH
18	Tiglate	Acetate	β-OH
19	2-Methylbutyrate	Isobutyrate	β-OH
20	2-Methylbutyrate	Acetate	β-OH
21	Acetate	Acetate	β-OH
22	Tiglate	H	β-OH
23	H	Acetate	β-OH
24	Tiglate	Isobutyrate	α-H
25	Tiglate	Acetate	α-H
26	2-Methylbutyrate	Acetate	α-H
27	H	Acetate	α-H

Fig. 1. Structures of phorbol (**1**), 4-deoxy-4α-phorbol (**2**) and esters of these diterpene alcohols (**3–27**) that have been isolated and spectroscopically characterized from extracts of croton oil

the purification of the toxic principles of croton oil took many years to achieve, some 25 esters of phorbol and the related diterpene, 4-deoxy-4α-phorbol (Fig. 1, **2**) have now been individually isolated and spectroscopically characterized from this plant source. In work conducted in the 1960's, the Hecker group at Heidelberg isolated 11 phorbol 12,13-diesters found to act as tumor promoters for mouse skin, which they termed A factors (Fig. 1, **3–6**) and B factors (Fig. 1, **7–13**), depending on whether the long-chain acyl substituent was affixed to C-12 or to C-13, respectively (Hecker and Schmidt 1974). Three non-naturally occurring phorbol diester products (Fig. 1, **14–16**) were also obtained by Hecker and co-workers, after the partial hydrolysis of a mixture of phorbol 12,13,20-triesters (Hecker and Schmidt 1974). Thus far, no native phorbol 12,13,20-triester has yet been characterized as a croton oil constituent. Recent work at the University of Illinois at Chicago has resulted in the isolation and characterization of a further 11 diterpene esters from croton oil, with compounds **17–23** (Fig. 1) and compounds **24–27** (Fig. 1) being short-chain esters of phorbol and 4-deoxy-4α-phorbol, respectively. All of these minor constituents were found to be diesters, with the exception of compounds **22**, **23**, and **27** (Fig. 1), which are monoesters (Marshall and Kinghorn 1984). Phorbol esters occur in a combined yield of over 1% w/w in croton oil, with the most abundant and also most potent biologically active compound in this series being 12-*O*-tetradecanoylphorbol 13-acetate (TPA, **3**) (Hecker and Schmidt 1974). The phorbol esters are widely used as biological tools in carcinogenesis and cellular biochemical experiments, and command a high price of some $2000 per gram as speciality chemicals (Balandrin et al. 1985).

Several different approaches have been made to the resolution of the phorbol ester constituents of croton oil, and a summary of these has been provided in another chapter of this Volume by Pieters and Vlietinck. These compounds have proven difficult to separate in the past because of their inherent lability, their tendency to occur in complex mixtures in low individual yields, and their potential severe toxicity to laboratory workers, due to their potent skin-irritant properties (Hecker and Schmidt 1974). Another complicating issue is that, when separated by either column chromatography or analytical TLC over silica gel, no significant resolution of the croton oil phorbol diesters generally results other than the separation of the long-chain C-12 phorbol diesters (A factors) and the long-chain C-13 phorbol diesters (B factors) as two discrete zones (Hecker and Schmidt 1974). However, we have found that the individual homologs within the separated A- and B-factor groups will separate by two-dimensional TLC, if a second reversed-phase phase-bonded silica gel layer is used (Erdelmeier et al. 1988). We have also developed a reversed-phase OPLC method suitable for the rapid separation of many of the known long- and short-chain phorbol esters of croton oil (Erdelmeier et al. 1988). These two new analytical methods for the croton oil phorbol esters will be described in turn in the next section.

3 Separation of the Phorbol Esters of Croton Oil

3.1 Two-Dimensional Thin Layer Chromatography

3.1.1 Method 1 — Use of Prepared Bilayer Plates

Separation may be effectively carried out on Whatman Multi-K dual-phase SC 5 plates (20 × 20 cm; 0.25 cm layer thickness) (Whatman Chemical Separations, Inc., Clifton, New Jersey, USA), which are composed of a narrow band of silica gel (normal phase, NP) and a wide band of octadecylsilyl phase-bonded silica gel (reversed-phase, RP). Prior to two-dimensional TLC (2D-TLC), a commercial sample of croton oil may be partially purified by dissolution in *n*-hexanes (20 ml) and successive partition with aliquots (4 × 10 ml) of methanol-water (20:3), to afford a hydrophilic residue on drying ("croton resin"). As may be seen from Fig. 2, a sample of this croton resin (1 mg, dissolved in methanol and streaked onto the NP region of the plate) is separated primarily into phorbol 12,13,20-triester, phorbol 12,13-diester B factor, and phorbol 12,13-diester A factor regions, on elution with chloroform-acetone (4:1). Subsequent RP development in the second dimension, using methanol-water (9:1) as eluant, then results in good resolution of the croton oil phorbol esters, which are visualized as rusty brown zones in visible light after spraying with ethanolic sulfuric acid (60% v/v) on heating for 10 min at 110 °C (Fig. 2). The times taken for each development were found to be 45 min for the NP development and 95 min for the RP development. In the commercial sample of croton oil that was available to us, the phorbol diesters were found to

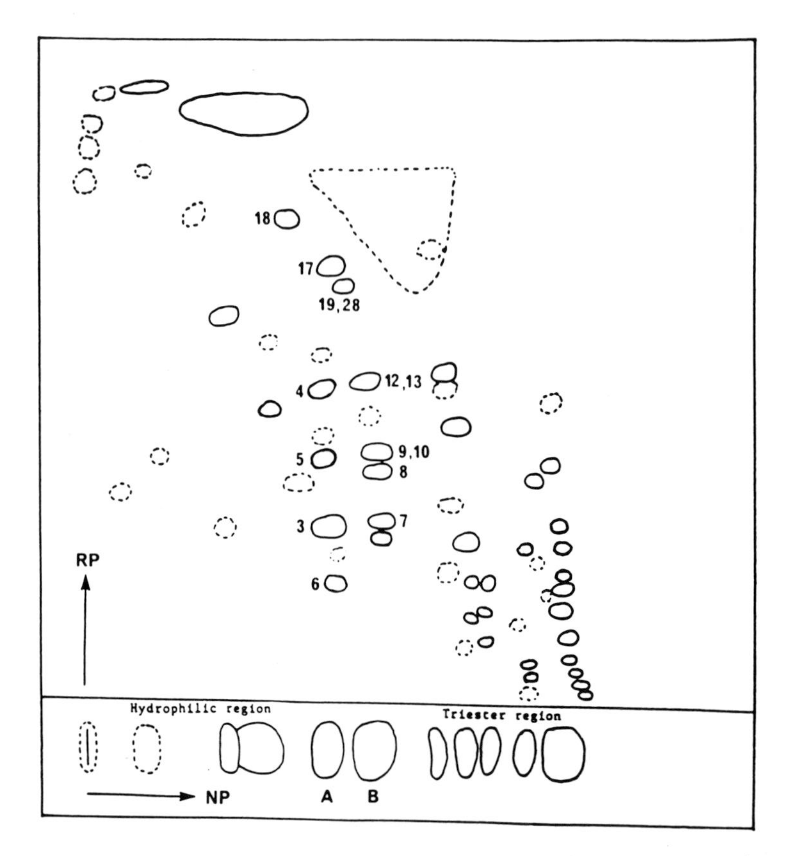

Fig. 2. Two-dimensional TLC chromatogram of the phorbol esters of croton resin. For the structures of compounds **3–10**, **12**, **13**, **17–21**, see Fig. 1 and Table 1. The tentative structure of compound **28** is shown in Table 1. Development modes: *NP* normal phase separation on silica gel; *RP* reversed-phase separation on silica gel. *A* and *B* represent phorbol 12,13-diesters A and B factors, having long-chain acyl functions at C-12 and at C-13, respectively. Identified phorbol 12,13-diesters and other TLC zones appearing as rusty brown zones in visible light are marked with *solid lines*, while the zones marked with *dotted lines* indicate presumed nonphorbol ester constituents of croton resin. For experimental conditions, see text

separate in order of their molecular weights (Table 1), and all of the known croton oil phorbol diester A factors (compounds **3–6**, Fig. 1), six of the known phorbol diester B factors (compounds **7–10** and **12–13**, Fig. 1), and three short-chain phorbol diesters (compounds **17–19**) were detectable using this method. These compounds could be identified by co-chromatography with authentic samples of phorbol esters obtained from our previous work on croton oil (Marshall 1985; Marshall and Kinghorn 1984). In general, the compounds are eluted in pure form, although this is not always the case, since several of the croton oil B-factors were obtained in the form of mixtures (Table 1). The homogeneity of the 12-*O*-tetradecanoylphorbol 13-acetate (TPA) (Fig. 1, **3**) zone was able to be confirmed by chemical ionization mass spectrometry (see Sect. 3.2.3).

3.1.2 Method 2 — Chromatogram Transfer Technique

In a modification of the method described in Section 3.1.1, similar resolution can be obtained after sequential separation on two separate chromatographic plates. Thus, the normal phase (NP) may be any commercially available aluminum-backed silica gel 60 GF254 plate (10 × 20 cm, 0.2 cm layer thickness), and the reversed phase (RP) can be any commercially available RP plate, as exemplified by the Whatman KC 18 F plate (20 × 20 cm, 0.25 cm layer thickness) (Whatman Chemical Separations Inc.). The same solvents are used for NP and RP elution as mentioned

Table 1. Separation of phorbol diesters of croton oil by 2D-TLC and RP-OPLC

Compound no.	R_1	R_2	Mol. wt.	R_f(2D-TLC)[a]	R_f(RP-OPLC)
3	Tetradecanoate	Acetate	616	0.35, 0.26[d]	0.24[d]
4	Decanoate	Acetate	560	0.35, 0.47[b]	—[c]
5	Dodecanoate	Acetate	588	0.35, 0.36[b]	0.39[b]
6	Hexadecanoate	Acetate	644	0.35, 0.17[b]	0.13[b]
7	2-Methylbutyrate	Dodecanoate	630	0.41, 0.27	0.24[e]
8	2-Methylbutyrate	Decanoate	602	0.41, 0.35[b]	0.34[d]
9	Tiglate	Decanoate	600	0.41, 0.38	0.40[e]
10	Acetate	Dodecanoate	588	0.41, 0.38	0.40[e]
12	Tiglate	Octanoate	572	0.41, 0.48	0.62[e]
13	Acetate	Decanoate	560	0.41, 0.48[b]	0.62[d]
17	Tiglate	Isobutyrate	516	0.35, 0.69[b]	0.82[d]
18	Tiglate	Acetate	488	0.27, 0.76[b]	—[c]
19	2-Methylbutyrate	Isobutyrate	518	0.35, 0.66[b]	0.76[b]
28[f]	Tiglate	2-Methylbutyrate	530	0.35, 0.66[b]	0.76[b]

[a]Values refer to NP and RP, respectively.
[b]Identity checked by co-chromatography.
[c]Not detected.
[d]Identity checked by co-chromatography and CI-MS.
[e]Molecular ion determined by CI-MS.
[f]Tentative identification.

for development with prepared bilayer plates, and the same visualization reagent is employed. Again, the solute (croton resin, 1 mg in a minimum volume of methanol) is streaked as a 1-cm band prior to elution in the first dimension (NP). After this initial TLC development, this NP plate is dried, and the track containing the whole chromatogram is then cut out to produce a silica gel strip of dimensions 1.5–2 × 20 cm. This strip is then dipped horizontally in methanol, until the chromatographed zones migrate to the top of the plate (see Fig. 3). The cut-out strip is then clamped face down to the RP (second dimension) plate, in the manner shown in Fig. 4. Transfer of the solute from the NP to RP plate is accomplished by inserting the clamped plates into methanol for a few seconds. After removing the clamp, the RP plate is then subjected to solvent development. The times taken for these two development are about 60 min (NP) and 75 min (RP). This latter technique has greater flexibility than the use of prepared bilayer plates, in that any aluminum- or glass-backed NP plate can be used in combination with any available RP plate.

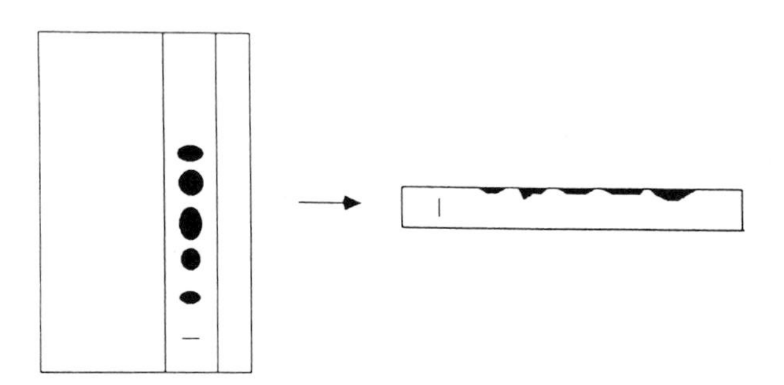

Fig. 3. Cut-out of normal-phase (NP) TLC-track containing partially separated components of croton resin

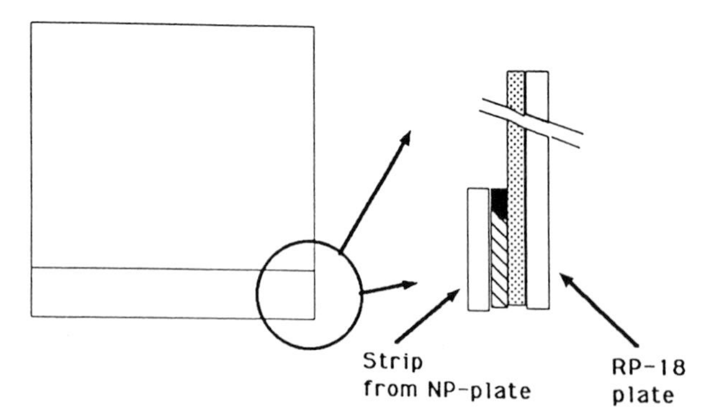

Fig. 4. Chromatogram-transfer technique for the components of croton resin from a normal-phase (*NP*) plate strip to a reversed-phase (*RP*) plate

3.2 Reversed-Phase Overpressure Layer Chromatography

3.2.1 Preliminary Separation of Croton Resin by Low-Pressure Column Chromatography

Croton resin is first prepared by solvent partition in the manner described in Section 3.1.1. From 20 g of croton oil originally subjected to solvent extraction, 1.12 g of "croton resin" could be obtained in our hands. A portion (800 mg) is then dissolved in chloroform-acetone (5:1, 2 ml), and subjected to low-pressure liquid chromatography, using this same binary solvent system, over silica gel H (45.7 × 2.54 cm i.d. glass column; flow rate 3.0 ml/min). Fractions (ca. 4 ml each) are collected, and it was found that fractions 73–81 were constituted by croton oil B factors (85 mg) and fractions 90–113 represented a crude mixture of croton oil A factors (55 mg).

3.2.2 Conditions Used for Reversed-Phase Overpressure Layer Chromatography

Separation of the resolved croton oil A and B factors by reversed-phase overpressure layer chromatography (RP-OPLC) may be conducted using the Chrompres 25 apparatus (Labor MIM, Budapest, Hungary) with the mobile phase (methanol-water, 9:1) delivered with a minipump at a flow rate of 0.1 ml/min (4–8 bar back pressure). Separation is effected on a RP-18 HPTLC plate (10 × 20 cm; 0.2 cm layer thickness) (E. Merck, Darmstadt, Germany), with plate-edge impregnation performed using solid paraffin, and a cushion pressure of 25 bar. Portions of the crude croton oil A factors (100 μg) and B factors (150 μg) obtained by low-pressure liquid column chromatography (Sect. 3.2.1) are then individually

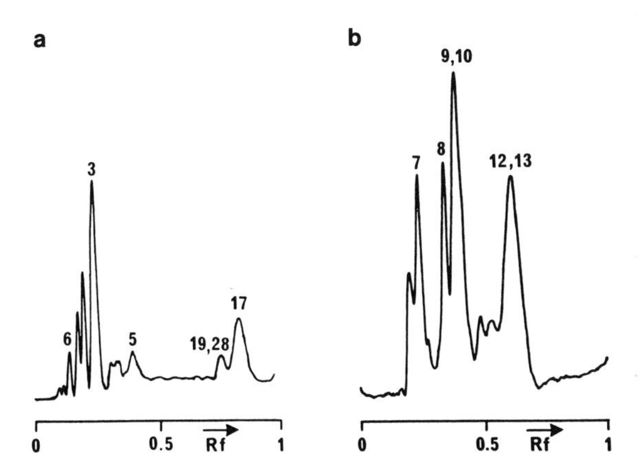

Fig. 5. Densitograms showing reversed-phase overpressure layer chromatographic (RP-OPLC) separations of croton oil phorbol 12,13-diesters. **a** Croton oil A factors. **b** Croton oil B factors. For the structures of compounds **3, 5–10, 12, 13, 17** and **19**, see Fig. 1 and Table 1, and for the tentative structure of compound **28**, see Table 1. Experimental conditions are given in the text

subjected to RP-OPLC (development time, 30 min each). Densitograms for each RP-OPLC separation can then be obtained after UV visualization (254 nm) on a Zeiss KM 3 chromatogram spectrophotometer, as shown in Fig. 5a (croton oil A factors) and 5b (croton oil B factors). Compound identifications can be made either by co-chromatography with authentic samples of the phorbol esters and/or by coupling with off-line chemical-ionization mass spectrometry (see Sect. 3.2.3).

If phorbol ester homologs are considered that vary by only one methylene group, such as the A-factors, compounds **3** and **6**, and the B-factors, compounds **7** and **8**, it may be seen from Table 1 that such compounds separate by 2D-TLC with ΔRf value of nearly 0.10. In addition to a shorter development time, the resolution of the croton oil A- and B-factors is much increased by RP-OPLC, when compared with 2D-TLC. However, such is the physico-chemical similarity between many of the phorbol esters of croton oil, that several of these compounds can not be individually resolved by RP-OPLC, even when high-performance TLC plates are employed for this purpose (see Table 1).

3.2.3 Chemical Ionization Mass Spectrometry

In order to assess the purity of the zones obtained after RP-OPLC, we have found positive-ion chemical-ionization mass spectrometry (CI-MS) to be a sensitive and specific analytical method for the detection of phorbol esters of croton oil. An instrument on which CI-MS may be performed is the Finnigan 4510 gas chromatograph-mass spectrometer (Finnigan-MAT, Sunnyvale, California, USA), equipped with a direct exposure probe system. Two reagent gases, methane and ammonia, will provide useful analytical data with which to identify and determine the purity of the separated croton oil phorbol-ester constituents. Separated zones after RP-OPLC are scraped and transferred to 4-ml vials. Methanol (0.5 ml) is then added to each vial, and the resultant suspensions are sonicated for a few seconds to homogenize the extracted materials. Each solute can then be individually extracted in a Pierce Reacti-Therm 18800 heating module (Pierce Chemical Co., Rockford, Illinois, USA) for 15 min at 45 °C. Samples are then filtered through a Pyrex glass filter (No. 36060) into Pierce vials. Solvent can be removed using a Reacti-Vap evaporating unit (Pierce), by flushing with nitrogen. Prior to CI-MS analysis, samples are redissolved in 20 μl of methanol, with 1-μl portions analyzed. As a result of this approach, it was found that the major zones from the RP-OPLC separations (compounds **3**, **7–10**, **12**, **13** and **17**) and zone 3 (TPA) from the 2D-TLC separation gave useful CI-MSdata. Quantities of the other separated compounds were too small for CI-MS analysis. When ammonia is used as reactant gas, these compounds are characterized by prominent $[M + NH_4]^+$ adduct ions, along with MH^+ protonated molecular ions of low relative intensity. Rather more detailed fragmentation pathways may be observed for these compounds when methane is used as the reactant gas, with the molecular ions of the phorbol esters typically observed via prominent $[M + H]^+$ protonated molecular ions and $[M + C_2H_5]+$ and $[M + C_3H_7]+$ adduct ions (Erdelmeier et al. 1988). CI-MS is preferred over electron-impact mass spectrometry not only because of its increased sensitivity, but also because it is a soft-ionization mass spectral technique, and therefore reveals more useful information on the molecular weights of the croton oil phorbol esters.

4 Summary and Conclusions

It has been shown that the 2D-TLC and RP-OPLC planar techniques rapidly separate the principal phorbol 12,13-diester constituents of the seed oil of *Croton tiglium* (croton oil), with excellent resolution. The 2D-TLC method may be performed with either a prepared bilayer plate or using the chromatogram transfer technique, and has the potential for use as a convenient general phytochemical screening procedure suitable for the detection and qualitative analysis of toxic phorbol ester-related diterpenoid constituents of other plants in the families Euphorbiaceae and Thymelaeaceae that also produce these compounds. In addition, the technique could be employed to show the absence of these compounds, and hence the safety of products manufactured from species that accumulate phorbol and biogenetically related diterpene esters. For example, the seeds of the Chinese tallow tree (*Sapium sebiferum* L.) are of considerable economic importance for the production of soap, candles, as well as illuminating and drying oils (Yang and Kinghorn 1985), but they also produce toxic phorbol esters (Brooks et al. 1987).

It is apparent from Fig. 2 that many phorbol ester constituents of croton oil remain to be characterized, since many similar-staining zones were apparent after 2D-TLC. Among the phorbol 12,13-diesters, one compound that has been detected by both 2D-TLC and RP-OPLC is 12-*O*-tiglyphorbol 13-(2-methyl)butyrate (**28**, Table 1), a compound which was only tentatively identified in a previous isolation study on croton oil, since the relative positions of ester substitution were not confirmed by selective hydrolysis (Marshall 1985). There is also great scope for the use of the techniques described in this chapter for the purification of the croton oil phorbol triesters or "cryptic cocarcinogens", particularly if used in combination with the rotation locular counter-current chromatographic procedure described by Pieters and Vlietinck (1986), which leads to little or no loss of such compounds during extraction. The RP-OPLC procedure is suitable for scale-up, and would allow the isolation of the presently uncharacterized phorbol di- and triester constituents of croton oil in milligram quantities in just a few hours.

References

Balandrin MF, Klocke JA, Wurtele ES, Bolinger WH (1985) Natural plant chemicals: sources of industrial and medicinal materials. Science 228:1154–1160

Brooks G, Morrice NA, Ellis C, Aitken A, Evans AT, Evans FJ (1987) Toxic phorbol esters from Chinese tallow stimulate protein kinase C. Toxicon 25:1229–1233

Dallenbach-Toelke K, Nyiredy S, Meier B, Sticher O (1986) Optimization of overpressured layer chromatography of polar, naturally occurring compounds by the "PRISMA" model. J Chromatogr 365:63–72

Erdelmeier CAJ, Erdelmeier I, Kinghorn AD, Farnsworth NR (1986) Use of overpressure layer chromatography (OPLC) for the separation of natural products with antineoplastic activity. J Nat Prod 49:1133–1137

Erdelmeier CAJ, Kinghorn AD, Farnsworth NR (1987) On-plate injection in the preparative separation of alkaloids and glycosides using overpressured layer chromatography. J Chromatogr 389:345–349

Erdelmeier CAJ, Van Leeuwen PAS, Kinghorn AD (1988) Phorbol diester constituents of croton oil: separation by two-dimensional TLC and rapid purification utilizing reversed-phase overpressure layer chromatography (RP-OPLC). Planta Med 54:71–75

Fullas F, Kim J, Compadre CM, Kinghorn AD (1989) Separation of natural product sweetening agents using overpressured layer chromatography. J Chromatogr 464:213–219

Hauck HE, Jost W (1983) Investigations and results obtained with overpressured thin-layer chromatography. J Chromatogr 262:113–120

Hecker E, Schmidt R (1974) Phorbol esters — the irritants and cocarcinogens of *Croton tiglium* L. Progr Chem Org Nat Prod 31:377–467

Marshall GT (1985) New phorbol ester constituents of croton oil. Thesis, University of Illinois at Chicago, Chicago, Illinois

Marshall GT, Kinghorn AD (1984) Short-chain phorbol ester constituents of croton oil. J Am Oil Chem Soc 61:1220–1225

Nyiredy S, Erdelmeier CAJ, Dallenbach-Toelke K, Nyiredy-Mikita K, Sticher O (1986) Preparative on-line overpressure-layer chromatography (OPLC): A new separation technique for natural products. J Nat Prod 49:885–891

Pieters LA, Vlietinck AJ (1986) Rotation locular counter-current chromatography and quantitative ^{1}H-nuclear magnetic resonance spectroscopy of the phorbol ester constituents of croton oil. Planta Med 52:465–468

Tyihák E (1987) Overpressured layer chromatography and its applicability in pharmaceutical and biomedical analysis. J Pharm Biomed Anal 5:191–203

Tyihák E, Mincsovics E, Kalász H (1979) New planar liquid chromatographic technique: overpressured thin-layer chromatography. J Chromatogr 174:75–81

Tyihák E, Mincsovics E, Kalász H, Nagy J (1981) Optimization of operating parameters in overpressured thin-layer chromatography. J Chromatogr 211:45–51

Witkiewicz Z, Bladek J (1986) Overpressured thin-layer chromatography. J Chromatogr 373:111–140

Yang P, Kinghorn AD (1985) Coumarin constituents of the Chinese tallow tree. J Nat Prod 48:373–375

Rotation Locular Countercurrent Chromatography Analysis of Croton Oil

L.A.C. PIETERS and A.J. VLIETINCK

1 Introduction

Continuous liquid-liquid partition methods not employing any solid support matrix are generally termed countercurrent chromatography (CCC). Consequently, all complications arising from the use of such solid supports, including tailing of solute peaks, adsorptive sample loss, deactivation, contamination, etc. are eliminated. Therefore CCC provides a valuable complementary method to adsorption chromatography, especially for the fractionation of complex mixtures of natural products. CCC techniques allow a mild isolation of natural products, and their significance as phytochemical methods is continuously increasing.

In the early 1970's, droplet CCC (DCCC) was introduced for preparative separations. A DCCC apparatus consists of a series of vertically arranged straight tubes connected to each other top-to-bottom. DCCC is carried out by passing droplets of the mobile phase through a column of stationary phase. After introduction into the column, the mobile phase forms multiple droplets in the stationary phase at regular intervals, each droplet serving as a partition unit. The mobile phase may be either heavier or lighter than the stationary phase. When heavier, the mobile phase is introduced at the top of the column, and when lighter, at the bottom. This system yields effective separations, but the time required for separation usually ranges from 1 to 5 days. In addition, the formation of suitable droplets of the mobile phase in the stationary phase depends on the interfacial tension, the viscosity of the solvents, the difference in density between the two phases, the internal diameter of the column, and the flow rate of the mobile phase. This crucial step limits the choice of the solvent system. Many DCCC separations that have been described in the past involve polar compounds, including many glycosides. However, the potential of DCCC for the separation of weakly polar or apolar compounds has increased enormously since the use of nonaqueous solvent systems.

When centrally perforated discs are placed into the columns at regular intervals, multiple partition units, locules, are created, and the column is called a locular column. The optimal retention of the stationary phase is achieved by column inclination, and efficient mixing of the two phases is attained by rotation of the locular column. With rotation locular countercurrent chromatography (RLCC) almost universal use of conventional two-phase solvent systems is possible. Although DCCC is the best-known method for the countercurrent chromatographic analysis of natural products at the moment, in cases where suitable solvent systems are not readily available, other techniques such as RLCC may provide a valuable alternative to DCCC. However, the resolution is mostly lower (Domon et al. 1982; Hostettmann et al. 1984; Snyder et al. 1984; Ito 1986; Martin et al. 1986).

2 Croton Oil, the Seed Oil of *Croton tiglium*

Croton oil, the seed oil of *Croton tiglium* L. (Euphorbiaceae), contains phorbol esters, which are responsible for the remarkable biological activity of this oil. These diterpene esters exhibit two important toxicological actions. After application to the skin, they produce an intense inflammation, and on continued application, for instance to mice, following a single subthreshold dose of a carcinogen, they show a tumor-promoting activity. They are biologically active in submicrogram doses, and in a situation of repeated or chronic exposure they are second-order carcinogenic risk factors.

The structure of phorbol, which is a polyhydroxylated tigliane derivative, is shown in Fig. 1. The toxic diterpenes of the plant families Euphorbiaceae and Thymelaeaceae are known as piglianes, daphnanes, and ingenanes. Chemically they all are partial esters of the tetracyclic or tricyclic diterpene parent alcohols, such as the prototypes phorbol, resiniferonol, and ingenol. The toxic components of croton oil, the first tigliane diterpenes to be isolated from plants, are phorbol 12,13-diesters, which contain one acid of carbon chain length 8 to 16, and one of 2 to 6. These are potent tumor-promoters. The phorbol 12,13-diesters include a series of esters in which a long-chain acyl derivate is attached to C-12 of the phorbol nucleus and a short-chain acyl derivate to C-13 (the A-series), or this may be reversed, as in the B-series of diesters. In both symmetrical and asymmetrical diesters the compounds with a combined chain length of 14 to 20 carbons usually have the highest biological activity. Phorbol 12,13-diesters possessing two short-chain ester functionalities, which are also present in croton oil, are not tumor-promoting. The presence of a free allylic hydroxyl group at C-20 is essential for high biological activity. Phorbol 12,13, 20-triesters have low promoting activity

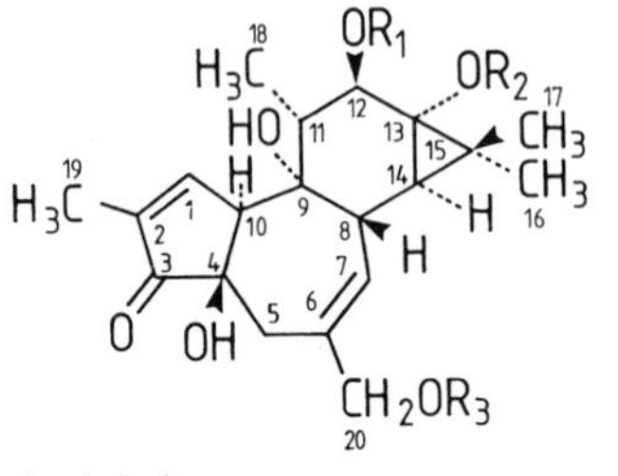

Acyl chains

R_1	R_2	R_3	
H	H	H	Phorbol
Long	short	H	A-series of phorbol diesters
Short	long	H	B-series of phorbol diesters
Short	short	H	Short-chain phorbol diesters
Acyl	acyl	acyl	Phorbol triesters

Long = carboxylic acid with carbon chain length 8 to 16
Short = carboxylic acid with carbon chain length 2 to 6

Fig. 1. Phorbol esters

unless the C-20 acyl group is removed by hydrolysis, e.g., by metabolic activation by esterases or lipases. Therefore phorbol triesters are known as cryptic tumor-promoters. The steric configuration at the ring junction of C-4 and C-10 is also critical for high promoting activity. In phorbol the C-10 hydrogen and C-4 hydroxy are *trans*, whereas in the epimer 4α-phorbol they are *cis*. 4α-Phorbol 12,13-diesters are inactive both as tumor-promoters and as irritants. Loss of the 4-hydroxyl group does not affect biological activity, for 4-deoxyphorbol diesters are biologically active, whereas 4-deoxy-4α-phorbol diesters are not. In general, a potent tumor-promoter must have both a highly lipophilic part of the molecule, and a hydrophilic part. Major alterations in the hydrophilic or lipophilic nature of the molecule as, for example, the lack of a lipophilic portion in phorbol itself, result in a substantial or even complete loss of biological activities. Croton oil has been used as a purgative and as a counterirritant to the skin, but it acts so powerfully that it is deemed unsafe for use in human medicine. The availability of the active components of croton oil and of semi-synthetic derivatives of phorbol has been of inestimable value in studying the biochemical mechanisms of tumor promotion (Slaga et al. 1976; Hecker 1978; Diamond et al. 1980; Hecker 1981; Evans and Taylor 1983).

3 RLCC Analysis of Croton Oil

3.1 Introduction

Many diterpene esters are unstable compounds and degrade during isolation procedures. Hydrolysis or transesterification reactions are known to occur during chromatographic separation. The compounds are sensitive to heat, light, oxygen, acid and alkaline conditions (Schmidt and Hecker 1975; Evans and Taylor 1983). As the diterpene esters are neutral lipid compounds purification generally requires sophisticated separation methods. This task is aggravated by the fact that in most cases complex mixtures of esters of the same diterpene nucleus are present. To date separation procedures are based upon a combination of chromatographic and partition techniques.

Originally liquid-liquid distributions of the O'Keeffe and Craig type were successfully used for the isolation of phorbol diesters from croton oil by Hecker and his coworkers (Hecker and Schmidt 1974). In the first stage of this fractionation procedure croton oil was separated by an O'Keeffe distribution into a hydrophobic and a hydrophilic portion. The hydrophilic portion contained the phorbol 12,13-diesters. Next the A- and B-series were separated by column chromatography, and individual phorbol diesters were resolved by repeated Craig distributions. After a carefully controlled acid-catalyzed transesterification of the phorbol 12,13–20 triesters present in the hydrophobic portion, an "activated" hydrophobic portion, containing a large amount of biologically active phorbol diesters, was obtained.

A dry-column technique has been described as well, but has not received universal acceptance (Ocken 1969). Other techniques that have been used for the isolation of phorbol derivatives from croton oil include high-performance liquid chromatography (HPLC) (Bauer et al. 1983), DCCC in combination with low-pressure liquid column chromatography and preparative thin-layer chromatography (Marshall and Kinghorn 1984), and reversed-phase overpressure layer chromatography (RP-OPLC) (Erdelmeier et al. 1988).

These investigations showed that croton oil contained not only many phorbol diesters and triesters, but also minor amounts of 4-deoxy-4α-phorbol esters and phorbol monoesters. However, all these methods have in common that the first stage of the fractionation procedure involves a simple partition of croton oil between a lipophilic layer (e.g., hexane) and a hydrophilic layer (e.g., methanol/water), or a simple methanolic extraction, to separate the oil into a hydrophobic fraction, containing mainly lipids and phorbol triesters, and a hydrophilic fraction, containing mainly phorbol diesters. The more sophisticated chromatographic techniques mentioned above are subsequently used to analyze only the hydrophilic fraction. On the contrary, we preferred to follow the original approach of Hecker and his coworkers (Hecker and Schmidt 1974) to apply a multiplicative liquid-liquid distribution method directly to crude croton oil, and we decided to use RLCC for this purpose, because of the high versatility of this method for choosing a solvent system (Pieters and Vlietinck 1986). In a recent review on RLCC for natural products isolation it was suggested that for the initial bioassay guided fractionation of plant extracts RLCC would be more effective than simple partitioning (Kubo et al. 1988). In addition, the greater resolution of an initial separation by RLCC can be achieved without any danger of forming emulsions, a problem frequently encountered in separatory funnel-type solvent partitions.

3.2 Experimental Details and Discussion

Although a wide range of solvent systems, also including binary systems, may be used for RLCC, in general ternary or quaternary systems are preferred for reasons of selectivity. After filling the locular column with stationary phase, it is inclined at an angle of about 30° from the horizontal position, and the mobile phase is continuously introduced. It displaces the stationary phase in each loculus to the level of the hole leading to the next one, and it is collected at the outlet of the column. In the ascending mode the lighter mobile phase is applied at the bottom of the column, in the descending mode the heavier mobile phase at the top of the column. In practice, several columns, which are interconnected with fine tubings, are mounted on a rotating shaft. The rotation speed lies typically between 60 and 90 rpm. Our apparatus consisted of 16 columns (glass and Teflon partition) with a length of 50 cm, an internal diameter of 11 mm and a locular distance of 12 mm. A schematic representation of an RLCC apparatus is shown in Fig. 2.

For the initial fractionation of croton oil into a hydrophobic and a hydrophilic portion, Hecker and his coworkers used an O'Keeffe distribution employing a petroleum ether-methanol-water solvent system. Initially we used a similar system for RLCC, but better results were obtained after adding a certain amount of diethyl ether. Figure 3 shows the elution chromatogram of about 500 mg of crude croton oil using a petroleum ether-diethyl ether-methanol-water system in a ratio

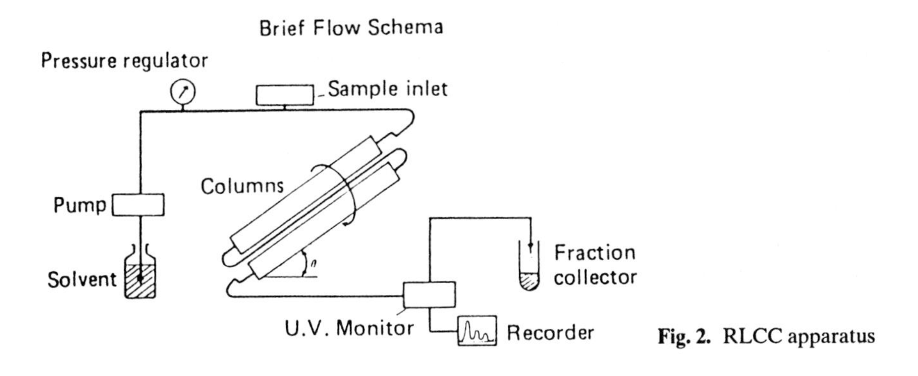

Fig. 2. RLCC apparatus

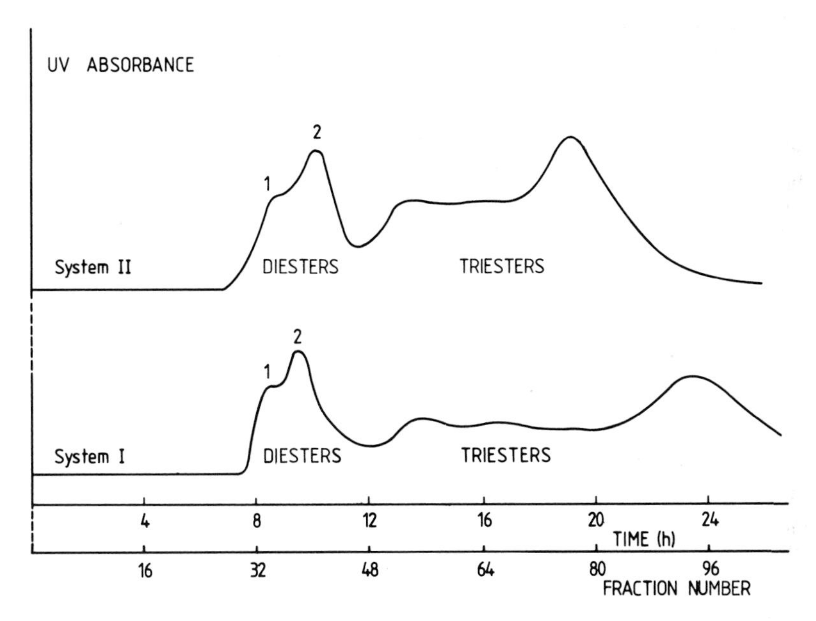

Fig. 3. Analytical RLCC elution chromatogram of croton oil (500 mg) (Pharmachemic, B-2610 Wilrijk, Belgium). Flow rate 0.7 ml/min, rotation speed 90 rpm, slope 30°, shortwave UV detection (254 nm), volume of each fraction about 10.5 ml. *Peak 1* short-chain phorbol diesters; *peak 2* A- and B-series of diesters

of 18–2–15–0.75 (I) and 18–1–15–0.3 (II), with the lipophilic layer as the stationary phase. The phorbol diesters could be separated from the triesters, but it was not possible to separate the different groups of diesters from each other using RLCC. The amount of diethyl ether and water in the solvent system had only little influence on the separation of the phorbol di- and triesters, but by using more diethyl ether the retention times of the lipophilic triesters were longer. The lipids of the oil, the triglycerides, were mainly retained by the stationary phase.

The RLCC fractions were characterized by thin-layer chromatography (TLC) and ^{1}H nuclear magnetic resonance spectroscopy (^{1}H NMR). The thin layer chromatogram corresponding to the RLCC elution chromatogram (system II) of Fig. 3 is shown in Fig. 4. The short-chain phorbol 12,13-diesters, the A-series and the B-series of diesters can be clearly distinguished by TLC. The last RLCC fractions containing phorbol triesters produced an additional spot with an Rf value of about 0.80, because of the presence of small amounts of triglycerides. Some fractions also produced a spot with Rf = 0.50, due to the free fatty acids of the oil. These were most easily detected on TLC using iodine vapors (as a yellow spot). In RLCC these acids were eluted just before the bulk of the phorbol triesters. They can be removed by a preliminary extraction with a 1% sodium carbonate solution.

^{1}H NMR was used as well for the characterization of the RLCC fractions. The most characteristic ^{1}H NMR signal of a phorbol ester is found at 7.6 ppm for H-1. This signal has the advantage of appearing at rather high field and thus of being far from "bulk signals". Comparison of the spectrum of a phorbol 12,13,20-triester with that of a phorbol 12,13-diester demonstrates that, when the C-20 acyl group of a triester is removed to give a C-20 primary hydroxyl group, there is an upfield shift in the allylic 2-proton signal of H-20, from 4.5 to 4.0 ppm. Phorbol 12,13-diesters can be distinguished from the corresponding triesters on this basis (Pieters and Vlietinck 1986).

The same RLCC procedure (solvent system II) was also used for the preparative analysis of about 2.5 g of croton oil. All the RLCC fractions (7.5 ml each with a flow rate of 0.5 ml/min) showing the same TLC spots were collected, and the amount of total phorbol esters, phorbol 12,13-diesters, and 12,13,20-triesters was determined by means of quantitative ^{1}H NMR spectroscopy, using the characteristic ^{1}H-NMR peaks mentioned earlier, with para-dinitrobenzene as an internal standard (Pieters and Vlietinck 1986). The results are shown in Table 1, giving the

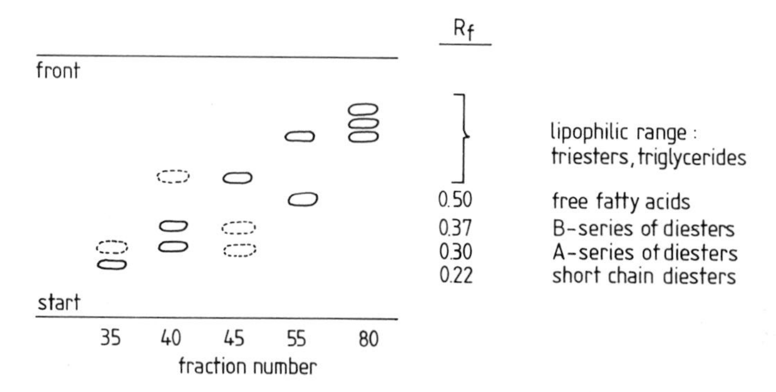

Fig. 4. TLC of phorbol esters containing fractions from croton oil on precoated silica gel 60 F_{254} plates, layer thickness 0.2 mm (Merck) with dichloromethane/acetone 3/1 as the mobile phase. For visualization of the spots two detection modes were used, UV light (at 254 nm, using plates with fluorescent indicator, and at 366 nm, after spraying with sulfuric acid/methanol 1:1 and heating for 15 min. at 110 °C), and iodine vapors (Synder et al. 1984; Pieters and Vlietinck 1986)

Table 1. Preparative RLCC isolation of the phorbol esters from croton oil. Results (% of the total phorbol ester weight)

	Fraction no.									
	1–37	38–44	45–48	49–54	55–64	65–80	81–91	92–100	101–140	Total
Phorbol diesters	–	0.94	2.98	7.28[a]	3.79	–	–	–	–	14.99
Phorbol triesters	–	–	0.29	1.62	7.62	31.87	33.29[a]	7.46	2.87	85.02

[a]Fractions with maximal contents.

relative amount of phorbol diesters as well as triesters in each sample vs the fraction number. The phorbol ester content was calculated as the diester factor Al (TPA, 12-O-tetradecanoylphorbol-13-acetate), being the major constituent of the diesters, with a molecular weight of 616.8 (Hecker and Schmidt 1974). Because molecular weights of the triesters were unknown, they were also calculated as diesters, which is their biologically active form.

RLCC appeared to be a good method to isolate phorbol esters from croton oil, and to separate at the same time the diesters from the triesters. When working on relatively large amounts of oil, the resolution obtained was rather low. The advantage of using RLCC for the isolation of phorbol esters from croton oil is that the diesters as well as the triesters are directly and completely extracted from the total oil, without any preliminary separation procedure, and that they are at the same time more or less separated from each other, depending on the weight and the concentration of the sample. No phorbol triesters are lost, and the samples contain only little triglycerides, if any. On the contrary, the initial partition of croton oil between a hydrophobic and a hydrophilic solvent (e.g., hexane and methanol/water) or a simple methanolic extraction of croton oil leads to a loss of about 50% of the phorbol triesters, which remain in the lipophilic residue of the oil (Pieters and Vlietinck 1986). Therefore, this preliminary partition or extraction should be avoided. Nevertheless, in many cases this is the first step in the analysis of croton oil, and mostly only the hydrophilic part of the oil is investigated (Ocken 1969; Bauer et al. 1983; Marshall and Kinghorn 1984; Erdelmeier et al. 1988). However, due to the cryptic nature of their tumor-promoting activity, the phorbol triesters are at least as important and even more dangerous than their corresponding biologically active diesters of the A- and B-series.

A disadvantage of this RLCC procedure is that the different groups of phorbol 12,13-diesters cannot be separated from each other. A vacuum liquid chromatographic (VLC) procedure to achieve this purpose, after a preliminary RLCC experiment, was described (Pieters and Vlietinck 1989). However, the resolving power of RLCC as well as VLC, especially for preparative work, is rather low, and for the isolation of individual phorbol ester factors or for the investigation of new phorbol esters from croton oil, other methods which have been mentioned earlier are to be preferred (Bauer et al. 1983; Marshall and Kinghorn 1984; Erdelmeier et al. 1988).

4 Summary and Conclusion

RLCC is a good method to isolate phorbol esters from croton oil on an analytical and preparative scale, and to separate at the same time the phorbol 12,13-diesters from the 12,13,20-triesters, although no base-line separation could be achieved. The diesters, as well as the triesters, can be extracted directly and completely from the oil, without any preliminary partition or extraction procedure, which causes a considerable loss of phorbol triesters (Pieters and Vlietinck 1986; Pieters 1988). For the isolation of individual phorbol esters from croton oil, other chromatographic methods with a higher resolving power should be used.

Acknowledgment. L.A.C. Pieters is a senior research assistant of the National Fund of Scientific Research (Belgium).

References

Bauer R, Tittel G, Wagner H (1983) HPLC — Nachweis und Isolierung von Phorbolestern aus Crotonöl. Planta Med 48:10–16

Diamond L, O'Brien TG, Baird WM (1980) Tumor-promoters and the mechanism of tumor-promotion. Adv Cancer Res 32:1–74

Domon B, Hostettmann K, Kovacevic K, Prelog V (1982) Separation of the enantiomers of (±)-norephedrine by rotation locular countercurrent chromatography. J Chromatogr 250:149–151

Erdelmeier CAJ, Van Leeuwen PAS, Kinghorn AD (1988) Phorbol diester constituents of croton oil: separation by two-dimensional TLC and rapid purification utilizing reversed-phase overpressure layer chromatography (RP-OPLC). Planta Med 54:71–75

Evans FJ, Taylor SE (1983) Pro-inflammatory, tumour-promoting and anti-tumour diterpenes of the plant families Euphorbiaceae and Thymelaeaceae. Progr Chem Org Nat Prod 44:1–99

Hecker E (1978) Structure-activity relationships in diterpene esters irritant and cocarcinogenic to mouse skin. In: Slaga TJ, Sivak A, Boutwell RK (eds) Mechanisms of tumour promotion and cocarcinogenesis. Raven, New York, pp 11–48

Hecker E (1981) Cocarcinogenesis and tumor-promoters of the diterpene ester type as possible carcinogenic risk factors. J Cancer Res Clin Oncol 99:103–124

Hecker E, Schmidt R (1974) Phorbol esters — the irritants and cocarcinogens of *Croton tiglium* L. Progr Chem Org Nat Prod 31:376–467

Hostettmann K, Hostettmann M, Marston A (1984) Isolation of natural products by droplet countercurrent chromatography and related methods. Nat Prod Rep 1:471–481

Ito Y (1986) Trends in countercurrent chromatography. Trends Anal Chem 5:142–147

Kubo I, Marshall GT, Hanke FJ (1988) Rotation locular countercurrent chromatography for natural products isolation. In: Mandava NB, Ito Y (eds) Countercurrent chromatography. Marcel Dekker, New York, Chap. 6

Marshall GT, Kinghorn AD (1984) Short-chain phorbol ester constituents of croton oil. J Am Oil Chem Soc 61:1220–1225

Martin DG, Biles C, Peltonen RE (1986) Countercurrent chromatography in the fractionation of natural products. Am Lab October 1986

Ocken PR (1969) Dry-column chromatographic isolation of fatty acid esters of phorbol from croton oil. J Lipid Res 10:460–462

Pieters LAC (1988) Possibilities and limitations of ^{1}H and ^{13}C nuclear magnetic resonance spectroscopy for the identification and the quantitative determination of some naturally occurring carcinogenic risk factors. PhD Thesis, University of Antwerp, chap. 5

Pieters LAC, Vlietinck AJ (1986) Rotation locular countercurrent chromatography and quantitative ^{1}H nuclear magnetic resonance of the phorbol ester constituents of croton oil. Planta Med 52:465–468

Pieters LAC, Vlietinck AJ (1989) Vacuum liquid chromatography and quantitative ^{1}H NMR spectroscpy of tumor-promoting diterpene esters. J Nat Prod 52:186–190

Schmidt R, Hecker E (1975) Autooxidation of phorbol esters under normal storage conditions. Cancer Res 35:1375–1377

Slaga TJ, Scribner JD, Thompson S, Viaje A (1976) Epidermal cell proliferation and promoting ability of phorbol esters. J Natl Cancer Inst 57:1145–1149

Snyder JK, Nakanishi K, Hostettmann K, Hostettmann M (1984) Applications of rotation locular countercurrent chromatography in natural products isolation. J Liquid Chromatogr 7:243–256

Oils and Waxes of Eucalypts Vacuum Distillation Method for Essential Oils

R.B. INMAN, P. DUNLOP, and J.F. JACKSON

1 Introduction

The genus *Eucalyptus* is restricted to the Australasian area and does not extend further northwest of the so-called Wallace Line into Asia (Pryor 1976). The genus does not ocurr naturally in New Zealand or New Caledonia to the east. *Eucalyptus* is a large genus with about 600 species which can be grouped together into subgenera *Blakella*, *Corymbia*, *Eudesmia*, *Gaubaea*, *Idiogenes*, *Monocalyptus* and the large subgenus *Symphyomyrtus*. A separate genus *Angophora* is generally considered "eucalyptoid" and contains seven species (Pryor 1976).

From the very earliest times of European settlement in Australia there was considerable interest in the essential oils of the indigenous Australian "vegetation" (Maiden 1895). Sandlewoods (*Fusanus* spp.) and their oils were exported to Asian centres for burning in temples, and the essential oils of *Melaleuca* species (tea tree) have long been of interest and known to contain cineole and terpinen-4-ol (Maiden 1895). Eucalyptus oils were also in demand for their medicinal value and were listed in the British Pharmacopoeia for 1885. Appearing under *Oleum Eucalypti*, it was defined as "the oil distilled from the fresh leaves of *Eucalyptus globulus*, Labill., *E. amygdalina*, Labill., and probably other species of eucalyptus". Under the heading Characters and Tests, eucalyptus oil was described as "colourless or pale straw-coloured becoming darker and thicker by exposure. It has an aromatic odour, and a spicy and pungent flavour, leaving a sensation of coldness in the mouth. It is neutral to litmus paper. Specific gravity about 0.90. Soluble in about an equal weight of alcohol."

2 Chemical Composition of *Eucalyptus* Oils and Waxes

The first comprehensive work published on the composition of the volatile oils of *Eucalyptus* is that of Baker and Smith (1920), and since then the advent of gas chromatography has resulted in a vastly expanded knowledge of the essential oils of eucalypts and other plant species (Penfold and Willis 1961; Guenther et al. 1975; Small 1977). One of the major components of many *Eucalyptus* species is cineole (1,8-cineole; 1,3,3-trimethyl-2-oxabicyclo[2,2,2]octane; C10H18O). Cineole, b.p. 174.4 °C, is generally considered to be a major component of commercial eucalyptus oil, and finds use in the pharmaceutical field, in confectionery, in stain removal and as a fuel additive. Other common components are α-pinene (2,6,6-trimethylbicyclo[3.1.1]hept-2-ene), b.p. 156 °C, limonene (1-methyl-4-(1-methylethenyl)-

cyclohexene), b.p. 176 °C, and p-cymene (1-methyl-4-(1-methylethyl)benzene) b.p. 177 °C. In such a large genus as *Eucalyptus* it is not surprising that many compositional differences can be found from one species to another, these include citronellal (in *E. citriodora*), nerolidol, ocimene, α-phellandrene, carvatonacetone and isovaleraldehyde (in *E. deglupta*), while methyl 3,4,5-trimethoxybenzoate is the major component of essential oils from *E. crenulata*, together with γ-terpinene, terpinolene and terpen-1-en-4-ol and others (McKern 1965).

Many of the components of eucalyptus oils are known to have insect repellent properties; cineole and p-cymene have been tested for ovipositional repellent activity against mosquito (*Aedes aegypti*), and although not as effective as the commercial insect repellent DEET (N,N,diethyl-m-toluamide) at the same concentration (Klocke et al. 1989), are nevertheless effective insect repellents at higher concentrations. Cineole is a better repellent than p-cymene.

The waxes of *Eucalyptus* are considered to be largely β-diketones, with carbon chain lengths of 29, 31 or 33, and diketone groups at 12, 14, 14, 16 and 16, 18 respectively (Tulloch 1976). In this context, the term "wax" is used here loosely and refers to "surface lipids" on leaves. The waxes consist of β-diketones, esters, flavanoids and others, with the β-diketones of general formula

$$R{-}(CH_2)_x{-}CH_2{-}(CH_2)_y{-}\underset{\underset{O}{\|}}{C}{-}CH_2{-}\underset{\underset{O}{\|}}{C}{-}(CH_2)_z{-}CH_3$$

These waxes have been investigated by Hallam and Chambers (1970), who found by electron microscopy that physically the leaf surface waxes were of two forms: "tube" waxes where β-diketones were the predominant chemical species present, or "plate" waxes where other waxes predominate. According to Thomas and Barber (1974) the dense layer of "tube" waxes on *E. urnigera* leaves are responsible for the increased frost hardiness of this species.

3 Methods of Analyses of *Eucalyptus* Waxes

Investigations into the waxes of *Eucalyptus* leaves have not been as numerous as that into the oils, and the present authors have little experience with these compounds. We refer the reader to the work of Hallam and Chambers (1970). These investigators, in addition to an electron microscope study of the leaf surface waxes, carried out wax isolation as follows: (1) reflux leaf samples in a reflux tube (2 in. by 2 ft.) over petroleum ether (b.p. 40–60 °C) for 1 h; (2) concentrate on a rotary evaporator, the residue being dissolved in a minimum volume of boiling acetone; (3) filter when hot and cool; (4) store at –5 °C for 12 h, over which time crystallization of the waxes may occur; (5) filter, and wash with cold acetone. Gas chromatography can be coupled with mass spectrometry for separation and chemical identification of components (Tulloch 1976).

4 Methods of Analysis of *Eucalyptus* Oils

It can be seen from the above discussion that eucalyptus oils were analyzed from early times by distillation from the leaves. However, a variety of tests were needed to further define the composition of these oils as can be gleaned from the British Pharmacopoeia of 1885 (quoted above). It was not until the advent of gas chromatography which could be coupled with mass spectrometry that definitive identification of individual components became easier. The reader is referred to Volume 3 in the series Modern Methods of Plant Analyses (Linskens and Jackson 1986) for several chapters dealing with gas chromatography and coupling with mass spectrometry for analysis of plant materials. This present chapter will concentrate on the extraction technique, since this appears to present many sources of error in analyzing quantitatively the oils of eucalyptus. Before considering the extraction procedures, however, it should be explained that many methods have been developed for determination of the essential oils of eucalyptus species following extraction, including for example, cineole determination by the o-cresol (o-methylphenol) method using a freezing point determination, or the congealing point method of Kleber and von Richenburg (see Ammon et al. 1985). Cineole can also be determined by the phosphoric acid method, or colorimetrically. In addition, cineole can be analyzed by proton magnetic resonance or infrared spectroscopy (Ammon et al. 1985). All of these methods have some defect leading to inaccuracy, and the method of choice remains gas chromatography, with the possibility of confirmation of results by gas chromatography coupled with mass spectrometry (Ammon et al. 1985; Linskens and Jackson 1986).

Concerning the extraction of eucalyptus oils, it seems that up until 1985 at least, all essential oil analyses of *Eucalyptus* species had been performed on oil obtained by steam distillation (Ammon et al. 1985). Several variations on the steam distillation apparatus have been used (Clevenger apparatus; McKern and Smith-White apparatus, see Ammon et al. 1985; Likens-Nickerson apparatus, see Likens and Nickerson 1964). According to Ammon et al. (1985), however, steam distillation can lead to losses in some terpenes, notably α-pinene. These workers compared solvent extraction of eucalypt leaves with steam distillation (Clevenger apparatus), and found lower results with steam distillation. Loss of α-pinene was particularly serious, sometimes as high as 30 to 35% with steam distillation compared to extraction with ethanol. Ammon and coworkers (1985) also maintain that the ethanol extraction procedures, besides giving reliably high yields, can be easily adapted to relatively small leaf samples (4 g) and can be carried out quickly (say 50 samples per week). Using the Clevenger apparatus, by way of contrast, Ammon et al. (1985) extracted 300 g leaves for each distillation, following the recommended procedure.

Although the solvent extraction procedure is simple and can be carried out on quite small samples, we wished to develop a distillation method that would yield pure oil not contaminated with nonvolatile material that could presumably be present in oil samples obtained by solvent extraction. We also hoped to avoid the problems associated with the steam distillation technique and aimed to develop a method that could be used with small amounts of leaves. The vacuum distillation method described below has been developed to meet these needs. The principle is quite simple, leaf powder is heated in a vacuum and volatile oils are condensed onto a cold finger within the distillation vessel. The method has been tested and shown to allow distillation of pinene with good yield and to require between 5–10 g of leaves.

5 Vacuum Distillation Method for Essential Oils

5.1 Preparation of Leaf Powder

Leaves are cut into 5-mm squares and then reduced to a fine powder after cooling in liquid nitrogen using a stainless steel Waring blender (Model no. SS110). Add liquid nitrogen to cover leaves and, immediately after evaporation of the nitrogen, blend for 10 s three times. Repeat liquid nitrogen/blending treatment, as above, three more times with a final blending of 30 s. Quickly transfer resulting leaf powder to the distillation tube (Fig. 1). If the distillation is not to be carried out immediately, the tube should be capped with a stopper, covered with Saran Wrap and stored at –20 °C until needed.

5.2 Distillation

The following procedure has been designed to allow a high yield distillation of pinene (a low boiling point oil). Add a filter paper baffle just above the powder (to minimize powder blowing about during the vacuum distillation). Connect up the system (Fig. 1) with the main vacuum isolation valve turned off.

1. Purge the distiller with dry air for about 2 min (open the air valve while the distiller stopper is left ajar). Tightly close the stopper and close the air valve.
2. Cool the distiller tube in a beaker of ethanol/dry ice to –60 °C and cool the cold finger bath with ethanol/dry ice to –75 °C and let distiller equilibrate for 15 min.
3. Open the main vacuum pump valve slowly and pump down till below 25 μ while the distiller tube is held at –60 °C. The ultimate vacuum achieved is usually about 15–30 μ.
4. Allow the distiller tube to slowly warm up while in the ethanol bath. The temperature should rise from –60 to +20 °C over a 20-min period.
5. Transfer distiller to the water bath at 25 °C and leave for 20 min. Raise the temperature to 50 °C for 20 min and finally raise temperature to 80 °C and leave for 240 min.
6. Remove distiller from the water bath and solid carbon dioxide from the cold finger cooling bath. Turn off the main isolation vacuum valve and open the dry air valve to allow the distiller to come to atmospheric pressure. Drain the ethanol from the cold finger cooling bath (to produce fast draining, place a stopper in the top of the

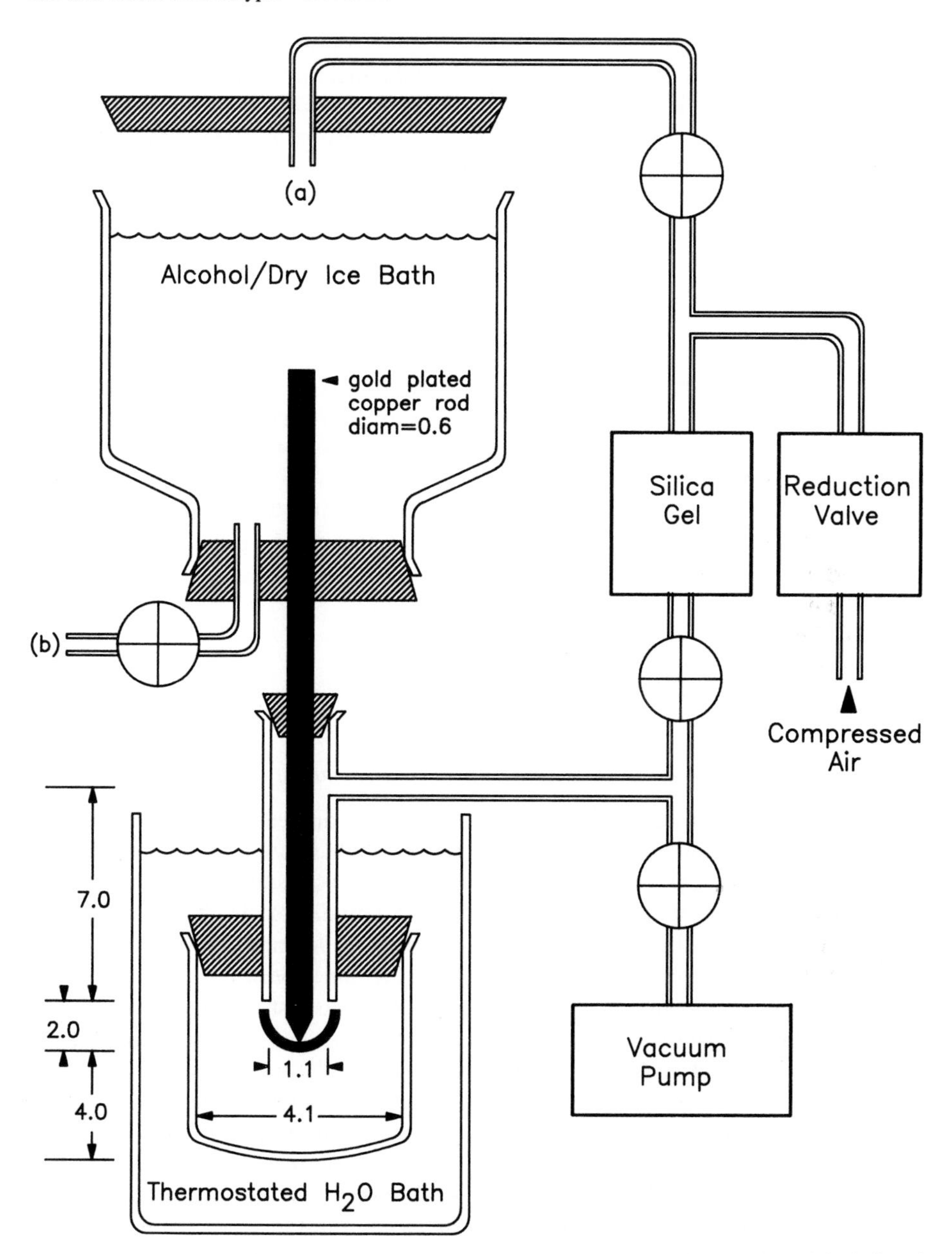

Fig. 1. Apparatus for the vacuum distillation of eucalyptus leaf powder. The equipment is designed for distillation of 5 to 10 g of powder. The figure shows dimensions in cm. The gold plated copper rod is cooled to –75 °C by the alcohol/dry ice bath situated above the distillation vessel. The condensate, which forms on the rod, runs down into the glass cup at the bottom of the cold finger when the cold finger is warmed to room temperature at the completion of distillation. The glass cup is glued to the cold finger with a small bead of silicon adhesive. The air inlet **a** is used to force out the alcohol from the cold finger cooling bath at the completion of the distillation (the stopper at **a** is inserted into the bath housing and the valve **b** opened). Initially, this apparatus was constructed with glass seals rather than the rubber seals shown in this figure (*hatched areas*), but it was felt that the greater vessel volume required by the glass seals was a disadvantage

cooling bath and apply air pressure from the air supply). Warm up the cold finger by placing a hair drier outlet hose over the cold finger for 10 min.

7. Open up the distiller and aspirate the oil and water with a bent capillary pipette and transfer to a glass conical centrifuge tube. Centrifuge at 10 000 rpm for 10 min. Aspirate the oil into a glass tube and record the weight of condensed water and oil.

5.3 Notes

Dead Volume

After a distillation, the condensed oil and water has to be aspirated from the cup and inevitably there will be a small amount of oil/water residue remaining. For accurate quantitation it is necessary to allow for the residue left in the cup and the aspiration pipettes. This was determined by simply adding a known weight of an oil/water mixture to the cup and then determining what weight could be aspirated. In this way it could be shown that the dead volume (the weight remaining in the cup and the pipettes after aspiration) for the apparatus in Fig. 1 corresponded to:

water	0.0162 ± 0.0092 g
oil	0.0363 ± 0.0058 g

Authentication and Reliability of the Distillation Procedure

1. Total oil contents are calculated on the basis of the dry leaf powder remaining at the end of the distillation. Thus the oil content, O/P (dry), is calculated as gram oil/gram residual powder; with the dead volume correction, described above, added to the weight of oil actually obtained. In a number of distillations involving nine different species, the overall uncertainty in oil content was found to be ± 8% (max. = 12%) between repeat distillations of the same species.
2. The vacuum distillation method described above was tested using pure oils with boiling points spanning the range found in *Eucalyptus*. The results in Table 1 show that these oils can be distilled as described above, without significant loss.
3. The distillation time of 4 h is somewhat arbitrary. As shown in Table 2, a distillation for 8 h yields a significant increase in oil yield but preliminary examination by GPC shows that the compositions of oils obtained at 4 and 8 h are not greatly different.
4. A possible disadvantage of this distillation method is that the moisture content of the leaves has to be reduced before distillation. For instance, with *camaldulensis*, if leaves are dried at 37 °C to produce a weight loss of 50%, then distillation results in an oil yield of 0.0217. Using leaves straight from the tree or leaves dried to a weight loss of 40% results in reduced oil yields (53% and 67% of the above value respectively).

   Furthermore, the powder resulting from undried leaves behaves badly during distillation; the powder tends to blow about and can end up on the cold finger and perhaps absorb condensed oil. We determined if the required

Table 1. Yields of various oils after vacuum distillation. Yields were calculated after correction for dead volume

Oil	b.p. (°C)	Oil added (g)	Yield (%)
α-Pinene	155	0.1241	92
d-Limonene	175	0.1420	85
d,l-Citronellol	225	0.1095	90

Table 2. Effect of Distillation time on yield of oil. Distillation was carried out at 80 °C

Species	Time (min) at 80 °C	O/P (dry)	Percent of 480 min data
globulus	480	0.0299	100
	240	0.0264	88
camaldulensis	480	0.0259	100
	240	0.0217	84
cloeziana	480	0.0094	100
	240	0.0080	85

drying at 37 °C results in evaporation of oil and found that it does not. For instance, if whole leaves (not powdered leaves) are put into the distiller and heated in a vacuum to 80 °C for 4 h then no oil is distilled. This experiment has been carried out with *grandis, globulus* and *socialis* which yield oil contents of 0.0167, 0.0268 and 0.0447, respectively, when leaf powder is used. If no significant oil evaporation takes place in a vacuum at 80 °C, then we assume that the required drying at 37°C and atmospheric pressure (see above) does not lead to oil evaporation. We find it of considerable interest that the oil glands in intact leaves were able to retain oil at 80 °C in a vacuum. This is particularly well illustrated with *socialis* (which has a high oil content); when viewed by microscopy, oil droplets can be immediately seen if the leaf surface is disturbed, however, as described above, the oil is retained at 80 °C in a vacuum. After heating in a vacuum, microscopy shows that these leaves still exude oil droplets if the surface is disturbed.

Comparison of the Various Methods

Inspection of the results obtained for *E. globulus* leaves by Ammon et al. (1985) show a mean value of O/P (wet) of 0.0146 and 0.0150 for solvent extraction and steam distillation respectively. The results obtained here with vacuum distillation (Table 2) compare favourably with the above values. For globulus, we obtain an oil content of 0.0299 after an 8hr distillation (0.0264 after a 4hr distillation), calculated with respect to the dry powder remaining after distillation. If these values are adjusted to a wet leaf basis (for our distillations the globulus leaves were dried to yield a 40–50% weight loss) then the oil contents would be between 0.0139–0.0167 for an 8hr distillation (0.0123–0.0148 for a 4hr distillation).

Ammon et al. (1985) reported rather large losses (30–35%) for a α-pinene with steam distillation methods. The vacuum distillation method described here does not show such losses, in fact, no more than 8 or 9% is lost by this method for the oils so far tested. The method is applicable to quite small amounts of material (5 to 10 g) and therefore is an improvement on the previously reported steam distillation procedures, which required approximately 300 g of material (Ammon et al. 1985). The method, therefore, has many advantages over the earlier steam distillation procedures and compares more than favourably (except for the time involved) with the solvent extraction method of Ammon et al. (1985).

We found that it was important to make a very fine powder before distillation. If the leaves rather than powdered leaves are distilled, then no oil is obtained after 4 h at 80 °C, whereas powdered leaves give good yields. It is possible that the oil glands in the leaves are essentially vacuum tight and remain tight after the leaves are taken from the tree. Presumably, these glands are made to open and close under the control of a regulatory system that requires some type of stimulus.

The stage of development of the leaf has been shown to be important in sampling for leaf volatile oils. There is quantitative variation in oil composition during leaf maturation in *Angophora costata* (Leach and Whiffin 1989), a result which may be broadly applicable to the closely related genus *Eucalyptus*. It appears that samples should be taken so as to exclude immature and aged leaves if meaningful comparisons are to be made between individual trees or species. The same investigators also showed that seasonal and diurnal variation in leaf volatile oils was less than genotypic variation between individuals in *Angophora costata*. They found that diurnal variation is so small as not to require any control sampling (Leach and Whiffen 1989).

References

Ammon DG, Bartin AFM, Clarke DA, Tjandra J (1985) Rapid and accurate determination of terpenes in the leaves of *Eucalyptus* species. Analyst 110:921–924

Baker RT, Smith HG (1920) Eucalypts especially with regard to their essential oils. Second Edition, Government Printer, Sydney

Guenther E, Gilbertson G, Koenig RT (1975) Essential oils and related products. Anal Chem 47:139–157

Hallam ND, Chambers TC (1970) The leaf waxes of the genus *Eucalyptus* L'Heritier. Aust J Bot 18:335–386

Klocke JA, Balandrin MF, Barnby MA, Yamasaki RB (1989) Liminoids, phenolics and furanocoumarins as insect antifeedants, repellents and growth inhibitory compounds. In: Arnason JT, Philogene BJR, Morand P (eds) Insecticides of plant origin. ACS Symp Ser 387, Am Chem Soc, Washington DC, pp 137–149

Leach GJ, Whiffin T (1989) Ontogenetic, seasonal and diurnal variation in leaf volatile oils and leaf phenolics of *Angophora costata*. Aust Syst Bot 2:99–111

Likens ST, Nickerson GB (1964) Detection of certain hop oil constituents in brewing products. Proc Am Soc Brew Chem 5–13

Linskens HF, Jackson JF (eds) (1986) Gas chromatography/mass spectrometry. Modern Methods of Plant Analyses, Vol 3. Springer, Berlin Heidelberg New York Tokyo

Maiden JH (1895) The chemistry of the Australian indigenous vegetation. Australian Association for the Advancement of Science, Proc, pp 167–209

McKern HHG (1965) Volatile oils and plant taxonomy. J R Soc NSW 98:1–10

Penfold AR, Willis JL (1961) The eucalypts: botany cultivation, chemistry and utilization. Leonard Harris, London, 245 pp

Pryor LD (1976) Biology of eucalypts. Studies in Biology No. 61. Edward Arnold, London

Small BEJ (1977) Assessing the Australian eucalyptus oil industry. For Timber 13:13

Thomas DA, Barber HN (1974) Studies on leaf characteristics of a cline of *Eucalyptus urnigera* from Mount Wellington, Tasmania. I. Water repellency and the freezing of leaves. Aust J Bot 22:501–512

Tulloch AP (1976) Chemistry of waxes of higher plants. In: Kolattukudy PE (ed) Chemistry and biochemistry of natural waxes. Elsevier, Amsterdam, pp 236–289

Analysis of Epicuticular Waxes

S. MISRA and A. GHOSH

1 Introduction

The plant surfaces which are exposed to the atmosphere, such as leaves, stems, fruits, and petals, are covered with a hydrophobic, water repellent substance called wax, of which the leaf waxes have received most attention. The outer surface of leaf epidermis is covered with a substance called cutin which is usually impregnated with wax; together they comprise the cuticle. The insoluble polymer cutin is composed of cross-linked hydroxy fatty acids (Kolattukudy 1975), which are released upon hydrolysis. The wax is a complex mixture of lipophilic substances such as hydrocarbons, wax esters, alcohols, and ketones (Tulloch 1976a). The cuticular wax plays an important role in preserving the water balance of the plant by reducing evaporation from the leaf surface. The hydrocarbons together with other waxy components serve as a barrier to the passage of water in and out of the cell, thus preventing water inundation or dehydration (Misra et al. 1984a; Weete et al. 1978). Other protective functions may include minimizing mechanical damage to leaf cells and inhibiting fungal and insect attack. The structural and functional roles of leaf epicuticular waxes along with their biosynthesis and some analytical aspects have been discussed in an excellent review by Eglinton and Hamilton (1967).

The most common constituent of leaf cuticular lipids is represented by the hydrocarbons, usually consisting of a mixture of n-alkanes of carbon chain lengths ranging between 25 to 35 carbon atoms. These hydrocarbons are predominated by the occurrence of odd carbon chain compounds. Branched chains of both iso- and anteiso-hydrocarbons and unsaturated hydrocarbons also occur, but only in trace amounts. Often one (or a few hydrocarbons) predominates and n-C_{29} or n-C_{31} is usually the dominant alkane in higher plants. For example, n-C_{29} alkane constitutes more than 90% of the alkanes of *Brassica olevacea* (Kolattukudy 1965), and an equally high proportion of n-C_{31} alkane is found in *Senecio odoris* (Kolattukudy 1968). Under exceptional circumstances, hydrocarbon components with even carbon chains predominate over the odd carbon chains (Misra et al. 1986, 1987, 1988). Iso-(2-methyl) and anteiso-(3-methyl) branched alkanes are found to occur in substantial amounts in many plants, e.g., tobacco (Mold et al. 1966; Kaneda 1967). Generally, odd numbers of carbon atoms are predominant in iso-alkanes, while, anteiso-alkanes have an even number of carbon atoms. Occasionally, hydrocarbons with internal and/or multiple branches, and unsaturated and cyclic structures (Kuksis 1964; Wollrab et al. 1965, 1967; Mold et al. 1966; Stransky et al. 1970) have been found to occur as minor components. Saturated n-alkane, n-C_{17} predominates in lower plants such as algae, however, unsaturated and internally branched components also occur (Youngblood et al. 1971).

Oxygenated alkane derivatives such as aliphatic alcohols and ketones occur in the cuticular wax of many plants. Aliphatic alcohols are found in the free form as components of cuticular wax lipids or more usually in the esterified form as wax esters (Kolattukudy 1975). These alcohols may be present with normal, branched, or unsaturated chains of various lengths and with primary, secondary, or rarely, with tertiary alcoholic functions. The major components of free primary alcohols generally have chain lengths of C_{14}-C_{32}. Secondary alcohols occur less frequently in plant waxes. Examples of secondary alcohols are 10- and 15-nonacosanol and 9- and 16-hentriacontanol (Tulloch 1976a).

The classes of ketones which are constituents of plant waxes may occur as mono- or β-diketones. Monoketones are of necessity odd-carbon numbered, e.g., nonacosane-15-one and hentriacontan-16-one are constituents of some plant cuticular waxes (Kolattukudy 1975). The other ketone class is represented in nature by the β-diketones which are constituents of plant waxes (Tulloch 1976b). These have an odd number of carbon atoms with chain lengths usually of C_{29}, C_{31}, and C_{33} carbons; the carbonyl groups are located on even positions, e.g., 12,14-, 14,16-, and 16,18- are the most common. Hydroxy and oxo β-diketones were found in many members of the Triticeae (Tulloch 1976b; Tulloch et al. 1980).

The most common constituents of the soluble lipids of the plant cuticle are wax esters, defined as fatty acid esters of fatty alcohols. Usually, wax esters are composed of n-alkanoic acids and n-alkan-l-ols with an even number of carbon atoms ranging from C_{12} to C_{32}. Carbon chain lengths of C_{20} to C_{24} tend to predominate in the acyl chain portion, while in the alcohol portion C_{24} to C_{28} carbon atoms are often the major components. The cuticular lipids of most plants contain free fatty alcohols and free fatty acids as significant components. Chain lengths of the free alcohols are generally similar to that of esterified alcohols, whereas substantially longer chain lengths are found in the free fatty acid moieties compared to those found in the wax esters. Generally, the unsaturated fatty acids of plant leaf surface waxes are composed of mono-, di-, and trienoic moieties (Misra et al. 1986), whereas the unsaturated alcohols are mostly monoenoic (Misra et al. 1987).

Several plants have been found to contain fatty aldehydes with chain lengths similar to those of alcohols (Schmid and Bandi 1969; Kolattukudy 1970). Other, less common components of cuticular lipids include alkane diols, ketols, terpenes, flavones, and aromatic hydrocarbons (Kolattukudy 1975; Martin and Juniper 1970). Recently, estolides, which are triacylglycerol components containing hydroxy fatty acids, the hydroxyl group of which is esterified to an additional fatty acid, have been found to occur in plant waxes (Tulloch and Bergter 1981).

Environmental factors which influence cuticle development have been reviewed by Hull et al. (1975). Plants can become conditioned to water stress by depositing more wax on the leaf surface, which may result in increased resistance to cuticular transpiration (Weete et al. 1978). Stress produced by inundation under tidal water results in the deposition of higher amounts of hydrocarbons and wax esters in the leaves (Misra et al. 1984a), with a high proportion of unsaturated fatty acids in the wax esters and higher levels of low molecular weight hydrocarbons in the inundated leaves (Misra et al. 1986, 1988). Methods for analyzing plant waxes have been reviewed earlier (Tulloch 1975).

2 Extraction of Epicuticular Wax

Various methods to extract waxes have been described using relatively nonpolar solvents. Because leaf epicuticular lipids are largely in hydrophobically associated form, neutral lipids may be extracted with nonpolar solvents like light petroleum ether, diethyl ether, chloroform, or benzene.

Epicuticular wax may be easily isolated from the plant tissue by dipping it in chloroform for a brief period of 10–30 s (Kolattukudy 1975). The extract is then filtered and concentrated by rotary evaporation. Wax was extracted from plants by stirring the plant pieces in light petroleum (40–60 °C) for about 10 s (Tulloch and Weenink 1969). The extraction of wax was also carried out by soaking the leaf discs for 16 h in petrol (40–60 °C) at room temperature (Cowlishaw et al. 1983); the extract was decanted and evaporated to dryness. To avoid the loss of wax, particularly in small-scale operations, the authors prefer the most widely used lipid extraction procedure of Bligh and Dyer (1959), a simplified version of the "classical" Folch et al. (1957) procedure. Plants are cut into small pieces and homogenized in methanol-chloroform (2:1), centrifuged; the residue is then rehomogenized with methanol-chloroform-water (2:1:0.8). After centrifugation, the residue is finally homogenized with the first solvent system and recentrifuged. The extract is then diluted with one volume each of chloroform and water to form the two-phased system, chloroform and methanol-water, and allowed to separate in a separatory funnel. The lower chloroform layer containing lipid is dried over sodium sulfate and concentrated by rotary evaporation at room temperature.

3 Fractionation of Wax Components

3.1 Column Chromatography

Fractionation of wax may be conveniently done by silicic acid column chromatography after activation of silicic acid at 120 °C for 18 h (Tulloch and Weenink 1969). The column was eluted with hexane containing increasing proportions of diethyl ether. Hydrocarbons were eluted with pure hexane; wax esters with hexane-diethyl ether (99:1); free alcohols with hexane-diethyl ether (49:1); less polar estolide and free fatty acids with hexane-diethyl ether (47:3 to 9:1); more polar estolides with hexane-diethyl ether (4:1 to 3:2); and diols with hexane-diethyl ether-ethanol (5:4:1) (Tulloch and Bergter 1981). During this fractionation, β-diketones and esters were eluted together and the fraction caused high UV absorption at 273 nm due to the presence of β-diketones (Tulloch and Hoffman 1979). Wax esters were purified by removal of admixed β-diketones first as the Cu complex (Horn et al. 1964) and then as the semicarbazone (Tulloch and Hoffman 1974). Hydroxy β-diketones which elute with free acid may also be removed as the Cu complex (Tulloch and Hoffman 1971).

When the total lipids from plant leaves are extracted by the method of Bligh and Dyer (1959), it is necessary to fractionate them into neutral, glyco-, and phospholipids using silicic acid column chromatography (Rouser et al. 1967). The neutral lipid fraction, which contains epicuticular wax components, is eluted by chloroform, whereas glyco- and phospholipids are eluted, respectively, by acetone and methanol. For small sample quantities column chromatography may be conveniently carried out using a Pasteur pipette filled with silicic acid (Kates 1986).

3.2 Thin Layer Chromatography (TLC)

For small quantities of total lipids TLC is a very convenient technique for the isolation of neutral lipids using a solvent system of chloroform-methanol (9:1), which moves the total neutral lipids to the solvent front leaving the polar lipids near the origin. The neutral lipid fraction is eluted out of the adsorbent using chloroform-methanol (1:1), as described by Kates (1986).

3.2.1 Fractionation of Neutral Lipids by TLC

For the separation of wax components in the neutral lipids by TLC, chloroform was the solvent of choice (Tulloch and Weenink 1969; Tulloch and Hoffman 1971). Leaf waxes of durum wheat were separated (Tulloch and Hoffman 1971) by using a solvent system of chloroform containing 1% ethanol by volume. Clear separation was obtained for hydrocarbons, esters, β-diketones, alcohols, hydroxy β-diketones, and acids. Benzene was used to separate the epicuticular waxes of *Chionochloa* and five fractions comprised of acids, alcohols, aldehydes, esters, and alkanes were separated (Cowlishaw et al. 1983).

For separation of neutral lipid components, the method described by Mangold (1969) has been found to be useful when all the components are of interest (Misra et al. 1988), using a solvent system of light petroleum ether-diethyl ether-acetic acid (80:20:1.0). We have observed that light petroleum ether in the b.p. ranges of 40–60 or 60–70 °C is equally good. Moreover, instead of light petroleum ether, n-hexane may also be used without loss of any resolution. According to this method, hydrocarbons, steryl esters, and wax esters have been found to overlap as shown in Fig. 1. To obtain these compounds in a pure state, the overlapped bands were rechromatographed using a solvent system of n-hexane-diethyl ether (49:1), whereby complete resolution was possible (Misra et al. 1986, 1987, 1988; Fig. 2). It should be noted here that elution sequences of steryl ester and wax ester (Fig. 1) have been reversed by using less polar solvent without acetic acid (Fig. 2). Bands were visualized by iodine vapor, which was then evaporated by leaving the TLC plate for sometime at room temperature; the compounds were then recovered.

4 Analysis of Wax Components

The various components of wax obtained by fractionation on TLC or by column chromatography and the techniques applied for their analysis, particularly by GLC of their appropriate derivatives, will be discussed in later sections.

4.1 Analysis of Hydrocarbon by Gas Liquid Chromatography (GLC)

Hydrocarbons are analyzed most conveniently by GLC without derivatization using nonpolar liquid phases, such as SE-30, OV-17, OV-1, etc. In the present

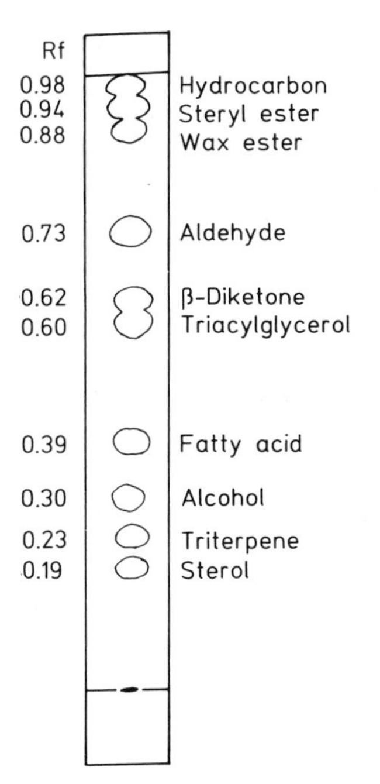

Fig. 1. Thin layer chromatogram of the neutral lipid fraction obtained from total leaf lipid of *P. coarctata* (Misra et al. 1988), using a solvent system of light petroleum ether (40–60 °C)-diethyl ether-acetic acid (80:20:1) according to Mangold (1969). Spots were visualized by exposing the TLC plate to iodine vapor

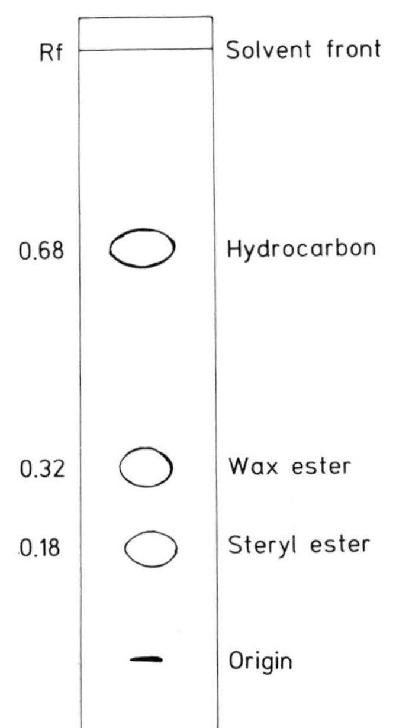

Fig. 2. Separation of hydrocarbon, steryl ester, and wax ester obtained by the fractionation of neutral lipids as shown in Fig. 1. Developing solvent was n-hexane-diethyl ether (49:1) according to Misra et al. (1986, 1987, 1988). TLC plate was visualized by iodine vapor

discussion only the conventional packed column, which is most widely used, will be considered. The column material, consisting of 1–3% of the liquid phase coated on solid supports such as Gas Chrom-P, Gas Chrom-Q, Gas Chrom-Z, or Chromosorb-W (HP), may be used. The particle size of the solid support depends on the size of the column, particularly the internal diameter. For example, for a column with 2–3 mm i.d. 80–100 mesh of the support is preferred, whereas for columns with 4–6 mm i.d. generally 60–80 mesh particles are preferred. The length of the column may vary from 4–12 ft. for general analytical purposes. For a general discussion on gas liquid chromatography, particularly applicable to lipids, the reader is referred to Christie (1982) and Kates (1986).

A 3% OV-17 liquid phase supported on Chromosorb-W (HP) 60–80 mesh, and packed into a 1.8 m × 6 mm stainless steel or glass column was used to analyze the plant leaf hydrocarbons (Misra et al. 1985, 1986, 1987, 1988) and the yeast hydrocarbons (Misra et al. 1984c). Hydrocarbons found were from C_{16}-C_{38} with normal, iso- and anteiso-chains. Initially, the temperature was kept isothermal around 220 °C for 8–12 min and was then programmed at a rate of 4 °C/min to a final temperature of about 320 °C. The carrier gas was nitrogen with a flow rate of 60 ml/min. For compounds with a wide range of molecular weights, a programmed temperature GLC run is absolutely necessary. Generally at the beginning, the isothermal run is carried out for some time to allow the low molecular weight compounds to produce good resolution, then programming is started at a convenient rate from a certain point, such that the high molecular weight compounds are also eluted at a reasonable time with good resolution.

As leaf hydrocarbons are composed of compounds with a wide range of molecular weights, with normal and branched chains like iso-(2-methyl) and anteiso-(3-methyl) compounds, they could be identified by two isothermal runs, one at a low temperature and the other at a higher temperature. For example, the hydrocarbons of *Porteresia coarctata* leaves were analyzed (Misra et al. 1988) on a 3% SE-30 column at two different temperatures under isothermal conditions, namely, 200 °C for identification of the low molecular weight components, and isothermally at 270 °C for high molecular weight compounds. In Fig. 3, the chromatogram obtained for the isothermal run at 200 °C is presented. The even-carbon chain components were identified by GLC of authentic even-carbon chain n-hydrocarbons under identical conditions. These authentic even-carbon chain n-hydrocarbons may be obtained from various commercial sources. Identification should be further confirmed by the addition of a single authentic hydrocarbon of known chain length to the sample followed by GLC of the mixture. For the identification of odd-carbon chain n-hydrocarbons, a semilogarithmic plot with the logarithm of the relative retention time (log RRT) against the carbon chain length is drawn (Fig. 3). The log RRT of the components in the chromatogram, besides the even-carbon chain compounds which fit in the linear plot, enables the identification of the odd-carbon chain compounds of the n-series. Since the nonpolar liquid phases like OV-17, SE-30, etc. are known to be boiling point separators, the iso-compounds, which have lower vapor pressures compared to those of the isomeric n-compounds, will be eluted before the n-compounds. For similar reasons the anteiso-compounds will be eluted before the corresponding iso-compounds.

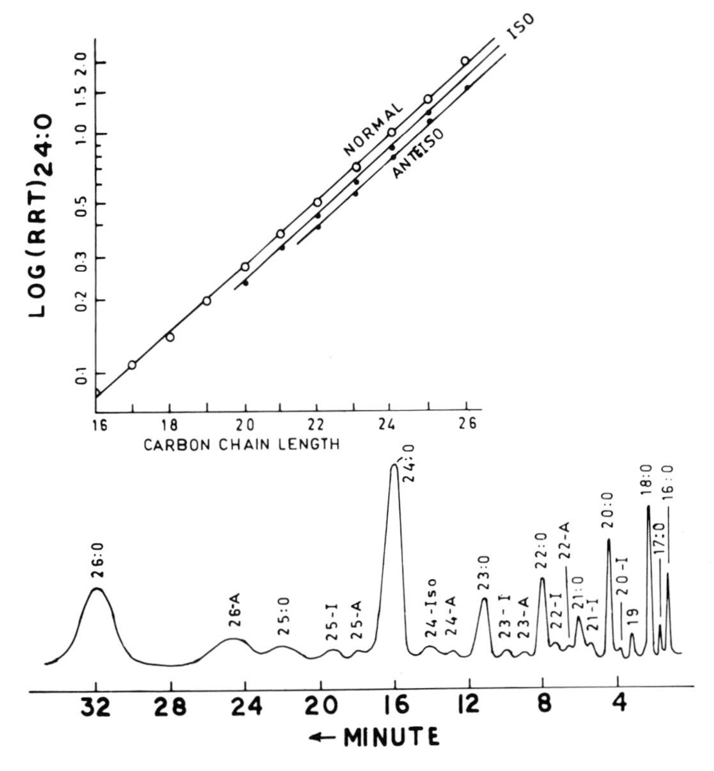

Fig. 3. Gas liquid chromatogram of the hydrocarbon fraction obtained by TLC, as in Fig. 2. Column was 3% SE-30 coated on 80–100 mesh Chromosorb W (HP) and packed into a 1.8 m × 3 mm i.d. glass column. Oven temperature was isothermal at 200 °C with detector and injection port temperatures of 370 and 250 °C, respectively: carrier gas was nitrogen with a flow rate of 60 ml/min. Relative retention times (*RRT*) were determined with respect to 24:0 hydrocarbon and plotted against carbon chain lengths. After the elution of 26:0, the oven temperature was raised to 350 °C and the high molecular components were eluted out. The instrument was a Pye Unicam 104 gas chromatograph with a dual column and dual flame ionization detector. The hydrocarbon sample was from the leaves of *P. coarctata* of Chuksar island (Misra et al. 1988). *Numbers* represent carbon chain lengths; *A* and *I* represent anteiso- and iso-hydrocarbons, respectively

Therefore, the peaks just preceding the peaks of n-alkanes are considered to be for iso-alkanes. A semilogarithmic plot (Fig. 3) confirms the identification of the iso-compounds. Similarly, the anteiso-compounds are also identified. The presence of any unsaturated compounds should be confirmed by reduction and rechromatography of the sample.

Plant hydrocarbons are also frequently unsaturated and, because the vapor pressures of the latter are lower than those of the corresponding saturated ones, they are eluted just before the n-alkanes, overlapping the peaks for branched chain alkanes. Under such circumstances, the presence of unsaturated and/or branched chain compounds should be confirmed by catalytic hydrogenation (Ghosh and Dutta 1972) of the sample and rechromatography. The disappearance or decrease in the peak area at the positions of the branched chain compounds with a proportionate increase in the n-alkane peak area will confirm the presence of unsaturated n-alkanes with particular chain lengths. Finally, after identification of the hydrocarbons in the lower and higher molecular weight regions, a chromatogram showing the presence of all the components can be obtained by a programmed temperature run (Fig. 4), which is necessary for the quantitation of all the hydrocarbon components.

4.1.1 Quantitation of Hydrocarbons

The relative proportions of hydrocarbon components are determined from the gas chromatogram by triangulation and determination of each peak area (Burchfield

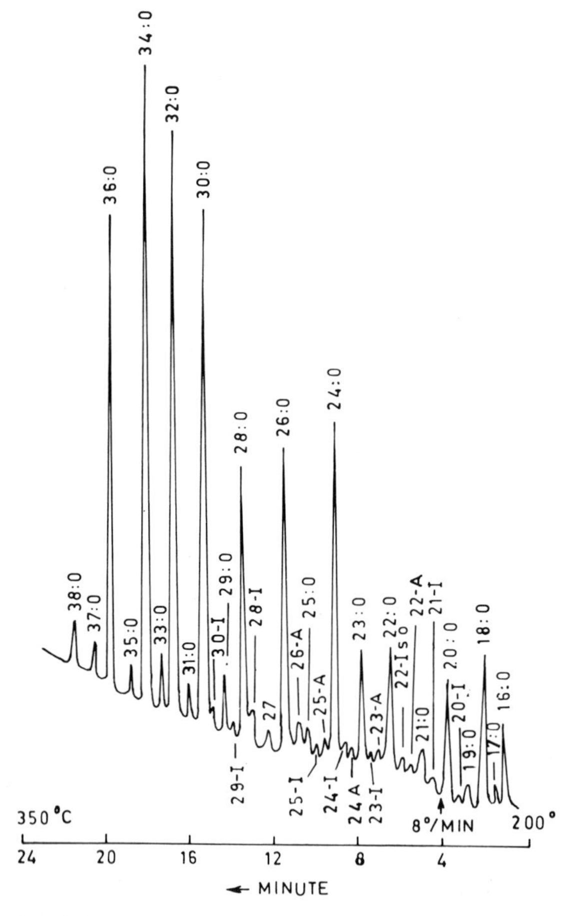

Fig. 4. Programmed temperature GLC chromatogram of the hydrocarbons of *P. coarctata* leaves (Misra et al. 1988). Oven temperature programming at the rate of 8 °C/min, started after 4 min of the isothermal run at 200 °C to a final temperature of 350 °C. Other GLC conditions were same as in Fig. 3

and Storrs 1962). Percentages of each component are then determined by internal normalization, i.e., the total peak area is considered to be 100%. In this calculation, it has been assumed that all the components of the mixture have been eluted from the GLC column and that no partial degradation has occurred during chromatography. Area percentages may also be conveniently determined by using an integrator.

For the determination of absolute amounts of each hydrocarbon components, gravimetry, i.e., weighing of the hydrocarbon fraction in a microanalytical balance, is the method of choice for relatively large samples. For microsamples the addition of a known amount of internal standards to the sample before GLC analysis is a very convenient way to determine the absolute amount of the fraction as well as of each component hydrocarbon in the fraction. The choice of internal standard should be made such that it elutes from the GLC column without overlapping any component of the mixture to which it has been added. Common internal standards used for estimating hydrocarbons are tetracosane (Tulloch 1981), triacontane, and p-dioctylbenzene (Tulloch 1973). It has been indicated that responses relative to these internal standards sometimes required correction (Tulloch 1973).

4.2 Analysis of Wax Esters

Since the wax esters are usually composed of long chain monohydric alcohols and long chain monocarboxylic fatty acids, there are two possible approaches to analyze them. Wax estes may be subjected to GLC analysis directly, whereby a separation according to carbon chain length is obtained, or they may be hydrolyzed to produce the component alcohols and fatty acids, which are separated, suitably derivatized, and analyzed by GLC.

4.2.1 Analysis of Wax Esters by Gas Liquid Chromatography According to Carbon Number

Wax esters are eluted from GLC columns containing nonpolar stationary phases under comparable conditions for the GLC analysis of diacylglycerol acetates which have similar molecular weights (Iyengar and Schlenk 1967; Litchfield 1972). The components were easily separated, depending on the differences in carbon number by one or two units, where the carbon atoms of the alcohols and fatty acid moieties have equal values. Separation of wax esters based on carbon number as well as on the degree of unsaturation on an Apolar 10C column was achieved with some resolution. However, better results were obtained when molecular species were separated according to unsaturations on silver nitrate-coated TLC, whereby the fractions were analyzed by GLC on the same column (Takagi et al. 1976). Thin layer chromatographic separation of wax esters based on their degree of unsaturation was carried out by developing with benzene-hexane (1:1) on a $AgNO_3$-silicic acid plate. Spots were detected under ultraviolet light after spraying with 2′,7′-dichlorofluorescein reagent. Short chain ($<C_{24}$) and long chain ($>C_{24}$) wax ester fractions can be separated by TLC on silicic acid plates, developed in petroleum ether-diethyl ether (95:5) and visualized under UV light after spraying with rhodamine 6G (Ackman et al. 1973). The short chain wax esters were recovered from the upper band, whereas the long chains from the lower band, which were subsequently analyzed by using a stainless steel, wall-coated, open-tubular GLC column, 46 m × 0.25 mm i.d. coated with DEGS (diethyleneglycol succinate), BDS (butanediol succinate), or Apizon-L columns, whereby excellent resolutions were obtained (Ackman et al. 1973). For separation of the wax esters, according to carbon chain lengths, the hydrogenated sample was examined by a 0.53-m-long, packed JXR column (Ackman et al. 1973).

The composition of leaf surface waxes of *Triticum* species were determined, without preliminary separation, by GLC using Dexsil 300 as liquid phase (Tulloch 1973). Five ml of chloroform solution containing about 1 mg each of p-dioctyl benzene and octadecyl octadenanoate was added to about 15–30 mg of wax. Chloroform was removed and the mixture acetylated with acetic anhydride and pyridine, 0.5 ml each, at 100 °C. Reagents were removed and the sample analyzed by GLC as a chloroform solution. To estimate methyl esters, part of the acetylated wax was treated with diazomethane and analyzed by GLC. Columns were 1 m × 3 mm stainless steel packed with 60–80 mesh acid washed and silanized Chromosorb-W coated with 1.5% Dexil 300. Temperature was programmed from 125–400 °C at a rate of 3 °C/min and helium was used as the carrier gas with a flow rate of 60 ml/min (Tulloch 1973). The Dexsil 300 liquid phase gave very sharp peaks and resolved the major components of hydrocarbons, free alcohols as acetates, free acids as methyl esters, long chain esters and diesters. Relative responses of β-diketones were variable and almost the complete disappearance of β-diketone peaks occurred after acetylation of the whole wax (Tulloch 1973). In these temperature programmed GLC analyses of the derivatized whole wax, identifications were made by constructing a plot of retention temperatures against the carbon numbers of synthetic authentic compounds, which was almost linear (Harris and Hapgood 1966; Tulloch and Weenink 1969).

Separation of wax esters of some mangrove plant leaves (Misra et al. 1987) was achieved according to carbon chain lengths on 3% SE-30 liquid phase supported on 80–100 mesh Chromosorb-W (HP) packed into a 1.8 m × 3 mm coiled glass column. Nitrogen was used as the carrier gas with a flow rate of 100 ml/min. Temperature was programmed from 230 to 350 °C. A typical gas chromatogram of the wax ester of *Avicennia officinalis* leaf wax has been presented in Fig. 5, the composition of which has already been reported (Misra et al. 1987). The wax ester was reduced catalytically using platinum oxide (Ghosh and Dutta 1972) prior to GLC for a clear separation. Further confirmation of the identifications was made by the addition of individual synthetic wax esters of known carbon chain lengths.

4.2.2 Quantitation of Wax Esters

Relative area percentages of the various wax ester components are determined from the chromatograms in a similar manner as discussed in Section 4.1.1. If the wax ester fraction is relatively large it may be conveniently weighed in a microanalytical balance and absolute amounts of each component may be determined from the area percentages. Addition of a known quantity of octadecyl octadecanoate to the wax ester fraction as internal standard before GLC has been commonly used (Tulloch 1973, 1981; Tulloch and Hoffman 1971; Cowlishaw et al. 1983).

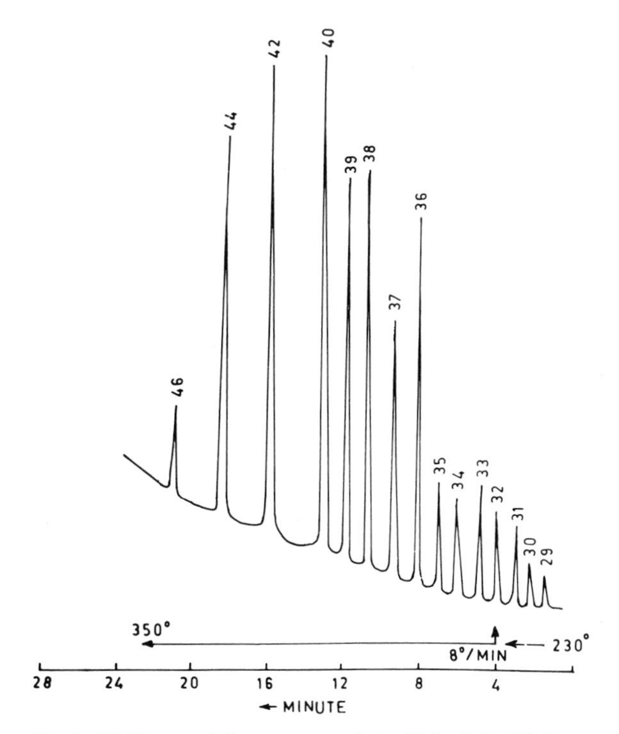

Fig. 5. GLC trace of the wax ester of *A. officinalis* leaf (Misra et al. 1987). The wax ester fraction was reduced by catalytic hydrogenation (Ghosh and Dutta 1972) before GLC. Column oven temperature programming at the rate of 8 °C/min started after the isothermal run at 230 °C for 4 min to a final temperature of 350 °C. Nitrogen flow rate was 100 ml/min. Other GLC conditions were same as in Fig. 3

It is desirable to check the percentage of C_{36} component in the sample before adding octadecyl octadecanoate to the wax ester sample. This may be done by withdrawing a known volume, e.g., 100 µl, from a 1-ml solution of wax esters. The GLC analysis is carried out using the 100-µl aliquot and components are identified by using the synthetic standard externally. The internal standard can then may be chosen such that the component with an identical carbon number to that of the standard does not exceed the 5% level, or preferably, it is absent in the sample. The appropriate internal standard may then be added to the residual 900-µl portion and after analysis the results are computed to 1 ml to obtain the absolute amounts of each component. A 10–20-mg portion of the internal standard is accurately weighed and diluted to a definitive volume; an aliquot of this solution containing 20–70 µg of the standard is withdrawn by a microliter syringe and added to the sample fraction.

4.2.3 Synthesis of Wax Esters

Individual wax esters of known carbon chain lengths may be synthesized by condensing acyl chloride of known chain length with primary n-alcohols of known chain length (Misra et al. 1983a). Dry alcohol is dissolved in distilled and dry, light petroleum ether (40–60 °C) or diethyl ether to which the acyl chloride is added gradually with continuous stirring. A few drops of pyridine is added to the mixture which is stirred at room temperature under anhydrous conditions for 12 h. The solvent is then removed in a rotary vacuum evaporator, and the residue dissolved in chloroform and subjected to preparative TLC using a solvent system of light petroleum ether-diethyl ether-acetic acid (80:20:1.0) according to Mangold (1969; Sect. 3.2.1). The bands are located by iodine vapor and recovered. The homogeneity of the wax ester should be checked by GLC. IR spectroscopy of the compound should have absorption bands at 1730–1750 cm^{-1} and 1150–1200 cm^{-1} characteristic for long chain esters. It is advisable to hydrolyze part of the synthetic wax ester. Moreover, the chain lengths of alcohols and fatty acids are determined to confirm the structure, which will be discussed in a later section.

4.2.3.1 Preparation of Acyl Chloride

A wide range of n-alcohols and n-fatty acids are available from various commercial sources. Acyl chloride may be conveniently prepared by refluxing the fatty acid in dry petroleum ether (40–60 °C) with an excess of oxalyl chloride under anhydrous conditions for 2–3 h. The solvent is removed in a rotary evaporator and the acyl chloride is redissolved in dry petroleum ether. This acyl chloride solution is used up as early as possible for the synthesis of wax esters.

4.2.4 Chemical Hydrolysis of Wax Esters and Analysis of Alcohols and Fatty Acids

Wax esters may be hydrolyzed by refluxing them with an excess of 1 M potassium hydroxide solution in 95% ethanol for 1 h (Christie 1982). Water is added to the cooled solution, which is extracted thoroughly with diethyl ether three to four times and dried over anhydrous sodium sulfate; alcohols are recovered by removal of the solvent. The aqueous layer is acidified with 6 M hydrochloric acid and free fatty acids are extracted three to four times with diethyl ether and dried over anhydrous sodium sulfate.

4.2.4.1 Acetylation of Alcohols for GLC Analysis

Alcohol are most conveniently analyzed by GLC as acetates, which are prepared by using acetic anhydride and pyridine (Privette and Nutter 1967). An alcohol sample of 1–50 mg is dissolved in 1 ml dry pyridine and 3 ml of distilled acetic anhydride is added to it. The mixture is taken in a vial with a PTFE lined screw cap and heated for 3 h in a boiling water bath, or may be left overnight at room temperature. After cooling, the solvent is evaporated under reduced pressure and the crude reaction mixture is partitioned in 60 ml of chloroform-methanol-water (8:4:3). The acetylated alcohols are dissolved in the chloroform layer, which is separated and dried over sodium sulfate. The alcohol acetates are purified by preparative TLC using a solvent system of light petroleum ether-diethyl ether-acetic acid (80:20:1.0) according to Mangold (1969; Sect. 3.2.1).

4.2.4.2 Methylation of Fatty Acid for GLC Analysis

Fatty acids are most conveniently methylated in quantitative yields by diazomethane. Diazomethane may be prepared by reacting an aqueous ethanolic solution of potassium hydroxide with N-methyl-N-nitroso-p-toluenesulfonamide (Diazald, Aldrich Chemical Company) in diethyl ether (Kates 1986). An ethereal solution of Diazald is taken in a round bottomed flask, which is connected to a receiver directly through a distillation head. Potassium hydroxide solution is added dropwise from a dropping funnel connected to the flask containing the Diazald solution, which is kept on an ice bath. Diazomethane gas is distilled by replacing the ice bath by a water bath at 30–40 °C and stirring the contents of the flask with a magnetic stirrer. Diazomethane is collected in the receiver containing dry diethyl ether cooled in an ice bath to give a bright yellow solution. Precautions should be taken by using a fume hood, furthermore, the diazomethane or the ethyl ether is never distilled completely. Diazomethane may also be generated in a similar way using N-nitroso-N-methyl urea which may be prepared in the laboratory (Vogel 1951) and stored for an indefinite period in a deep freezer at –20 °C. For methylation, to a solution of fatty acids in dry ethyl ether containing a few drops of dry methanol, the diazomethane solution is added dropwise until the yellow color persists. The mixture is left for 15 min at room temperature and the solvent is evaporated in a stream of nitrogen. The methyl ester thus prepared is dissolved in chloroform and preferably purified by preparative TLC according to Mangold (1969) before GLC analysis.

4.2.4.3 Hydrogenation of Wax Esters, Alcohols and Fatty Acid

Catalytic hydrogenation using platinum oxide (PtO_2) is commonly used to reduce the double bonds of wax esters and their hydrolysis products, viz., alcohols and fatty acids. Catalytic reduction of wax esters may be done directly, whereas acetates and methyl esters of alcohols and fatty acids, respectively, are preferred for hydrogenation, because of their greater solubilities in organic solvents. For a microscale hydrogenation, the following method is suitable for most purposes (Ghosh and Dutta 1972).

A small conical flask of 5–10 ml capacity is fitted with inlet and outlet tubes through a ground glass joint. The outlet is connected to a Pasteur pipette which is dipped into a conical flask containing mercury. One mg of Adam's catalyst is placed in the flask containing a magnetic fly and 1 ml dry methanol is added. Hydrogen is passed for 1–2 min and then stirring is started until the catalyst turns black and coagulates into lumps. At this stage the hydrogen is turned off and 1–2 mg of the sample in 1 ml of dry methanol is introduced into the flask, air is removed by hydrogen and then the sample is stirred for 2 h at room temperature, maintaining a low flow and a slight positive pressure. The solution is filtered and the catalyst is washed with methanol and ethyl ether. After removal of the solvent the sample is redissolved in hexane or ethyl ether for GLC analysis. More than one sample may be hydrogenated simultaneously by connecting the outlet of the first to the inlet of the second, etc., the outlet of the last flask being dipped into mercury through a Pasteur pipette. All the flasks are placed on one magnetic stirrer.

4.2.4.4 Analysis of Alcohol Acetates by GLC

Alcohol acetates are conveniently analyzed on nonpolar phases, such as Apiezon-L or M (Emery and Gear 1969; Farquhar 1962) or OV-101 (Nicolaides et al. 1970), as well as on polar phases such as BDS, EGA (Farquhar 1962). Alcohol acetates were also analyzed on a silicone SE-30 column (Tulloch and Weenick 1969), Dexsil 300 (Tulloch 1973), and a silicone OV-17 column (Misra et al. 1983a, 1986, 1987; Pakrashi et al. 1989). Because of the occurrence of long chain alcohols in natural waxes, generally GLC analysis is done at elevated temperatures using nonpolar liquid phases, the polar phases being inadequate to withstand such temperatures.

A 3% OV-17 liquid phase supported on Chromosorb-W (HP), 80–100 mesh, and packed into a glass column (1.8 m × 3 mm) may be used for the analysis of alcohol acetates. The nitrogen flow rate should be around 50–60 ml/min and temperature programming is necessary because of the presence of compounds with a wide range of carbon numbers in natural samples. Temperature programming from 200 to 350 °C gives a good resolution of alcohol acetates from C_{14}-C_{36} (Misra et al. 1987), i.e., saturated and monoenoic components, which elute just before the saturated homologues. Hydrogenation and reanalysis by GLC is essential to identify unsaturated components and also to examine the presence of branched chain components (Sect. 4.1). Identifications are made by using acetates of authentic alcohols externally and internally (Sect. 4.1). Semilogarithmic plots of relative retention times (RRT) against carbon chain lengths at two different isothermal temperatures are also helpful for identifications when limited numbers of authentic standards are available. Comparison of separation factors with published values are also useful in identification of the components (Kates 1986). A typical chromatogram of an alcohol acetate sample is presented in Fig. 6.

Quantitation procedures for alcohol acetates are similar to those discussed previously for hydrocarbon and wax esters with the choice of proper internal standards.

4.2.4.5 Analysis of Fatty Acid Methyl Esters (FAME) by GLC

Nonpolar liquid phases such as SE-30, OV-1, JXR, QF-1, or Apiezon-L permit the separation of fatty acid esters mainly on the basis of their molecular weights, although there can be some separation of unsaturated from saturated fatty acids of the same chain lengths. These silicone and high molecular weight hydrocarbon liquid phases allow one to elute the unsaturated components before the saturated compounds of the same chain length, but in packed columns there is very little separation. Such phases are now only used for the analysis of oxygenated or high molecular weight fatty acids.

Clear separations of esters of the same chain lengths, but with zero to six double bonds, and unsaturated components eluting after the corresponding saturated ones, are achieved by using polar polyester phases, which are much more suitable to fatty acid analysis. These polar phases can be subdivided into three groups: (1) highly polar phases consisting of polymeric ethyleneglycol succinate

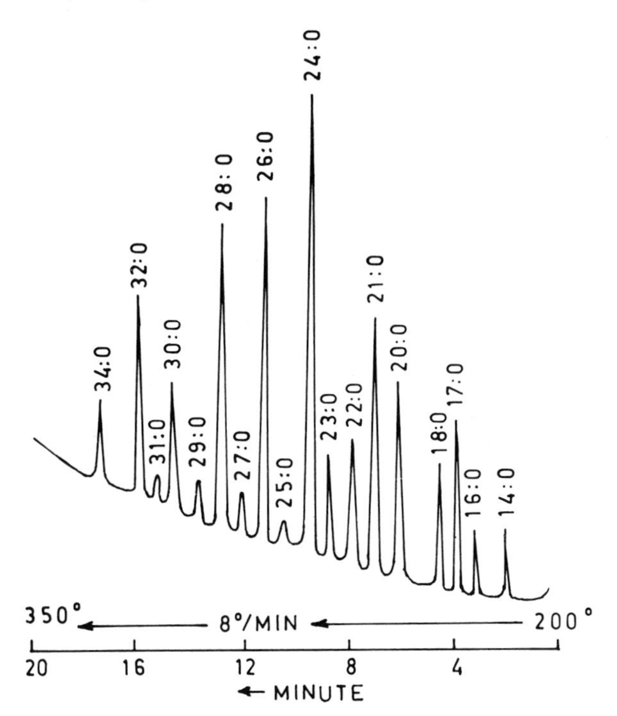

Fig. 6. GLC tracing of the alcohol acetates obtained by the lipolysis (Misra et al. 1983a) of the wax esters of *A. officinalis* leaves (Misra et al. 1987). Oven temperature was programmed from 200 to 350 °C at the rate of 8 °C/min. Other GLC conditions were the same as in Fig. 3

(EGS), diethyleneglycol succinate (DEGS), and EGSS-X, the later being a copolymer of EGS with a methyl silicone; (2) medium polarity phases, such as polyethyleneglycol adipate (PEGA), butanediol succinate (BDS), and EGSS-Y, the later being a copolymer of EGS with a higher proportion of the methyl silicone than in EGSS-X; (3) low polarity phases, such as neopentylglycol succinate (NPGS) and EGSP-Z. Low polarity phases are principally used in open-tubular and support-coated, open-tubular (SCOT) columns, since when they are used in packed columns, resolutions are poor for saturated and monoenoic components of the same chain length. Recently, a range of new stationary phases, which are alkylpolysiloxanes containing various polar substituents, especially nitrile groups, are commercially available with a wide range of polarities. The most useful of these liquid phases are more polar than EGSS-X and are sold under various trade names, such as Silar 10C, Silar 9CP, SP2340, and OV275. These phases are stable at relatively higher temperatures and afford excellent separations of polyunsaturated fatty acids (Myher et al. 1974; Heckers et al. 1977).

The quality of the separations can be influenced by the nature of the support material used for the stationary phase. Acid washed and silanized support materials are commercially available and have been found to be most suitable for analytical columns.

For identification of fatty acids analyzed as methyl esters, standard mixtures containing accurately known amounts of methyl esters of saturated, monoenoic, and polyenoic fatty acids are commercially available from a number of reputable biochemical suppliers. For the provisional identification of fatty acids and also for checking the quantification procedure used, these standards are invaluable. Identifications are done by direct comparison of the retention times of the standards with those of the unknown esters on the same columns under identical isothermal conditions (Fig. 7). Secondary external standards consisting of a natural fatty acid mixture of a known composition may also be used for identification and for the preparation of plots of retention parameters which are used for identification (Ackman and Burgher 1965) of fatty acids. Relative retention times (RRT) of fatty acid esters relative to those of a chosen standard, such as 16:0, 18:0, or 18:1ω9, which are commonly occurring components, are determined under isothermal conditions. When the logarithms of the retention times of fatty acid esters are plotted against the number of carbon atoms, curves approaching straight lines are obtained for members of different homologous series (James et al. 1952; James and Martin 1956; Lipsky and Landowne 1958). This relationship is linear over a wide range of chain lengths, but departure from linearity has been observed at shorter chain lengths. For fatty acid methyl esters, plots of logarithms of RRT against carbon chain lengths for saturated, monoenoic, dienoic, and trienoic esters are parallel straight lines. For a particular chain length, viz., for 18 carbon chains, the plot for saturated esters occupies the lowest and the trienoic the highest positions (Fig. 8), in accordance to their retention times on a polar column. For un-

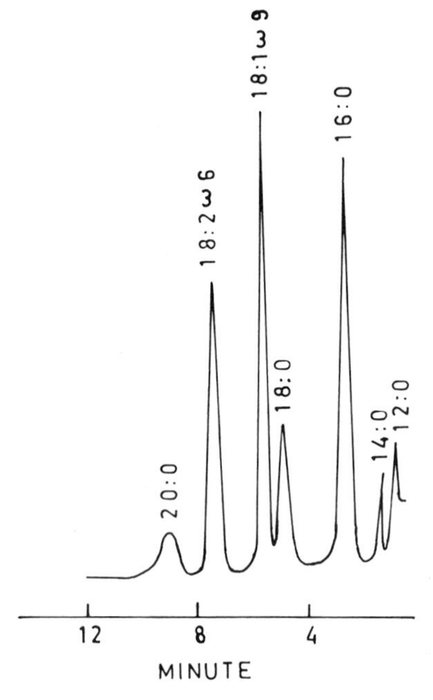

Fig. 7. GLC tracing of the fatty acid methyl esters obtained by the lipolysis (Misra et al. 1983) of the wax esters of *A. officinalis* leaves (Misra et al. 1987). Column was 15% DEGS coated on 80–100 mesh Chromosorb W (HP) and packed into a 2.1 m × 2 mm i.d. glass column. Oven temperature was isothermal at 180 °C. Injection port and detector temperatures were 230 and 250 °C, respectively. Nitrogen carrier gas was used with a flow rate of 60 ml/min

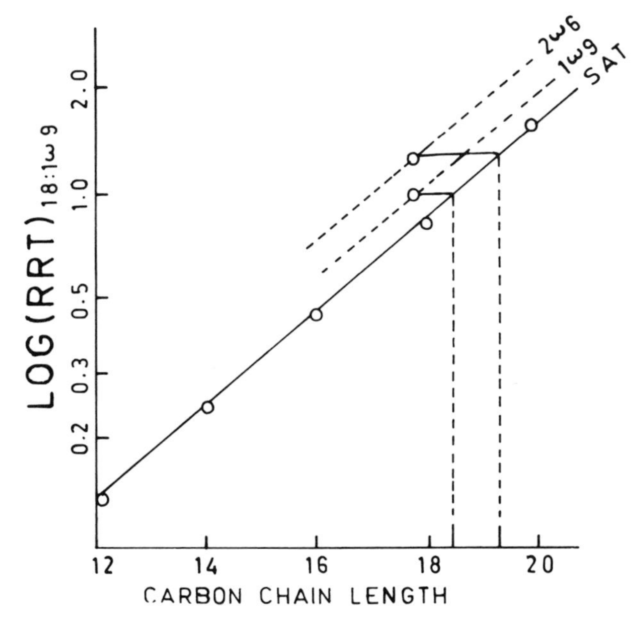

Fig. 8. Semilogarithmic plot of relative retention time (*RRT*) of the saturated (*SAT*) monoenoic (18:*1*ω9) and dienoic (18:*2*ω6) fatty acid methyl esters from wax esters of *A. officinalis* (Fig. 7). Equivalent chain length (ECL) values of the unsaturated esters were determined by fitting the RRT of the same into the semilogarithmic plot of the saturated esters and extrapolating to the carbon chain length axis. The ECL values of 18:1ω9 and 18:2ω6 determined from the plot were 18.5 and 19.4, respectively

saturated esters with a particular chain length and degree of unsaturation, three isomers may be found which are members of three different homologous series, viz., (n-3) or –ω3, (n-6) or –ω6 and (n-9) or –ω9 series, of which the latter is mostly confined to monoenoic acids. The –ω values are used for designating unsaturated acids with methylene-interrupted cis- double bonds and are defined as the number of carbon atoms from the center of the double bond farthest from the carboxyl group, to the end methyl group. Examples of such isomers are 18:3ω3 and 18:3ω6, the isomeric linolenic acids; 20:3ω3 and 20:3ω6 are the other examples. The –ω3 and –ω6 acids with different carbon numbers will yield parallel, semilogarithmic plots for the two different homologous series (Ackman 1969). Semilogarithmic plots are of value in identifications, thus predicting retention times of compounds which are not available for comparison and also for discovering compounds, if they belong to the same homologous series. The application of equivalent chain length (ECL) values expressing the elution sequence of esters is of considerable use (Miwa et al. 1960), and are also referred to as carbon numbers (Woodford and Van Gent 1960). ECL values can be calculated from an equation similar to that for Kovat's indices (Jamieson 1970), but are usually found by plotting the logarithms of RT or RRT against the carbon numbers of n-saturated fatty acid methyl esters (Fig. 8). The retention times of the unknown acids are measured under identical operational conditions and the ECL values are read directly from the graph.

Hydrogenation of fatty acid methyl esters has been proved to be very useful in the confirmation of the unsaturated components and may be conveniently carried out by following the procedure described in Section 4.2.4.3. The hydrogenated samples also should be run consecutively under identical conditions of GLC for the nonhydrogenated sample.

In conclusion, the identification of the fatty acid methyl esters is made by (1) the comparison of retention times of sample components with those of authentic standards; (2) the determination of the relative retention times of the components and a comparison of the same in the literature; (3) the construction of semi-logarithmic plots using authentic saturated and unsaturated standards and fitting the logarithm of RRT of the sample components in the plot; (4) the construction of a semilogarithmic plot by using authentic n-saturated methyl esters, the determination of ECL values of the sample components from the plot, and the comparison of the same with the literature; and (5) the identification of the saturated esters obtained by the hydrogenation and GLC of the sample.

Relative proportions of fatty acids are determined from area percentages (see Sect. 4.1.1), where absolute amounts of each component are determined by adding a known quantity of methyl pentadecanoate (15:0) as internal standard prior to GLC analysis (Misra et al. 1983b).

Application of GLC in the identification of fatty acids according to RRT and ECL values have been described elaborately in the excellent articles by Ackman (1969), Jamieson (1970), Christie (1982) and Kates (1986).

The composition of individual wax esters according to carbon number may be calculated from the compositions of fatty acids and alcohols, assuming a random distribution of the acid and alcohol moieties (Kolattukudy 1970). The results may be confirmed by GLC or GLC-mass spectrometry of intact hydrogenated wax esters.

4.2.5 Analysis of Wax Esters by Lipolysis on TLC Plates

Wax esters may be hydrolyzed on TLC plates coated by silica gel using porcine pancreatic lipase liberating free fatty acids and alcohols (Misra et al. 1983a) which were separated on the same plate. After recovery of the hydrolysates from the TLC plates, alcohols were converted to acetates and fatty acids to methyl esters and were analyzed by GLC. The method was found to be advantageous since only microquantities (1–2 mg) of wax ester are required, i.e., three to four samples can be lipolyzed on a 20 × 20 cm TLC plate, and the method is less time-consuming than the conventional chemical hydrolysis method.

4.2.5.1 Lipolysis of Wax Esters on TLC Plates Using Porcine Pancreatic Lipase

Silica gel G plates, 0.25 or 0.5 mm thick, were used after activation for 1–2 and 3–4 mg samples, respectively. The sample was applied in 4–5 cm wide lanes, on the same plate. Any good quality porcine pancreatic lipase may be used but should be freed from contaminated lipids before use, by shaking vigorously six times each with diethyl ether and acetone, respectively, centrifuging each time (Misra et al. 1984b) and then discarding the supernatants. The lipase pellet was dried under vacuum at 30 °C for 1 h and a saturated solution was prepared in 1 M Tris buffer, pH 8.2. After centrifugation the supernatant was used for the determination of lipase activity and soluble protein and then for lipolysis.

Protein in the lipase solution and wax esters to be used should be in the ratio of 1:1 (w/w). About 0.05–0.1 ml of lipase solution was applied evenly as bands at the origin (2–3 cm from the edge) of each lane on the TLC plate. The band was partially dried by a gentle stream of air and wax ester solutions in n-hexane were applied on the enzyme bands. The plates were incubated for 90–100 min in a chamber saturated with water vapor at 45 °C. The plates were then developed three times by diethyl ether in a TLC tank, each time to a height of 2 cm above the reaction zone to extract the reaction products. The plate was then developed up to 14 cm from the origin which was 2 cm above the reaction zone. The developing solvent was n-hexane-diethyl ether-acetic acid (80:20:1). Bands were located by iodine vapor and were identified by comparison with the R_f values of standards (Fig. 9). The reaction products, viz., fatty acids and alcohols, along with the unreacted wax esters, were recovered from the plate. Alcohols were converted to acetates and analyzed by GLC as described in Sects. 4.2.4.1 and 4.2.4.4, respectively. Fatty acids were converted to methyl esters and were analyzed by GLC as described in Sects. 4.2.4.2 and 4.2.4.5, respectively.

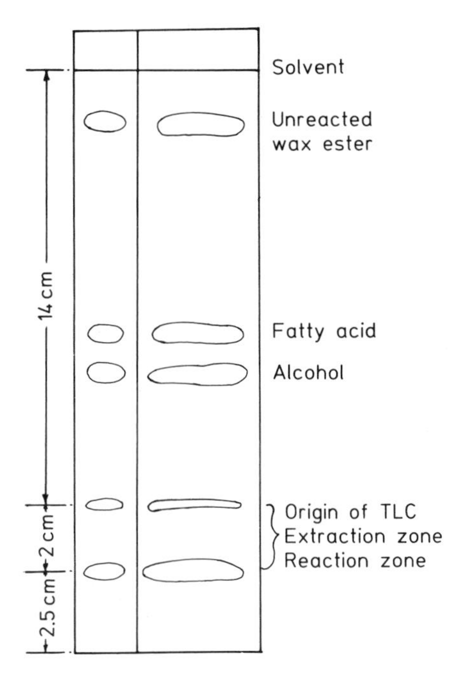

Fig. 9. Lipolysis of wax ester on a TLC plate (Misra et al. 1983a) A wax ester sample in hexane was applied on the lipase band at the reaction zone and incubated. Reaction products were eluted out from the reaction zone to the origin of TLC by developing the plate three times in the diethyl ether chamber and then developed to a height of 14 cm using light-petroleum ether-diethyl ether-acetic acid (80:20:1). Bands were visualized by exposure to iodine vapor. Fatty acid and alcohol were recovered, derivatized, and analyzed by GLC. Synthetic octadecyl oleate was treated similarly in the *left lane*. For experimental details, see text

4.3 Analysis of Free Alcohols and Fatty Acids

Free alcohols and fatty acids of waxes may be isolated by column chromatography (Sect. 3.1) or TLC (Sect. 3.2). Alcohols are acetylated (Sect. 4.2.4.1) and analyzed by GLC (Sect. 4.2.4.4). Fatty acids are methylated (Sect. 4.2.4.2) and analyzed by GLC (Sect. 4.2.4.5).

4.4 Analysis of Aldehydes

Long chain aldehydes with ten or more carbon atoms can be analyzed by GLC, without prior derivatization. The instability of free aldehydes over a period of time, during which they undergo condensation and polymerization, is the major disadvantage. Alternatively, it is advantageous to convert them to stable volatile derivatives. Long chain aldehydes may be readily converted to volatile dimethyl acetals in quantitative yields, which are very stable at neutral and alkaline pH.

4.4.1 Preparation of Dimethyl Acetal Derivatives

Dimethyl acetal derivatives of the aldehydes are prepared by refluxing with 2% anhydrous methanolic hydrogen chloride (1:20 w/v) for 2 h under anhydrous conditions. After cooling the mixture, hydrogen chloride is neutralized with a slight excess of anhydrous sodium carbonate. Light petroleum ether (40–60 °C) and water (1:1) are added to the mixture and shaken; the petroleum ether layer containing the dimethyl acetal is then removed. The aqueous methanolic phase is extracted further with light petroleum ether. Combined petroleum ether extract is shaken vigorously for a few minutes with a saturated solution of sodium metabisulfite to remove any traces of unreacted aldehyde, washed twice with water, and dried over anhydrous sodium sulfate. For microquantities of aldehydes, the metabisulfite washing step can be safely left out, since the yield of acetals is quantitative (Gray 1976).

4.4.2 Conversion to Fatty Acid Methyl Esters

For structural identification of aldehydes, oxidation to corresponding acids may be carried out according to Gray (1976). The method is applicable strictly only to saturated aldehydes, since unsaturated aldehydes may be oxidised also at the double bonds.

The solution of aldehyde (1–10 mg/ml) in glacial acetic acid is warmed slightly (40 °C) and small amounts of concentrated solution of chromium trioxide in glacial acetic acid are added with stirring until an excess is achieved. The solution is then poured into water (2 vol) and the aqueous layer is extracted three to four times with benzene. Combined extracts are dried over anhydrous sodium sulfate. After removal of benzene in a stream of nitrogen, the fatty acids are dissolved in dry ether containing a few drops of dry methanol and methylated and purified by TLC as described in Sect. 4.2.4.2. If the aldehydes are in dimethyl acetal forms, they are released by dissolving 1–20 mg of the sample in 2 ml of 90% acetic acid containing 0.1 ml of 2.5% methanolic hydrochloric acid and heating at 90–100 °C for 1 h. The solution is diluted with 2–3 ml of water and the acids are extracted with several portions of light petroleum ether (Kates 1986). Combined extracts are concentrated to yield the acid, which is methylated as described in Sect. 4.2.4.2.

4.4.3 Conversion to Alcohol Acetate

Free aldehydes (1–10 mg) are dissolved in 1 ml of dry ethyl ether, cooled to –20 °C and 3 ml of a 3% solution of lithium aluminum hydride (LAH) in dry ethyl ether is added with magnetic stirring. After 2 h at –20 °C, LAH is decomposed by adding 90% methanol-water dropwise to decompose the excess reagent (Farquhar 1962). The precipitate is centrifuged and washed several times with ether. Supernatants are combined and evaporated to dryness in vacuum after dilution with benzene (Kates 1986). The alcohols are converted to acetates as described in Sect. 4.2.4.1.

4.4.4 Analysis of Aldehydes and Derivatives by GLC

Dimethyl acetal derivatives of aldehydes may be analyzed by GLC on 10% Apiezon L columns but the support must be treated with alkali before it is coated with the liquid phase (Gray 1976). Polar phases such as BDS, EGA, and SP-2330, etc. may also be used to analyze the dimethyl acetal derivatives, but it is important that no traces of the acid catalyst used in the synthesis of these phases remain (Kates 1986). The above mentioned polar phases, and also and Apiezon L column that has not been treated with alkali, may be used for the analysis of free aldehydes. Fatty acid methyl esters derived by the oxidation of aldehydes are analyzed as described in Sect. 4.2.4.5, whereas the analysis of alcohol acetates obtained by the reduction of aldehydes has been described in Sect. 4.2.4.4.

4.5 Analysis of β-Diketones

Leaf waxes may contain β-diketones and oxygenated β-diketones, e.g., hydroxy- and oxo β-diketones. These β-diketones may be estimated in a mixture using UV spectrophotometry, and measuring the absorption at 273 nm (Horn et al. 1964). Upon alkaline hydrolysis of a single β-diketone, viz., hentriacontane-14-16-dione, equal amounts of pentadecan-2-one and heptadecan-2-one along with myristic (14:0) and palmitic (16:0) acids were obtained, which were analyzed by GLC to determine the structure of the parent compound (Tulloch and Weenink 1969). Plant leaf waxes normally contain β-diketones of various chain lengths in different proportions, which give poor responses in GLC (Tulloch 1973). When hydroxy β-diketone are mixtures of isomers (Tulloch and Hoffman 1973; Tulloch 1976b), they cannot be identified by GLC alone. The occurrence of oxo β-diketones along with hydroxy β-diketones (Tulloch and Hoffman 1976) also cannot be distinguished by GLC.

A very useful method for analyzing β-diketones and oxygenated β-diketones, which were better resolved as their trimethylsilyl (TMS) derivatives and were more conveniently identified by GC-MS, was developed by Tulloch and Hogge (1978). During silylation of natural waxes, free acids and alcohols present were also derivatized and separated from each other and from hydrocarbons; all the components were identified by GC-MS (Tulloch and Hogge 1978).

5 Conclusion

In the present chapter various chromatographic techniques have been discussed for the analysis of plant epicuticular waxes. Classes of wax components such as hydrocarbon, wax ester, etc. were obtained by chromatographic fractionation; these were identified by comparing Rf values with those of authentic standards. Each class was then analyzed by GLC of underivatized or derivatized fractions; components were identified by comparing retention times with those of standards. GLC identifications were also made by semilogarithmic plots of retention times and also by hydrogenation techniques. The methods described are suitable for tentative identification of components in a mixture and also indicate the occurrence of unusual components. Confirmation of identification, particularly for unusual components, should be done by GLC-MS.

References

Ackman RG (1969) Gas-liquid chromatography of fatty acids and esters. In: Lowenstein JM (ed) Methods in enzymology, vol 14. Academic Press, New York, pp 329–381

Ackman RG, Burgher RD (1965) Cod liver oil fatty acids as reference standards in the GLC of polyunsaturated fatty acids of animal origin: analysis of a dermal oil of the Atlantic leatherback turtle. J Am Oil Chem Soc 42:38–42

Ackman RG, Sipos JC, Eaton CA, Hilaman BL, Litchfield C (1973) Molecular species of wax esters in jaw fat of Atlantic bottlenose dolphin *Tursiops truncatus*. Lipids 8:661–667

Bligh EG, Dyer WJ (1959) A rapid method of total lipid extraction and purification. Can J Biochem Physiol 37:911–917

Burchfield HP, Storrs EE (1962) Biochemical applications of gas chromatography. Academic Press, New York

Christie WW (1982) Lipid analysis, 2nd edn. Pergamon, Oxford

Cowlishaw MG, Bickerstaffe R, Young H (1983) Epicuticular wax of four species of *Chionochloa*. Phytochemistry 22:119–124

Eglinton G, Hamilton RJ (1967) Leaf epicuticular waxes. Science 156:1322–1335

Emery EE, Gear JR (1969) Long chain esters in clover wax. Can J Biochem 47:1195–1197

Farquhar JW (1982) Identification and gas-liquid chromatographic behavior of plasmalogen aldehydes and their acetals, alcohol and acetylated alcohol derivatives. J Lipid Res 3:21–30

Folch J, Lees M, Stanley GHS (1957) A simple method for the isolation and purification of total lipids from animal tissues. J Biol Chem 226:497–509

Ghosh A, Dutta J (1972) A multi-hydrogenator for micro-scale hydrogenation of fatty acids. Trans Bose Res Inst 35:13–15

Gray GM (1976) Gas chromatography of the long-chain aldehydes. In: Marinetti GV (ed) Lipid chromatographic analysis, 2nd edn, vol 3. Marcel Dekker, New York, pp 897–923

Harris WE, Hapgood HW (1966) Programmed temperature gas chromatography. Wiley, New York, pp 141–168

Heckers H, Dittmar K, Melcher FW, Kalinowski HO (1977) Silar 10C, Silar 9CP, SP2340 and OV-275 in the gas-liquid chromatography of fatty acid methyl esters on packed columns. Chromatographic characteristics and molecular structures. J Chromatogr 135:93–107

Horn DHS, Kranz ZH, Lamberton JA (1964) The composition of *Eucalyptus* and some other leaf waxes. Aust J Chem 17:464–476

Hull HM, Morton HL, Wharrie JR (1975) Environmental influences on cuticle development and resultant foliar penetration. Bot Rev 41:421–452

Iyengar R, Schlenk H (1967) Wax esters of mullet (*Mugil cephalus*) roe oil. Biochemistry 6:396–402

James AT, Martin AJP (1956) Gas-liquid chromatography: the separation and identification of the methyl esters of saturated and unsaturated acids from formic acid to n-octadecanoic acid. Biochem J 63:144–152

James AT, Martin AJP, Smith GH (1952) Gas liquid partition chromatography: the separation and microestimation of ammonia and the methyl amines. Biochem J 52:238–242

Jamieson GR (1970) Structure determination of fatty acid esters by gas liquid chromatography. In: Gunstone FD (ed) Topics in lipid chemistry, vol 1. Logos, London, pp 107–159

Kaneda T (1967) Biosynthesis of long chain hydrocarbons: I Incorporation of L-valine, L-threonine, L-isoleucine and L-leucine into specific branched-chain hydrocarbons in tobacco. Biochemistry 6:2023–2032

Kates M (1986) Techniques of lipidology. In: Burdon RH, van Knippenberg PH (eds) Laboratory techniques in biochemistry and molecular biology, 2nd edn. Elsevier, Amsterdam

Kolattukudy PE (1965) Biosynthesis of wax in *Brassica oleracea*. Biochemistry 4:1844–1855

Kolattukudy PE (1968) Further evidence for an elongation-decarboxylation mechanism in the biosynthesis of paraffins in leaves. Plant Physiol 43:375–383

Kolattukudy PE (1970) Composition of the surface lipids of pea leaves (*Pisum sativam*). Lipids 5:398–402

Kolattukudy PE (1975) Biochemistry of cutin, suberin and waxes, the lipid barriers on plants. In: Galliard T, Mercer EI (eds) Recent advances in the chemistry and biochemistry of plant lipids. Academic Press, New York, pp 203–246

Kuksis A (1964) Hydrocarbon composition of some crude and refined edible seed oils. Biochemistry 3:1086–1093

Lipsky SR, Landowne RA (1958) A new partition agent for use in the rapid separation of fatty acid esters by gas-liquid chromatography. Biochim Biophys Acta 27:666–667

Litchfield C (1972) Analysis of triglycerides. Academic Press, New York

Mangold HK (1969) Aliphatic lipids. In: Stahl E (ed) Thin layer chromatography. Springer, Berlin Heidelberg New York, pp 363–421

Martin JT, Juniper BE (1970) The cuticles of plants. St Martins, New York

Misra S, Choudhury A, Dutta AK, Ghosh A, Dutta J (1983a) Enzymatic reactions on thin-layer chromatographic plates IV. Lipolysis of wax esters and separation of products on a single plate. J Chromatogr 280:313–320

Misra S, Dutta AK, Dhar T, Ghosh A, Choudhury A, Dutta J (1983b) Fatty acids of the mud skipper *Boleophthalmus boddaerti*. J Sci Food Agric 34:1413–1418

Misra S, Choudhury A, Ghosh A, Dutta J (1984a) The role of hydrophobic substances in leaves in adaptation of plants to periodic submersion by tidal water in a mangrove ecosystem. J Ecol 72:621–625

Misra S, Choudhury A, Dutta AK, Dutta J, Ghosh A (1984b) Lipid contaminants in commercial lipases. Lipids 19:302–303

Misra S, Ghosh A, Dutta J (1984c) Production and composition of microbial fat from *Rhodotorula glutinis*. J Sci Food Agric 35:59–65

Misra S, Choudhury A, Dutta J, Ghosh A (1985) Hydrocarbon and wax ester compositions of the leaves of some flooded and emergent plants of Sunderban mangrove forest. In: Krishnamurthy V (ed) Marine plants, their biology, chemistry and utilisation. Proc All India Symp marine plants, Dona Paula, Goa, India, pp 273–276

Misra S, Choudhury A, Pal PK, Ghosh A (1986) Effect on the leaf lipids of three species of mangrove of periodic submergence in tidal water. Phytochemistry 25:1083–1087

Misra S, Dutta AK, Chattopadhyay S, Choudhury A, Ghosh A (1987) Hydrocarbons and wax esters for seven species of mangrove leaves. Phytochemistry 26:3265–3268

Misra S, Choudhury A, Chattopadhyay S, Ghosh A (1988) Lipid composition of *Porteresia coarctata* from two different mangrove habitats in India. Phytochemistry 27:361–364

Mold JD, Means RE, Ruth JM (1966) The higher fatty acids of flue-cured tobacco. Phytochemistry 5:59–66

Miwa TK, Mikolajczak KL, Earle FR, Wolff IA (1960) Gas chromatographic characterisation of fatty acids. Identification constants for mono- and dicarboxylic methyl esters. Anal Chem 32:1739–1742

Myher JJ, Marai L, Kuksis A (1974) Identification of fatty acids by GC-MS using polar siloxane phase. Anal Biochem 62:188–203

Nicolaides N, Fu HC, Ansari MNA (1970) Diester waxes in surface lipids of animal skin. Lipids 5:299–307

Pakrashi SC, Dutta PK, Achari B, Misra S, Choudhuri A, Chattopadhyay S, Ghosh A (1989) Lipids and fatty acids of the horseshoe crabs *Tachypleus gigas* and *Carcinoscorpius rotundicauda*. Lipids 24:443–447

Privette OS, Nutter LJ (1967) Determination of the structure of lecithins via the formation of acetylated 1,2-diglycerides. Lipids 2:149–154

Rouser G, Kritchevsky G, Yamamoto A (1967) Column chromatographic and associated procedures for separation and determination of phosphatides and glycolipids. In: Marinetti GV (ed) Lipid chromatographic analysis, vol 1. Marcel Dekker, New York, pp 99–162

Schmid HHO, Bandi PC (1969) n-Triacontanal and other long chain aldehydes in the surface lipids of plants. Hoppe-Seyler's Z Physiol Chem 350S:462–466

Stransky K, Streibl M, Kubelka V (1970) Natural waxes XV. Hydrocarbon constituents of the leaf wax from the walnut tree (*Juglans regia*). Coll Czech Chem Commun 35:882–891

Takagi T, Itabashi Y, Ota T, Hayashi K (1976) Gas chromatographic separation of wax esters based on the degree of unsaturation. Lipids 11:354–356

Tulloch AP (1973) Composition of leaf surface waxes of *Triticum* species: variation with age and tissue. Phytochemistry 12:2225–2232

Tulloch AP (1975) Chromatographic analysis of natural waxes. J Chromatogr Sci 13:403–407

Tulloch AP (1976a) Chemistry of waxes of higher plants. In: Kolattukudy PE (ed) Chemistry and biochemistry of natural waxes. Elsevier, Amsterdam, pp 235–287

Tulloch AP (1976b) Epicuticular wax of *Agropyron smithii* leaves. Phytochemistry 15:1153–1156

Tulloch AP (1981) Composition of epicuticular waxes from 28 genera of Gramineae: differences between subfamilies. Can J Bot 59:1213–1221

Tulloch AP, Bergter L (1981) Epicuticular wax of *Juniperus scopulorum*. Phytochemistry 20:2711–2716

Tulloch AP, Hoffman LL (1971) Leaf wax of durum wheat. Phytochemistry 10:871–876

Tulloch AP, Hoffman LL (1973) Leaf wax of oats. Lipids 8:617–622

Tulloch AP, Hoffman LL (1974) Epicuticular waxes of *Secale cereale* and *Triticale* hexaploid leaves. Phytochemistry 13:2535–2540

Tulloch AP, Hoffman LL (1976) Epiculicular wax of *Agropyron intermedium*. Phytochemistry 15:1145–1151

Tulloch AP, Hoffman LL (1979) Epicuticular waxes of *Andropogon hallii* and *A. scoparius*. Phytochemistry 18:267–271

Tulloch AP, Hogge LR (1978) Gas chromatographic-mass spectrometric analysis of β-diketone-containing plant waxes. J Chromatogr 157:291–296

Tulloch AP, Weenink RO (1969) Composition of the leaf wax of Little Club Wheat. Can J Chem 47:3119–3126

Tulloch AP, Baum BR, Hoffman LL (1980) A survey of epicuticular waxes among genera of Triticeae. 2. Chemistry. Can J Bot 58:2602–2615

Vogel AI (1951) Text book of practical organic chemistry, 2nd edn. Longmans, London, pp 843–844

Weete JD, Leek GL, Peterson CM, Currie HE, Branch WD (1978) Lipid and surface wax synthesis in water-stressed cotton. Plant Physiol 62:675–677
Wollrab V, Streibl M, Sorm F (1965) Plant substances XXI. Composition of hydrocarbons from rose petal leaf wax. Coll Czech Chem Commun 30:1654–1659
Wollrab V, Streibl M, Sorm F (1967) Iso- and anteiso-alkanes in natural waxes. Chem Ind 1872–1873
Woodford FP, Van Gent CM (1960) Gas-liquid chromatography of fatty acid methyl esters: the carbon number as a parameter for comparison of columns. J Lipid Res 1:188–190
Youngblood WW, Blumer M, Guillard RL, Fiore F (1971) Saturated and unsaturated hydrocarbons in marine benthic algae. Mar Biol 8:190–201

Analysis of Flower and Pollen Volatiles

H.E.M. DOBSON

1 Introduction

Flowers have evolved complex olfactory and visual displays to attract animals for pollination. Many pollinator insects show high specificity to flower volatiles (Dodson et al. 1969; Williams and Whitten 1983; Pham-Delegue et al. 1986, 1989; Borg-Karlson 1990), and variations in their chemical composition could lead to reproductive isolation and speciation in plants. Since the pioneer studies of flower fragrance chemistry versus pollinators carried out on orchids (Kullenberg 1961; Dodson et al. 1969; Hills et al. 1972; Kullenberg and Bergström 1976; Bergström 1978), there has been a growing interest in the biological significance and systematic value of flower volatiles in diverse plant groups (e.g., Thien et al. 1975, 1985; Nilsson 1978; Gregg 1983; Williams 1983; Williams and Whitten 1983; Borg-Karlson and Groth 1986; Bergström 1987; Groth et al. 1987; Robacker et al. 1988; Pham-Delegue et al. 1989; Borg-Karlson 1990; Dobson et al. 1990).

Different parts of the flower may participate in volatile production, and petals are frequently the principal source of emission (Gregg 1983; Borg-Karlson 1990; Dobson et al. 1990). In some species, the androecium (staminodia, stamens, pollen) may make significant contributions, and pollen, in particular, often has distinctive odors that are used in orientation by flower visitors (von Frisch 1923; Porsch 1956; von Aufsess 1960; Dobson 1987). In contrast to studies of flowers, work on pollen volatiles is recent and only few methods have been tried (Dobson et al. 1987, 1990; Dobson 1988; J. Knudsen and L. Tollsten unpubl.).

2 Flower Volatile Chemistry

Flower volatiles generally comprise a blend of aliphatic, terpenoid, and aromatic compounds in differing representations. Some flowers include nitrogen- or sulfur-containing compounds (Kaiser and Lamparsky 1982; Nilsson et al. 1985; Joulain 1987; Burger et al. 1988; Robacker et al. 1988), which may be important in providing distinctive aromas.

The number and composition of volatiles emitted by flowers vary widely. In some cases the volatile profile is clearly dominated by certain compounds, with numerous other constituents present in minor quantities (e.g., Buttery et al. 1982, 1984, 1986; Matile and Altenburger 1988; Patt et al. 1988; Dobson et al. 1990; Loughrin et al. 1990). The characteristic features of flower fragrances as detected by humans are often due to only a portion of the total volatiles comprised of both major and minor components (Lindeman et al. 1982; Joulain 1987; Matile and Altenburger 1988).

Wide variation in flower volatiles has been reported between individual plants at the intraspecific level, both within and between populations (Gregg 1983; Groth et al. 1987; Burger et al. 1988; Bergström and Bergström 1989; Pham-Delegue et al. 1989; Tollsten and Bergström 1989), stressing the need to collect multiple samples. In addition, the chemical picture may be complicated by the presence of flower morphs with distinct volatile profiles within a species (Gregg 1983; Groth et al. 1987) and by hybridization (Loper 1976).

Many, and perhaps most, flowers emit only small quantities of volatiles, but in some species volatile production can reach high levels. Inflorescences of *Bartsia alpina.* (Scrophulariaceae) averaged 100 ng/day (Bergström and Bergström 1989), while in contrast flowers of *Hoya carnosa* (Asclepiadaceae) emitted 40–50 μg/day (Matile and Altenburger 1988) and inflorescences of *Platanthera stricta* (Orchidaceae) up to 50 μg/h (Patt et al. 1988).

3 Overview of Methodology

A preliminary evaluation of floral fragrance can be made using the human nose, most easily by first placing flowers in a closed container to concentrate the odors. For precise and objective chemical studies, a variety of methods have been used for collecting and analyzing flower volatiles. New and improved approaches are continually being developed and adopted in parallel with more sensitive instrumentation. Methodologies applied during the 1960's and 1970's are discussed by Bergström et al. (1980), Williams (1983), and Williams and Whitten (1983); the present review emphasizes techniques used during the 1980's.

3.1 Volatile Collection

Collection of volatiles from flowers has been achieved by solvent extraction, steam distillation, headspace cold-trapping, direct headspace sampling, and headspace sorption (Bergström et al. 1980; Williams 1983). While the headspace approaches involve the collection of volatiles released into the air, extraction and distillation processes remove substances from the floral tissues, making them generally more crude techniques that yield samples containing both fragrance and nonfragrance compounds. Previously the most commonly used methods for collecting volatiles, extraction and distillation have now been largely replaced by headspace techniques, which are more sensitive to air-borne volatiles and compatible with current analytical instrumentation. Headspace approaches have the additional advantage of being nondestructive to the flower material and therefore permitting the collection of replicate samples. Volatile collection by headspace sorption is particularly favorable in that it involves the accumulation of volatiles over the sampling period and thus yields enriched samples.

Choice of methods depends partly on the availability of equipment, type of flower material (e.g., whole flowers versus individual flower parts or pollen, laboratory versus field plants), and amount as well as general chemical composition of emitted volatiles. Methods involving separation steps to isolate the volatile compounds for analysis run a greater risk of loss of volatiles and although no method gives a complete picture of the actual volatile mixture emitted by the flower, headspace sorption clearly comes closest.

3.1.1 Extraction and Distillation

Volatile samples obtained by the more traditional approaches of solvent extraction and distillation yield both greater quantities and a larger number of compounds (frequently 70–100) compared with headspace approaches (e.g., Buttery et al. 1982; Flath et al. 1983; Etievant et al. 1984; Borg-Karlson et al. 1985a, 1987; Borg-Karlson 1990). However, their greater uptake of high boiling-low volatility compounds results in a relative under-representation, and even lack of detection, of the highly volatile compounds found in the headspace, some of which are critical in defining a fragrance (Kaiser and Lamparsky 1982; Toulemonde and Richard 1983; Lamparsky 1985; Joulain 1987). One advantage of extraction is that it requires simple equipment, making it easy to use in the field; the large amount of extraneous chemicals, including waxes and other lipid substances, extracted with the volatiles call for separation steps prior to analysis.

The appropriateness of extraction and distillation for descriptive studies of flower volatiles is further questioned by evidence which suggests that scent production in flowers is a dynamic and discrete excretory process rather than merely a matter of biosynthesis and evaporation. Differing abundances of specific volatiles in headspace compared to extraction samples were found to be not simply a function of the vapor pressures of individual compounds (Altenburger and Matile 1990). Additionally, in some orchids the volatiles are actively released from the surface of the flower with no accumulation in the floral tissues, such that solvent extraction and steam distillation fail to yield any fragrance volatiles (Williams 1983), and species displaying rhythmic volatile emissions in headspace samples showed no comparable periodicities in petal extract samples (Altenburger and Matile 1990).

3.1.2 Headspace Cold-Trapping

The collection of headspace volatiles by condensation at low temperatures in a cold trap yields samples with compositions that are similar to those obtained by direct headspace sampling (Borg-Karlson et al. 1987) and have odors similar to living flowers (Bergström et al. 1980; Joulain 1986). It has the disadvantage that water is collected along with the volatiles, causing possible ice formation in the equipment and requiring separation steps to remove water from the sample. This method is not very applicable to work in the field.

3.1.3 Direct Headspace Sampling

This approach consists of taking a sample of the air above the flower and therefore gives potentially the most true picture of the flower's volatile emission. However, its application is restricted to flowers with strong odors that secrete sufficient quantities of volatiles for detection in an air sample during analysis. Even in these cases the samples are generally weak, which places limitations on the amount of chemical work that can be done for identifying constituent compounds. This problem is aggravated by the fact that the entire sample is used up in a single analysis.

This technique was the principal method applied during the early, extensive work on fragrances from orchids pollinated by euglossine bees (Williams and Whitten 1983). It has now been largely replaced by the more effective and sensitive headspace sorption methods.

3.1.4 Headspace Sorption

The fragrance concentration problem presented by direct headspace sampling is eliminated in headspace sorption by the accumulative trapping of volatiles over a time period onto an adsorbent. Headspace sorption is undoubtedly the most efficient technique for collecting volatiles from flowers, especially those which release only small amounts. It is also the principal method that has been used for pollen odors. The popularity of headspace sorption lies in the efficacy of volatile collection, the reliable and accurate representation of actual volatile emissions, and the relative ease of applicability in both laboratory and field. It is given emphasis here.

3.2 Volatile Analysis

Separation of volatile compounds in a sample is accomplished using gas chromatography (GC). This also allows preliminary identification of selected constituents. Chemical identities of compounds are established using gas chromatography coupled with mass spectrometry (GC-MS). For more information on the identity and structure of a compound, additional approaches can be applied, such as Fourier transform infrared spectroscopy (FT-IR) or nuclear magnetic resonance spectrometry (NMR), provided that sufficient material is available.

Conditions used for the analysis by GC and GC-MS depend greatly on the instrument, column characteristics, and chemistry of the particular sample. On-column injection of samples is preferable to split/splitless injection for obtaining quantitative data for constituents covering a wide range of boiling points, since there is no discrimination between compounds due to selective vaporization. Dual detection using an N/P detector and flame ionization detector (FID) is useful for the analysis of samples that include nitrogen- and phosphorus-containing compounds (Wassgren and Bergström 1984). Most critical in the analysis is the type of GC column, since the polarity of the coating material influences how effectively fragrance components are separated. Optimally, samples are first analyzed on both polar and nonpolar columns to obtain a broad view of the range of constituents.

In general, columns of moderate polarity, such as those coated with polyethylene glycol (PEG; Carbowax, Superox FA, OV-351, etc.) are best for the broad gamet of flower volatiles, as they are sensitive to a wide range of compounds, including amines. Columns coated with methyl silicone (such as OV-101) have a nonpolar surface, which may be preferred for certain flowers. Appropriate column length is 25–30 m, with 0.25 mm inside diameter (0.15 mm for GC-MS with vacuum pressure). Temperature programming is a function of the column and the kinds of components to be separated. A wide temperature range allows the detection of compounds with a broad range of volatilities, which may be desirable or essential for flower samples. On a PEG column, low initial temperatures of 30–50 °C allow the resolution of compounds with high volatility, while high final temperatures over 250 °C are preferred for the separation of high boiling components. Rates of increase generally fall within the range of 4–8 °C/min; multiple programming can increase the resolution of certain compound groups.

3.3 Cautionary Notes

Studies of flower and particularly pollen volatiles involve very small quantities of chemicals. It is therefore critical to minimize contaminations arising from either solvents or equipment during sample collection and analysis. Even minor contaminants can mask from detection volatiles present in the flower sample.

Solvents must be of high purity (checked by GC), which often requires redistillation of even high-grade solvents. Glassware should be carefully cleaned with detergent, solvents, and heat. Glass equipment may be silanized after cleaning, which deactivates the glass surface and decreases its interaction with volatiles. Williams (1983) effectively decontaminated Plexiglas by washing with hot soapy water, followed by heating in an oven at 70 °C overnight. Any materials with oil base, including corks, wax, paraffin, most plastics, and rubber must be avoided and kept out of direct or indirect contact with the volatile sample, including exposure to solvents and air flow used in collecting volatiles or cleaning adsorbents; Teflon is a safe substitute.

4 Details of Collection Methods

4.1 Extraction and Distillation

4.1.1 Flowers

Sufficient material for GC analysis can be extracted from as few as three florets of *Achillea collina* (Compositae) using 20μl hexane for 48 h in darkness at room temperature (Cernaj et al. 1983). Borg-Karlson et al. (1985a, 1987) prepared extracts of *Ophrys* flowers and isolated labella (lip petals) using methanol, pentane, and hexane at room temperature for 3 days. The use of methanol had the disadvantage of requiring that fragrance compounds be transferred to a nonpolar solvent (such as pentane) prior to analysis and producing significant quantities of artifactual methyl esters.

More conventional approaches involve large-scale extractions of up to 200 kg flowers using hexane, pentane, or dichloromethane, which require many separative steps for the isolation of volatile extracts (Etievant et al. 1984; Harada and Mihara 1984; Kumar and Motto 1986). Volatile oils can also be obtained by vacuum steam-distilling flowers for 1 h followed by extraction of the distillate with dichloromethane (Erickson et al. 1987), diethyl ether (Andersen 1987), or ether-pentane (Lindeman et al. 1982).

High-pressure Soxhlet extraction using CO_2 as a solvent has been tried on flowers by Takeoka et al. (1985). Carbon dioxide has the advantages of being relatively inexpensive, chemically stable, inert, nontoxic, and also has a low boiling point, which facilitates its removal from the volatile extract. It was found to be equally or more effective for extracting plant volatiles in a Soxhlet apparatus than dichloromethane, and yielded considerably more compounds than either direct headspace sampling or steam distillation/extraction.

4.1.2 Pollen

The oily coating of pollen (i.e., pollenkitt) which contains volatiles can be removed by brief extraction with pentane or diethyl ether (Dobson 1988; Dobson et al. 1987). To do this, 5–10 mg pollen are placed on a Millipore filter (type AA, pore size 0.8 or 5.0 µm) fitted in a syringe and 0.5 ml solvent is rapidly passed through the pollen sample; several replicate samples can be combined to yield sufficient extract. After concentration, the extract is analyzed by GC, preferably fitted with a precolumn to remove low-volatile lipids present in the extract.

4.2 Headspace Cold-Trapping

Headspace fragrance compounds are collected in a liquid form by passing air over the sample and through a cold vessel in which the volatiles are condensed. The cold trap may be kept at 0 °C, -70 °C, or -196 °C using ice, solid CO_2-ethanol, or liquid nitrogen, respectively. During the condensation water is also accumulated, which can lead to the formation of ice plugs in the glass tubes and also requires that more material be collected due to loss incurred during subsequent water-removal extraction steps (Bergström et al. 1980).

In the procedures used by Bergström et al. (1980), a glass spiral was attached directly to the vessel containing the flowers and the air surrounding the flowers was pumped through it at a rate of 5–10 ml/min. After complete condensation the cold trap was rinsed three times with 1 ml methanol (to remove water and water-soluble substances) and diethyl ether or pentane. Joulain (1986) placed 100 g of freshly picked flowers in a round bottom flask and applied a 0.1 Torr vacuum for

30–45 min, during which the volatiles were condensed in a liquid nitrogen trap. The condensates were recovered at atmospheric pressure by reflux into a 100-ml flask containing 40 ml dichloromethane. The separated water was extracted with 20 ml solvent and the combined organic phases filtered and concentrated.

Cryogenic cold trapping, which involves the condensation of the air itself, has been successfully used on plant seedlings (Hibbard and Bjostad 1988) and may have potential value for flower fragrances containing compounds of very high volatility. The set-up is simple: the plant material is placed in a glass tube closed at one end and connected to a similar sample tube with Teflon tubing. Both tubes are immersed in a liquid nitrogen bath and, as air condenses in the sample tube, a vacuum is created that pulls air from the first tube through the plant material. The sample tube is then immediately placed into a precooled vessel at -183 °C, where only nitrogen and oxygen are boiled away, leaving behind plant volatiles.

4.3 Direct Headspace Sampling

4.3.1 Whole Plants

In a typical set-up (Hills et al. 1972), whole plants are placed inside a Plexiglas chamber during the peak period of fragrance production and an equilibration time of at least 30 min is allowed for the odor to saturate the chamber. A sample of air is taken by inserting a gas-tight syringe through a hole in the chamber wall (fitted with a septum), then 10 ml are immediately injected into the GC. To prevent contamination, used syringes are rinsed with acetone, washed with detergent, and rinsed again, followed by drying and storage in a 50 °C oven.

4.3.2 Flowers

Volatiles from orchid flowers (Gregg 1983) and fig synconia (Barker 1985) have been successfully concentrated within glass jars for sampling. Gregg (1983) also used isolated orchid petals by placing several labella into glass vials (7 × 2 cm); although she reported no evidence of injury volatiles resulting from the excision of the labella, altered fragrance profiles increased when volatile samples were collected after more than 8 h.

4.3.3 Direct Purging of Flowers

Air samples can also be collected by putting flowers or flower parts into glass tubes attached directly to the GC inlet. Volatiles are driven off the flowers and into the GC with carrier gas by heating the injection tube to about 150 °C for 10 min. The advantage here lies in the completeness of the volatile collection, but there are risks of artifact formation and breakdown of volatiles during heating (Bergström et al. 1980). This is a potentially good approach for small quantities of flower material having weak volatile emissions. Loper and Berdel (1978) followed a similar procedure by placing alfalfa flowers into a 100 cc syringe, equilibrating for

15 min at 32–40 °C, and transferring the volatiles to a pre-column cold trap, using helium gas and heat, prior to analysis.

4.4 Headspace Sorption

Fresh flower material is placed within a chamber having two small openings through which air is flowed. The openings are preferably situated such that the air flows up and over the flowers. The air exits the chamber through a cartridge that contains an adsorbent and in which volatiles mixed with the air become trapped (Fig. 1). The adsorbent cartridges are optimally attached directly to the chamber. Volatile collection is usually carried out at ambient temperature; Tatsuka et al. (1988) and Loughrin et al. (1990) warmed the flower chambers to 27–30 °C, but such manipulations should be done carefully and heating of the adsorbent cartridge avoided. At the termination of collection, the adsorbed volatiles are desorbed from the cartridge using heat (for charcoal and Tenax GC only) or solvents and the resulting sample analyzed by GC. In the case of liquid (solvent) samples, concentration of volatiles through evaporation of solvent may be necessary prior to analysis.

An alternative approach that is gaining increased popularity for the trapping of volatiles from plants and insects is the "closed-loop-stripping" technique (Boland et al. 1984; Lorbeer et al. 1984). In this set-up the air flow system is closed such that the same air is passed repeatedly over the sample and through the adsorbent cartridge. This decreases the in-flow of contaminants and minimizes disturbance of the organism. The method is applicable in field conditions and offers promise for studies of flower volatiles.

4.4.1 Air-Flow Parameters

Natural air is most commonly used as the entrainment gas. Hamilton-Kemp et al. (1989) found that plant volatiles collected with nitrogen gas differed both qualita-

Fig. 1. General schematic set-up for collecting flower volatiles by headspace sorption. Air flow is generated by either pressure (*A*) or suction (*B*)

tively and quantitatively from those collected with air; more particularly, use of nitrogen resulted in more alcohols and esters, the formation of which may be enhanced by the increased anaerobic conditions. Air flow through the headspace sorption system is achieved by either pressure (gas cylinder) or suction (vacuum pump or air ejector). Pressure flow results in some increased air pressure within the flower chamber, suction in reduced air pressure.

Air coming into the flower chamber should be as clean as possible to minimize sample contamination. This requires passing it first through a filter. Activated charcoal is frequently used, as it is very effective at adsorbing a wide range of compounds and is relatively inexpensive; it also has a high porosity and therefore a low resistance to air flow, allowing the air pressure within the flower chamber to be kept equal to that on the outside during sampling by suction. Alternatively, the filter cartrige can contain the same adsorbent as that employed to trap volatiles. Any tubing used to carry air between the pre-chamber filter and the volatile collecting cartridge should be made of unreactive material.

Air and flower chambers invariably contribute some contaminants to the sample, making it essential to collect control samples from the experimental set-up minus flowers. Simultaneously collected control samples become especially critical when the incoming air is not prefiltered, as is often the case in field conditions.

Determination of air flow rate involves consideration of several variables: (1) volatile concentration, (2) chamber volume, (3) duration of collection, and (4) breakthrough volume of volatile compounds in the adsorbent cartridge (see Sect. 4.4.5.1). Principally, the air flow must be high enough to carry as many volatiles as possible into the collection cartridge, but slow enough so that they are not lost from the cartridge. A turnover of the air volume in the chamber every 10 min is a good rule of thumb, but must be modified for very large or very small chambers, where air flow rate will be too high or too low, respectively. It is most cautious to perform preliminary tests on the particular set-up and flower material. Flow rates usually range from 100 ml to several liters per minute (Crisp 1980).

For chambers of 1–2l, flow rates of 50–150 ml/min are common. When using adsorbent cartridges 150 mg Porapak Q, flow rates of 100–150 ml/min have been considered appropriate for that contain periods of up to 24 h (Groth et al. 1987; Tollsten and Bergström 1989) and slightly lower rates of 50–100 ml/min when cartridges contain 70–100 mg (Borg-Karlson and Groth 1986). Rates of 85 ml/min were applied for 6 h using cartridges with 100 mg charcoal or 30 mg Tenax GC by Patt et al. (1988), and 50–100 ml/min for 1 h with 200 mg Tenax GC by Lewis et al. (1988). Using thin 5-mg charcoal filters for adsorption, Matile and Altenburger (1988) effectively trapped flower volatiles during 3-h sampling periods with a flow rate of 1 l/min.

4.4.2 Flowers

Collecting volatiles from flowers still attached to the plant minimizes the possible release of injury-related compounds and the interruption of normal secretory processes. However, when this is not feasible, especially if sampling several flowers or inflorescences from separate plants, cut flowers still on the stem which can be kept in water are next best. Rapid water loss from cut flowers without stems can be decreased by covering the severed surfaces with aluminum foil or Teflon tape (paraffin is a more effective sealer, but can contribute hydrocarbon contaminants).

When stem and leaves are included with the flower material, separate control samples should be collected simultaneously from all-green plant parts to identify the nonflower volatiles present in the experimental sample. Subtraction of "green" volatiles from the flower sample is most reliable when volatiles are collected from the same individual or group of plants in order to eliminate the effect of inter-individual variation (Pellmyr et al. 1987).

4.4.2.1 Containment of Sample

The flower sample is placed in a chamber of appropriate size made of glass (cylinder or flask, preferably two-piece) or Plexiglas (box) and provided with openings for air flow. Water vases for cut stems are covered with aluminum foil to minimize evaporation. When collecting volatiles from flowers or inflorescences still attached to the rooted plant, bags or open-ended glass cylinders can be used to enclose the samples (Fig. 2). Field set-ups in which the cylinder is open at the bottom, and the incoming air is therefore unfiltered, have been successfully used by Burger et al. (1988) and Tollsten and Bergström (1989) for flowers with strong

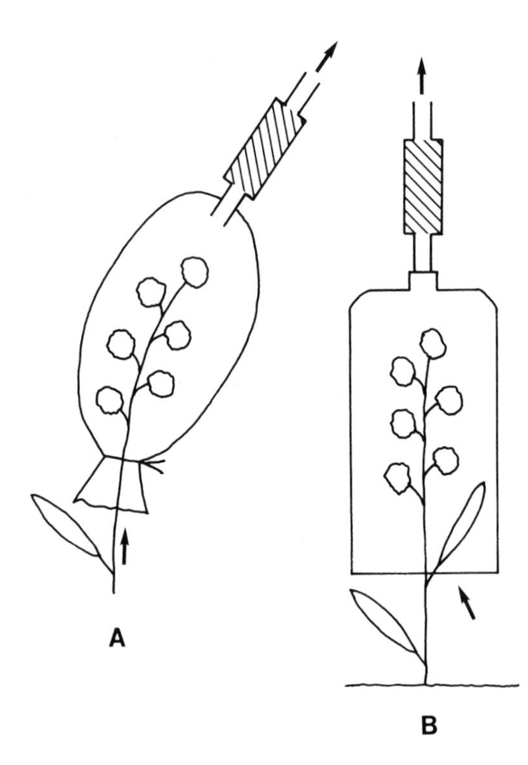

Fig. 2. Field set-up: containment of flowers attached to rooted plants for headspace sorption. **A** Transparent polyacetate bag. **B** Glass cylinder; a stand is used to hold it up above the ground

odors. In the case of flowers with weak volatile emissions it is necessary to filter the incoming air, such as done by Borg-Karlson et al. (1985a), who packed the bottom of the cylinder with activated charcoal held in a piece of cloth that was snuggly fitted around the plant stalk.

The use of bags requires previous testing to find a kind that does not release or adsorb volatiles. Workers in Sweden have found transparent polyacetate bags (for cooking food in ovens) to work well for this purpose (Borg-Karlson et al. 1985a; Dobson et al. 1987); Pham-Delegue et al. (1989) employed bags made of polyvinyl fluoride film (Tedlar, Dupont de Nemours). The bags are placed over the flowers and simply tied closed around the stem. Air enters the system through the bottom of the bag (around the stem) and is drawn out by suction through a hole into which is inserted an adsorbent cartridge.

4.4.2.2 Timing of Sampling

The length of time required to trap enough volatiles for analysis depends on the plant species. Collection is preferentially carried out when volatile emission by the flower is at its highest levels, but in cases where it shows no clear periodicities, or information on this is lacking, volatiles are collected over 12 or 24 h. For flowers of *Ophrys* (Orchidaceae), at least 12–24 h was necessary to concentrate relatively small amounts of volatiles, while for other plants, especially those with strong odors, only 1–6 h may be sufficient (Bergström et al. 1980; Williams and Whitten 1983).

However, volatile production can vary widely between plants, and timing may need adjustment accordingly. Patt et al. (1988) found that at any given time of day total floral emissions from *Platanthera stricta* (Orchidaceae) varied over tenfold (for Tenax GC) among individual plants, in spite of the fact that all inflorescences were of similar size and stage of maturity. Volatile composition and emission may be influenced by flower age and developmental stage (Lewis et al. 1988; Pham-Delegue et al. 1989; Tollsten and Bergström 1989), as well as by environmental conditions (Chang et al. 1988; Robacker et al. 1988).

In plants showing periodicities of volatile emission, the rhythmic patterns may be under the control of endogenous circadian clocks or may depend on environmental variables, particularly photoperiod (Loper and Lapioli 1971; Altenburger and Matile 1988, 1990), a factor to be considered when sampling from greenhouse plants. Furthermore, in some species, rhythmic volatile production is expressed only when the flower is attached to the plant (Matile and Altenburger 1988).

Short sampling periods are necessary to recognize any cyclic patterns in volatile emission (Loper and Lapioli 1971; Matile and Altenburger 1988) or in volatile composition, such as differences between day and night (Nilsson 1978). Matile and Altenburger (1988) devised the set-up shown in Fig. 3 to study volatile rhythmicities. Using electrically controlled valves, air flow was directed through one of a dozen adsorbent cartridges for each 3-h sampling period and then switched to another. When they sampled from excised flowers, the peduncle was kept in a 0.05 M glucose solution.

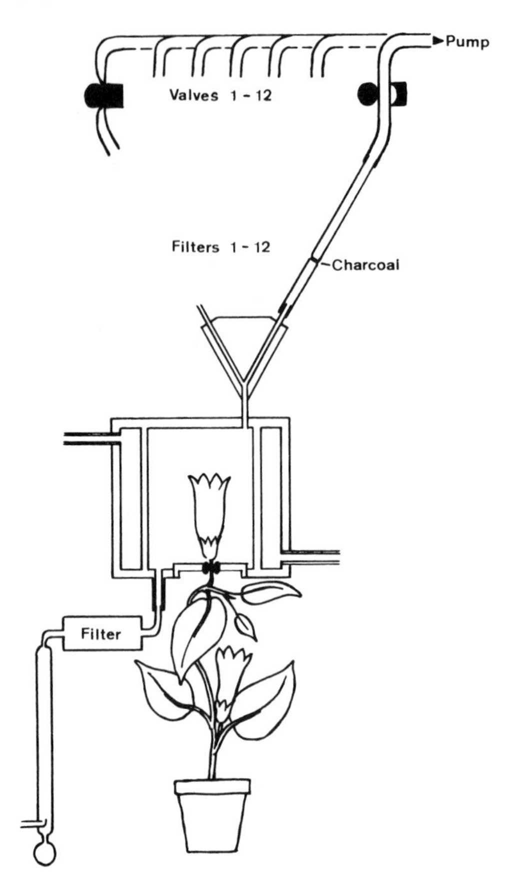

Fig. 3. Layout used for collecting volatiles from individual flowers at specific time intervals. (Matile and Altenburger 1988)

4.4.3 Flower Parts

Volatile emissions from separate flower parts can be studied by carefully isolating petals, sepals, stamens, or other parts from the flower and placing them into glass cylinders of the appropriate size (Dobson et al. 1990). Since some parts may produce only small quantities of volatiles, it is best to use pre-filtered air for volatile uptake. The methods are applicable in both field and laboratory conditions.

It should be noted that removal of parts from the flower body may disrupt metabolic processes, thus making it important that the procedure be timed to coincide with maximum volatile emission. In addition, some damage may be incurred to the sample, with the possible release of wound-related substances. Orchid labellar petals have been observed to produce aliphatic aldehydes following their excission from the flower (Borg-Karlson unpubl.), and compositional changes in volatiles from isolated labella of other orchids reported by Gregg (1983) may similarly represent responses to wounding. No clear evidence of such chemical responses were found by Dobson et al. (1990) in their studies of *Rosa* flower parts.

4.4.4 Pollen

Collection of pollen volatiles presents problems due to (1) small amounts of emitted volatiles, (2) containment of volatiles in the oily pollenkitt that includes triglycerides and waxes, which may slow the release of volatile compounds, and (3) difficulty of obtaining fresh pollen in sizable quantities. Since the pollenkitt is deposited onto the pollen grains from the anther tapetum, pollen must be harvested within a short time after the anthers have opened in order to obtain fresh pollenkitt with maximum odor.

For strong-smelling pollen, 0.1 g samples may be sufficient (Dobson et al. 1987, 1990). There is, however, no upper limit on sample size and for many pollens larger samples appear to be required to collect sufficient quantities of volatiles. Harvested pollen has been stored in closed vials at -5 °C before volatile collection, although effects of storage have not been evaluated. To collect volatiles, pollen is placed into a small glass tube (15 × 5 mm) and held in with a minimal amount of silanized glass wool. Adsorbent cartridges are attached to both ends of the tube (one to filter air, one to collect volatiles) and sampling is carried out with an air flow of 10 ml/min for up to 24 h at room temperature.

Among over a dozen different pollens sampled using this procedure, several with weak odors have failed to provide sufficient amounts of volatiles for their detection by GC (Dobson unpubl.), and background contaminants often mask the pollen compounds. Improved volatile collection might be obtained by using isolated pollenkitt (applied to filter paper) rather than pollen grains, which would increase the surface area in contact with the air flow. Alternate methods to be considered include GC analysis of pollenkitt extracts (Sect. 4.1) and direct headspace purging of pollen (Sect. 4.3).

4.4.5 Adsorbents

4.4.5.1 Characteristics

The properties of various adsorbents that can be used in volatile collection are described by Crisp (1980) and Schaefer (1981). Discussion of the differing applicability of selected adsorbents in studies of flower volatiles, however, is more limited (Bergström et al. 1980; Chang et al. 1988; Patt et al. 1988). The most popular adsorbents currently used for flower volatiles are activated charcoal, and the porous polymers Tenax GC and Porapak Q. These three materials differ slightly in their adsorption properties and in the procedures used for volatile desorption, both of which influence final volatile recovery. Uniformity of methodology is essential in comparative studies.

During volatile uptake, a cartridge filled with adsorbent functions similarly to a chromatographic column: adsorbed volatiles are moved with the air flow along the cartridge at a rate that is a function of both the compound and the adsorbent. Of importance in determining appropriate air-flow rates and duration of volatile collection is the amount of air needed until a compound starts to escape from the cartridge, namely the breakthrough volume. In general, compounds of higher volatility have smaller breakthrough volumes, and the retention time of a compound in the cartridge increases with its chain length.

When flowers have weak odors it is desirable to collect volatiles for as long a period as possible, in which case breakthrough volume becomes critical for deciding on the duration of sampling. The retention and adsorption capacity of Porapak Q is generally higher than for Tenax GC, making it preferable for these, and probably for most, flower species (Schaefer 1981). When the flower fragrance is strong, less time is needed for volatile uptake.

Charcoal is the most efficient adsorbent for trapping a wide spectrum of volatiles, but effectivity of volatile desorption is of major concern and in this respect Tenax GC performs better than activated charcoal, to which some compounds bind tightly (Schaefer 1981). In comparisons of charcoal with Tenax GC, Patt et al. (1988) found that neither adsorbent alone effectively trapped/desorbed the full range of volatile compounds emitted by *Platanthera* (Orchidaceae) flowers. Bergström et al. (1980) obtained similar adsorptions from flowers using either Porapak Q or activated charcoal, whereas Tenax GC showed a greater affinity for polar compounds and, according to Williams (1983), does not easily adsorb (i.e., easily desorbs) several monoterpenes. However, Tenax GC and Porapak Q in general appear to perform equally effectively for most flowers (I. Groth pers. commun.). The adsorption efficiency of Porapak Q for flower volatiles has been reported to be over 99% (150 mg, 24 h, 150 ml/min) (L. Tollsten unpubl.), and for some floral monoterpenes 74–99% (8 h, 0.5 l/min) (Chang et al. 1988). Porapak Q was also shown to have a higher trapping efficiency for leaf volatiles than either Tenax GC or charcoal (Schaefer 1981). It is interesting to note that in the instances where Porapak Q is given very high ratings, cleaning and desorption are done exclusively with solvents. Poropak Q conditioned with heat (which can cause breakdown of the polymer) but desorbed with solvents does not appear to perform as well (Cole 1980). This suggests that treating Porapak Q with heat at any time decreases its adsorption capacity, although this has not been rigorously examined. Because of the incomplete trapping of all fragrance compounds by any single adsorbent, Williams (1983) suggests the combined use of charcoal with either Tenax GC or Porapak Q.

Mesh sizes commonly used for Porapak Q are 50–80 and 80–100, for Tenax GC 60–80; the higher mesh size has smaller beads and thus a greater total effective surface area, but it offers more resistance to air flow.

4.4.5.2 Cartridges

Adsorbents are commonly packed into small glass tubes and held in with silanized glass wool on either side. The cartridges are then covered with aluminum foil to keep out light. Long narrow cartridges adsorb more effectively than short wide ones, since it takes longer for compounds to move through the adsorbent column (but resistance to air is increased).

The amount of adsorbent needed depends on the size of the plant sample, rate of air flow, and duration of sampling. Typical quantities range from 70–200 mg Porapak Q for field and laboratory studies involving only one or a few plants sampled over 12–24 h. Cartridges approximately 30–50 × 5 mm serve well for 150 mg Porapak Q (Borg-Karlson et al. 1985a; Dobson et al. 1987; Bergström and Bergström 1989). Pham-Delegue et al. (1989) used large quantities of Porapak Q, 2.2 g within 10 × 1 cm tubes, for collecting sunflower volatiles. Cartridges of Tenax GC may contain only 30 mg (Patt et al. 1988), 75 mg within 40 × 4 mm tubes (Borg-Karlson et al. 1985b), or 200 mg in 150 × 6 mm tubes (Lewis et al. 1988). Buttery et al. (1982, 1984, 1986) and Loughrin et al. (1990) used around 1.5 g Tenax GC, within 1.3 × 7 cm cartridges, and in one case as much as 10 g, for large flower samples weighing 200 g. Activated charcoal can be likewise packed into cartridges, such as in quantities of 100 mg (Patt et al. 1988). Alternatively, one can use small manufactured charcoal filters (5 mg) that are embedded between two grids (Bender & Hobein AG, Zürich, or Brechbühler AG, Urdorf, Switzerland); these appear to be very efficient (Matile and Altenburger 1988) and are presently being used in several laboratories. They may prove to be especially useful for flower or pollen samples with weak odor emissions. Burger et al. (1988) prepared open tubular traps by melting carbon particles onto the inside surface of glass capillaries.

Different adsorbents can be combined by placing them in the same cartridge, either separated with glass wool or blended together, or by connecting separate cartridges in succession. If different adsorbents are used in the same cartridge, methods for desorption and cleaning (see sections below) must be compatible with both materials. Williams (1983) prefers cartridges 75 × 9 mm filled with Tenax GC and activated charcoal that are separated with glass wool; during sampling, air passes first through the Tenax, but during desorption the cartridge is reversed such that the nitrogen gas first contacts charcoal, a necessary measure to prevent the irreversible adsorption of Tenax-held compounds onto the charcoal.

4.4.5.3 Cleaning

Adsorbents must be treated to remove contaminants prior to their first use (i.e., conditioned) and subsequently cleaned after each fragrance desorption; the same methods are often applied for both processes. This can be done with either heat or solvents, or both (Crisp 1980). Conditioning adsorbents with heat under a flow of nitrogen gas is a popular method, especially for Tenax GC and charcoal, both of which can sustain high temperatures without damage. Porapak Q is sensitive to heat and should be conditioned and cleaned exclusively with solvents.

Heat is very effective for the cleaning and desorption of Tenax GC and charcoal. Tenax can be initially conditioned by heating to 200–250 °C with a nitrogen gas flow for up to 24 h within the adsorbent cartridges (Bergström et al. 1980); Buttery et al. (1982) eluted Tenax GC with diethyl ether prior to heating. Williams (1983) conditions his Tenax-charcoal cartridges at 275 °C. For solvent elution from Tenax GC, diethyl ether and pentane are commonly used; dichloromethane must be avoided since it dissolves the polymer. Matile and Altenburger (1988) treated their charcoal filters exclusively with solvents by eluting 30 times with the sequence of hot ethanol (60 °C), dichloromethane, and carbon disulfide.

Porapak Q has a much lower heat limit than Tenax GC and charcoal, and temperatures higher than 100 °C can produce artifacts arising from breakdown of the polymer (Krumperman 1972; Williams et al. 1978; G. Birgersson pers. commun.). This heat lability and the unclear damage caused by even lower temperatures to the adsorptive capacity of Porapak Q make solvent extraction the safest approach. Workers at the Department of Chemical Ecology in Göteborg have found solvents to be sufficient for decontaminating Porapak Q (I. Groth pers. commun.). Accordingly, for the initial cleaning, Porapak Q can be Soxhlet extracted with dichloromethane for 6 h, followed by a similar extraction with pentane; after each extraction, the solvents are driven off with nitrogen gas. The Porapak Q is then packed into cartridges and washed again briefly with each solvent. It is also possible to do the complete treatment with already packed cartridges. For cleaning after each fragrance desorption, cartridges are washed with 2 ml quantities of dichloromethane, followed by diethyl ether and pentane, and dried with nitrogen gas.

4.4.5.4 Desorption

Fragrance volatiles are desorbed with either heat or solvents. The choice of procedures depends partly on the adsorbent, following some of the same limitations discussed above for cleaning, and partly on the sample form (gas or liquid) desired for analysis.

Volatiles can be heat-desorbed directly into the GC. In this case the entire sample may be used in one injection, making it impossible to perform repeated injections or further chemical analyses on the particular sample. However, this method allows the detection of very small amounts of volatiles, since no dilution is involved. Use of high temperatures runs the risk of thermal breakdown of certain volatiles, such as oxygenated monoterpenes. Alternatively, all or a portion of the heat-desorbed volatiles may be collected by condensation in a cold trap and mixed with solvent to produce a liquid sample, which is easier to store and offers the possibility of replicate analyses from a single sample.

Desorption with solvents is simple, avoids any thermal modification of volatiles, and provides the advantages of a liquid sample. It does, however, introduce any impurities present in the solvent and dilutes the sample, which must often be concentrated prior to analysis. Concentration invariably results in the loss of some highly volatile compounds and it is thus desirable to use a minimum amount of solvent during desorption. Elution of volatiles with solvent has gained increasing popularity in studies of flowers and is presently the method of choice used by N. Williams (pers. commun.) and by workers in Sweden.

Heat Desorption. For direct desorption into the GC, the volatiles are driven off the desorbent by heating (requires special modification of the GC injection port; commercial thermal desorption units are available for attachment to a GC) and cryofocused on the column, using solid CO_2 or liquid nitrogen, prior to analysis (Bergström et al. 1980; Williams 1983). Lewis et al. (1988) and Tatsuka et al. (1988) followed the same principles to desorb volatiles from Tenax GC and Burger et al. (1988) from charcoal.

Williams and Whitten (1983) developed an alternative method for collecting liquid samples from heat-desorbed volatiles. For heat desorption, the adsorbent cartridge is placed in a device made of copper tubing, with one end connected to a source of nitrogen gas (flow rate 30 ml/min). Gas carrying the desorbed volatiles is passed through a 30-cm-long glass capillary tube (1 mm diameter) fitted inside an aluminum block that is provided with a copper cold finger cooled with liquid nitrogen. A temperature gradient is thus established along the block and the volatiles condense inside the glass capillary tube. The condensed compounds are eluted with 1 ml pentane (or hexane).

Solvent Desorption. The most effective solvent for charcoal is carbon disulfide (Crisp 1980). Matile and Altenburger (1988) eluted volatiles from 5-mg charcoal filters by applying 10 µl carbon disulfide with a syringe and carefully moving it up and down ten times across the filter. Patt et al. (1988) desorbed cartridges containing 100 mg charcoal with 100 µl dichloromethane. For Tenax GC, Nilsson (1981) and Patt et al. (1988) used diethyl ether, the latter by applying as little as 400 µl to cartridges with 30 mg polymer; pentane can also be used. Pentane and diethyl ether are best for Porapak Q, especially if the sample is to be concentrated (Borg-Karlson et al. 1985a; Dobson et al. 1987; Tollsten and Bergström 1989; Borg-Karlson 1990). Solvent quantities of 2 ml have been found to suffice for eluting volatiles trapped on 150 mg Porapak Q. Pham-Delegue et al. (1989) desorbed volatiles with Freon 11 by percolation of 2 × 25 ml; freon requires the maintenance of low temperature conditions.

4.4.5.5 Final Sample Preparation and Storage

Volatiles are usually desorbed from the cartridges soon after completion of sample collection. During any waiting periods the cartridges are wrapped in aluminum foil (avoiding tape and sticky labels, which can produce contaminations) and kept in cool, dry, dark conditions. Patt et al. (1988) stored Tenax GC and charcoal cartridges with trapped volatiles at -20 °C for 1 month before desorption. In cases of long-distance field work, Porapak Q cartridges have been placed in envelopes and mailed to the laboratory; both highly volatile and heat-sensitive compounds were successfully desorbed after such handling (I. Groth pers. commun.).

Following elution, the resulting liquid samples should be stored at low temperature in tightly closed vials. Any concentration steps are preferentially carried out just prior to GC analysis. The extent of concentration required for analysis depends on the strength of the sample. Volatiles collected from some euglossine-pollinated orchid flowers and eluted in 1 ml solvent can often be analyzed directly (N. Williams pers. commun.), whereas eluted pollen volatiles usually need to be concentrated to 10µl (Dobson et al. 1987). Samples comprising approximately 2 ml can be concentrated carefully in a water bath, preferentially in special tapered vials with very narrow (2–3 mm wide) tubes that minimize the loss of fragrance volatiles during solvent evaporation (von Klimetzek et al. 1989), or just at room temperature. Recent developments of retention-gap pre-columns, which allow the injection of up to 200 µl samples into the GC (McCabe et al. 1989), offer promise in eliminating the need for sample concentration.

4.5 Passive Sorption

Modified methods have been developed to collect flower volatiles onto adsorbents in the absence of air flow, i.e., passively, which can be of special value for field work. In this case, volatile uptake is determined by diffusion and is thus more dependent on the concentration of individual compounds, such that different volatile profiles may be obtained using air flow and passive sorption (Lewis et al. 1988).

Early methods include placing strips of glass fiber paper impregnated with Convalex-10 pump oil (dissolved in ether) (Holman and Heimermann 1973) or silica gel impregnated with silicone oil (Bergström et al. 1980) into chambers containing the flowers. The trapped volatiles are desorbed by heating directly into the GC or first into a cold trap followed by solvent elution (Williams 1983). Although ease of sample handling makes these methods practical for the field, preparation of the sorption materials is tedious.

A simpler approach consists of trapping volatiles onto Tenax GC or Porapak Q that is spread out on filter paper or aluminum foil to maximize the adsorptive surface. Using this principle, Armbruster et al. (1989) collected volatiles by placing Tenax GC, held in a folded filter paper, into a jar with flower material; after collection the Tenax GC was sealed in a glass ampule. Bergström et al. (1980) have placed flowers or flower parts directly onto Porapak Q beads for several hours to several days, which is similar to the traditional enfleurage technique of the perfume industry. This method tends to favor the uptake of low-volatile compounds (I. Groth, pers. commun.). In both cases, the volatiles are desorbed using same procedures described for adsorbent cartridges.

Acknowledgments. Thanks are given to L. Ågren, G. Bergström, G. Birgersson, A.-K. Borg-Karlson, I. Groth, L. Tollsten, A.-B. Wassgren, and N.H. Williams for helpful comments and discussions.

References

Altenburger R, Matile P (1988) Circadian rhythmicity of fragrance emission in flowers of *Hoya carnosa* R. Br. Planta 174:248–252

Altenburger R, Matile P (1990) Further observations on rhythmic emission of fragrance in flowers. Planta 180:194–197

Andersen JF (1987) Composition of the floral odor of *Cucurbita maxima* Duchesne (Cucurbitaceae). J Agric Food Chem 35:60–62

Armbruster WS, Keller S, Matsuki M, Clausen TP (1989) Pollination of *Dalechampia magnoliifolia* (Euphorbiaceae) by male euglossine bees. Am J Bot 76:1279–1285

Barker NP (1985) Evidence of a volatile attractant in *Ficus ingens* (Moraceae). Bothalia 15:607–611

Bergström G (1978) Role of volatile chemicals in *Ophrys*-pollinator interactions. In: Harborne G (ed) Biochemical aspects of plant and animal coevolution. Academic Press, London, pp 207–231

Bergström G (1987) On the role of volatile chemical signals in the evolution and speciation of plants and insects: why do flowers smell and why do they smell differently? In: Labeyrie V, Fabres G, Lachaise D (eds) Insect-plants. W Junk, Dordrecht, pp 321–327

Bergström G, Appelgren M, Borg-Karlson A-K, Groth I, Strömberg S, Strömberg S (1980) Studies on natural odoriferous compounds XXII. Techniques for the isolation/enrichment of plant volatiles in the analyses of *Ophrys* orchids (Orchidaceae). Chem Scr 16:173–180

Bergström J, Bergström G (1989) Floral scents of *Bartsia alpina* (Scrophulariaceae): chemical composition and variation between individual plants. Nord J Bot 9:363–365

Boland W, Ney P, Jaenicke L, Gassmann G (1984) A "closed-loop-stripping" technique as a versatile tool for metabolic studies of volatiles. In: Schreier P (ed) Analysis of volatiles, methods and applications. de Gruyter, Berlin, pp 371–380

Borg-Karlson A-K (1990) Chemical and ethological studies of pollination in the genus *Ophrys* (Orchidaceae), review article. Phytochemistry 29:1359–1387

Borg-Karlson A-K, Groth I (1986) Volatiles from the flowers of four species in the sections Arachnitiformes and Araneiferae of the genus *Ophrys* as insect mimetic attractants. Phytochemistry 25:1297–1299

Borg-Karlson A-K, Bergström G, Groth I (1985a) Chemical basis for the relationship between *Ophrys* orchids and their pollinators I. Volatile compounds of *Ophrys lutea* and *O. fusca* as mimetic attractants/excitants. Chem Scr 25:283–294

Borg-Karlson A-K, Eidmann HH, Lindström M, Norin T, Wiersma N (1985b) Odoriferous compounds from the flowers of the conifers *Picea abies*, *Pinus sylvestris* and *Larix sibirica*. Phytochemistry 24:455–456

Borg-Karlson A-K, Bergström G, Kullenberg B (1987) Chemical basis for the relationship between *Ophrys* orchids and their pollinators II. Volatile compounds of *O. insectifera* and *O. speculum* as insect mimetic attractants/excitants. Chem Scr 27:303–311

Burger BV, Munro ZM, Visser JH (1988) Determination of plant volatiles 1: analysis of the insect-attracting allomone of the parasitic plant *Hydnora africana* using Grob-Habich activated charcoal traps. J High Resolut Chromatogr Chromatogr Commun 11:496–499

Buttery RG, Kamm JA, Ling LC (1982) Volatile components of alfalfa flowers and pods. J Agric Food Chem 30:739–742

Buttery RG, Kamm JA, Ling LC (1984) Volatile components of red clover leaves, flowers, and seed pods: possible insect attractants. J Agric Food Chem 32:254–256

Buttery RG, Maddox DM, Light DM, Ling LC (1986) Volatile components of yellow starthistle. J Agric Food Chem 34:786–788

Černaj P, Repčák M, Tesařik K, Hončariv R (1983) Terpenoid compounds from different parts of *Achillea collina* Becker inflorescences. Biol Plant 25:221–224

Chang JF, Benedict JH, Payne TL, Camp BJ (1988) Volatile monoterpenes collected from the air surrounding flower buds of seven cotton genotypes. Crop Sci 28:685–688

Cole RA (1980) The use of porous polymers for the collection of plant volatiles. J Sci Food Agric 31:1242–1249

Crisp S (1980) Solid sorbent gas samplers. Ann Occup Hyg 23:47–76

Dobson HEM (1987) Role of flower and pollen aromas in host-plant recognition by solitary bees. Oecologia 72:618–623

Dobson HEM (1988) Survey of pollen and pollenkitt lipids — chemical cues to flower visitors? Am J Bot 75:170–182

Dobson HEM, Bergström J, Bergström G, Groth I (1987) Pollen and flower volatiles in two *Rosa* species. Phytochemistry 26:3171–3173

Dobson HEM, Bergström G, Groth I (1990) Differences in fragrance chemistry between flower parts of *Rosa rugosa* (Rosaceae). Isr J Bot 39:143–156

Dodson CH, Dressler RL, Hills HG, Adams RM, Williams NH (1969) Biologically active compounds in orchid fragrances. Science 164:1243–1249

Erickson BJ, Young AM, Strand MA, Erickson EH (1987) Pollination biology of *Theobroma* and *Herrania* (Sterculiaceae) II. Analyses of floral oils. Insect Sci Appl 8:301–310

Etievant PX, Azar M, Pham-Delegue MH, Masson CJ (1984) Isolation and identification of volatile constituents of sunflowers (*Helianthus annuus* L.). J Agric Food Chem 32:503–509

Flath RA, Mon TR, Lorenz G, Whitten CJ, Mackley JW (1983) Volatile components of *Acacia* sp. blossoms. J Agric Food Chem 31:1167–1170

Gregg KB (1983) Variation in floral fragrances and morphology: incipient speciation in *Cycnoches*? Bot Gaz 144:566–576

Groth I, Bergström G, Pellmyr O (1987) Floral fragrances in *Cimicifuga*: chemical polymorphism and incipient speciation in *Cimicifuga simplex*. Biochem Syst Ecol 15:441–444

Hamilton-Kemp TR, Rodriguez JG, Archbold DD, Andersen RA, Loughrin JH, Patterson CG, Lowry SR (1989) Strawberry resistance to *Tetranychus urticae* Koch: effects of flower, fruit, and foliage removal—comparisons of air- vs. nitrogen-entrained volatile compounds. J Chem Ecol 15:1465–1473

Harada K, Mihara S (1984) The volatile constituents of Freesia flower (*Freesia hybrida* Hort.) Agric Biol Chem 48:2843–2845

Hibbard BE, Bjostad LB (1988) Behavioral responses of western corn rootworm larvae to volatile semiochemicals from corn seedlings. J Chem Ecol 14:1523–1539

Hills HG, Williams NH, Dodson CH (1972) Floral fragrances and isolating mechanisms in the genus *Catasetum* (Orchidaceae). Biotropica 4:61–76

Holman RT, Heimermann WH (1973) Identification of components of orchid fragrances by gas chromatography-mass spectrometry. Am Orchid Soc Bull 42:678–682

Joulain D (1986) Study of the fragrance given off by certain springtime flowers. In: Brunkel E-J (ed) Progress in essential oil research. Walter de Gruyter, Berlin, pp 57–67

Joulain D (1987) The composition of the headspace from fragrant flowers: further results. Flav Fragr J 2:149–155

Kaiser R, Lamparsky D (1982) Constituants azotés en trace de quelques absolues de fleurs et leurs headspaces correspondants. In: VIIIe Congrès Intern Huiles Essen 1980 Annales Techn, Fedarom Ed, Grasse, pp 287–294

Krumperman PH (1972) Erroneous peaks from Porapak-Q traps. J Agric Food Chem 20:909

Kullenberg B (1961) Studies in *Ophrys* pollination. Zool Bidr Upps 34:1–340

Kullenberg B, Bergström G (1976) Hymenoptera aculeata males as pollinators of *Ophrys* orchids. Zool Scr 5:13–23

Kumar N, Motto MG (1986) Volatile constituents of *Peony* flowers. Phytochemistry 25:250–253

Lamparsky D (1985) Headspace techniques as a versatile complementary tool to increase knowledge about constituents of domestic or exotic flowers and fruits. In: Svendsen AB, Scheffer JJC (eds) Essential oils and aromatic plants. W Junk, Dordrecht, pp 79–92

Lewis JA, Moore CJ, Fletcher MT, Drew RA, Kitching W (1988) Volatile compounds from the flowers of *Spathiphyllum cannaefolium*. Phytochemistry 27:2755–2757

Lindeman A, Jounela-Eriksson P, Lounasmaa M (1982) The aroma composition of the flower of meadowsweet (*Filipendula ulmaria* (L.)Maxim.) Lebensm Wiss Technol 15:286–289

Loper GM (1976) Differences in alfalfa flower volatiles among parent and F1 plants. Crop Sci 16:107–110

Loper GM, Berdel RL (1978) Seasonal emanation of ocimene from alfalfa flowers with three irrigation treatments. Crop Sci 18:447–452

Loper GM, Lapioli AM (1971) Photoperiodic effects on the emanation of volatiles from alfalfa (*Medicago sativa* L.) florets. Plant Physiol 49:729–732

Lorbeer E, Mayr M, Hausmann B, Kratzl K (1984) Zur Identifizierung flüchtiger Substanzen aus biologischem Material mit Hilfe der CLSA (Closed Loop Stripping Apparatus). Monatsh Chem 115:1107–1112

Loughrin JH, Hamilton-Kemp TR, Andersen RA, Hildebrand DF (1990) Headspace compounds from flowers of *Nicotiana tabacum* and related species. J Agric Food Chem 38:455–460

Matile P, Altenburger R (1988) Rhythms of fragrance emission in flowers. Planta 174:242–247

McCabe T, Hiller JF, Morabito PL (1989) An automated large volume on-column injection technique for capillary gas chromatography. J High Resolut Chromatogr 12:517–521

Nilsson LA (1978) Pollination ecology and adaptation in *Platanthera chlorantha* (Orchidaceae). Bot Not 131:35–51

Nilsson LA (1981) The pollination ecology of *Listera ovata* (Orchidaceae). Nord J Bot 1:461–480

Nilsson LA (1983) Anthecology of *Orchis morio* (Orchidaceae) at its outpost in the north. Nova Acta Reg Soc Sci Upsaliensis Ser V:C 3:167–179

Nilsson LA, Jonsson L, Rason L, Randrianjohany E (1985) Monophily and pollination mechanisms in *Angraecum arachnites* Schltr. (Orchidaceae) in a guild of long-tongued hawk-moths (Sphingidae) in Madagascar. Biol J Linn Soc 26:1–19

Patt JM, Rhoades DF, Corkill JA (1988) Analysis of the floral fragrance of *Platanthera stricta*. Phytochemistry 27:91–95

Pellmyr O, Bergström G, Groth I (1987) Floral fragrances in *Actaea*, using differential chromatograms to discern between floral and vegetative volatiles. Phytochemistry 26:1603–1606

Pham-Delegue MH, Masson C, Etievant P, Azar M (1986) Selective olfactory choices of the honeybee among sunflower aromas: a study of combined olfactory conditioning and chemical analysis. J Chem Ecol 12:781–793

Pham-Delegue MH, Etievant P, Guichard E, Masson C (1989) Sunflower volatiles involved in honeybee discrimination among genotypes and flowering stages. J Chem Ecol 15:329–343

Porsch O (1956) Windpollen und Blumeninsekt. Österr Bot Z 103:1–18

Robacker DC, Meeuse BJD, Erickson EH (1988) Floral aroma: how far will plants go to attract pollinators? BioScience 38:390–398

Schaefer J (1981) Comparison of adsorbents in head space sampling. In: Schreier P (ed) Flavour '81. Third Weurman Symp Proc Intern Conf., Walter de Gruyter, Berlin, pp 301–313

Takeoka G, Ebeler S, Jennings W (1985) Capillary gas chromatography analysis of volatile flavor compounds. In: Bills DD, Mussinan CJ (eds) Characterization and measurement of flavor compounds. Am Chem Soc Symp Ser 289, Washington, DC, pp 95–108

Tatsuka K, Kohama M, Suekane S (1988) Floral fragrance components of *Zygopetalum mackayi* (Orchidaceae). Agric Biol Chem 52:1599–1600

Thien LB, Heimermann WH, Holman RT (1975) Floral odors and quantitative taxonomy of *Magnolia* and *Liriodendron*. Taxon 24:557–568

Thien LB, Bernhardt P, Gibbs GW, Pellmyr O, Bergström G, Groth I, McPherson G (1985) The pollination of *Zygogynum* (Winteraceae) by a moth, *Sabatinca* (Micropterigidae): an ancient association? Science 227:540–543

Tollsten L, Bergström J (1989) Variation and post-pollination changes in floral odours released by *Platanthera bifolia* (Orchidaceae). Nord J Bot 9:359–362

Toulemonde B, Richard HMJ (1983) Volatile constituents of dry elder (*Sambucus nigra* L.) flowers. J Agric Food Chem 31:365–370

von Aufsess A (1960) Geruchliche Nahorientierung der Biene bei entomophilen und ornithophilen Blüten. Z Vergl Physiol 43:469–498

von Frisch K (1923) Über die "Sprache" der Bienen. Zool Jahrb Abt Allg Zool Physiol 40:1–186

von Klimetzek D, Köhler J, Krohn S, Francke W (1989) Das Pheromon-System des Waldreben-Borkenkäfers, *Xylocleptes bispinus* Duft. (Col., Scolytidae). J Appl Entomol 107:304–309

Wassgren A-B, Bergström G (1984) A simple effluent splitter for capillary gas chromatography. J High Resolut Chromatogr Chromatogr Commun 7:154–155

Williams AA, May HV, Tucknott OG (1978) Observations of the use of porous polymers for collecting volatiles from synthetic mixtures reminiscent of fermented ciders. J Sci Food Agric 29:1041–1054

Williams NH (1983) Floral fragrances as cues in animal behavior. In: Jones EC, Little RJ (eds) Handbook of experimental pollination biology. Sci Acad Eds Van Nostrand Reinhold, New York, pp 50–72

Williams NH, Whitten M (1983) Orchid floral fragrances and male euglossine bees: methods and advances in the last sesquidecade. Biol Bull 164:355–395

Bioactivities of Diterpenoids from Marine Algae

C. TRINGALI

1 Introduction

In the last decades, interest in marine organisms has risen sharply. As a result, notwithstanding that only a small proportion of the over 200 000 species thought to be present in the oceans has been examined to date, a great mass of biological and chemical data has been acquired. Particular attention has been given to secondary metabolites: indeed, their chemical structures are frequently unprecedented in terrestrial sources, and their pharmacological properties seem to be promising (Baker 1976; Kaul and Sinderman 1978; Fautin 1988). Furthermore, since secondary metabolites are involved in chemoreception in the marine ecosystem (Mackie and Grant 1974), the study of marine natural products is of interest not only in bioorganic chemistry and in the biomedical field, but also in chemical ecology.

To date, the chemical work has revealed over 3000 new metabolites from marine organisms (Faulkner 1977, 1984a,b, 1986, 1987, 1988), most of which have been obtained from algae; the filetic distribution within this group shows that the great majority of data refer to seaweeds (Chlorophyta, Phaeophyta, and Rhodophyta) (Ireland et al. 1988). Notwithstanding that many species of macroalgae have been examined in relation to their pharmacological properties (Bhakuni and Silva 1974; Naqvi et al. 1980; Hoppe et al. 1979; Hoppe and Levring 1982), the percentage of algal metabolites isolated through biological screening or subjected to bioactivity tests following their purification is rather low. This is partly due to the insufficient diffusion of the knowledge of the methods for bioactivity analysis.

Thus, this chapter is devoted to the methods for determining bioactivity of diterpenoids (diterpenes and compounds of mixed biogenesis with a diterpenic moiety) obtained from seaweeds; the bioassays herewith described can be run with relative ease even in a laboratory normally dedicated to phytochemical research.

Diterpenoids account for about 70% of the compounds obtained from algae and are therefore an amply representative class of the various biological activities reported for algal metabolites. The results of bioassays on algal diterpenoids are also reported. The activities taken into account are of interest in the fields of both pharmacology (e.g., antimicrobial, cytotoxic, molluscicidal), and chemical ecology (e.g., antialgal or ichthyotoxic).

The literature up to 1988 has been reviewed, and a total of 103 bioactive compounds are reported, whose biological properties are discussed in separate sections of this chapter and summarized in Table 8.

2 Antimicrobial Activity

2.1 Background

The term antimicrobial will be used here in the restrictive, but widely accepted meaning of "active against bacteria or fungi". Activity against microalgae or viruses will be considered elsewhere. Since the pioneering work of Pratt et al. (1951), several researchers have investigated the antimicrobial properties of extracts from marine algae (among others: Burkholder et al. 1960; Olesen et al. 1964; Henriquez et al. 1979; Caccamese et al. 1980; Hodgson 1984; Pesando and Caram 1984; see also literature cited therein). The early diffusion and popularity of screening tests employing bacteria and/or fungi as target organisms to detect bioactivities in crude extracts or in pure compounds from marine algae is probably due to the simplicity of the methods normally employed, when compared with other pharmacological assays. To date, antimicrobial compounds isolated from marine organisms have never displayed activities competing with those of the classic antibiotics from microorganisms; anyway, it should be mentioned that many biologically active compounds show antimicrobial properties, and antimicrobial activity may be used as a "correlative assay" for other bioactivities whose direct detection would be more difficult, when determined against a broad spectrum of selected microorganisms (Betina 1983). Antimicrobial data may also be useful in marine ecology studies, particularly if the target organism is a marine isolate, as in several cases reported in Tables 1 and 2.

2.2 Description of the Methods

Tests against bacteria and fungi are similar and will be discussed jointly; the main differences will be specified in the relevant paragraphs. These antimicrobial assays are generally referred to as "sensitivity" or "susceptibility" tests, in indicating the response of pathogenic microorganisms to known chemiotherapics; their application to the evaluation of new natural products against selected microbial strains requires only minor adjustments. A spectrum of test microorganisms is frequently employed, including at least a Gram-positive bacterium (usually *Staphylococcus aureus* or *Bacillus subtilis*), a Gram-negative bacterium (usually *Escherichia coli*) and one or more fungi (*Candida albicans* is the most commonly used). Studies aimed at the assessment of the ecological role of antimicrobial algal metabolites have been carried out mainly by Fenical and coworkers using marine bacteria and fungi. (Gerwick 1981; Norris and Fenical 1982; Paul 1985; Paul and Fenical 1987).

Only a limited number of experimental details concerning antimicrobial screening can be found in the chemical literature on diterpenoids from marine algae; this is probably due to the abundance of books and manuals dealing with this subject (see, for instance, Lennette et al. 1985). The following description is aimed at covering only the essential aspects of the methods most frequently used.

Three principal techniques for antimicrobial testing are employed: agar diffusion, agar dilution, and broth dilution. These terms, although widely used, are slightly misleading because the antimicrobial agent, rather than the medium—agar or broth—is diluted. The principle of these methods is to administer known concentrations of the sample under investigation to a microorganism inoculated in a medium, and to observe, after an appropriate incubation period, the growth inhibition in comparison with a blank. Susceptibility tests with antifungal agents are usually more difficult than tests with antibacterial agents due mainly to more specific growing conditions (temperature and incubation time) of fungi in comparison with bacteria. All these methods, normally employed in routine clinical analysis, may be carried out in plant research laboratories on condition that minimal specialist equipment and some personnel trained in microbiological techniques are available. This is particularly recommended if human pathogenic microorganisms are to be employed.

"Bioautographic" techniques are modifications of the sensitivity methods which have proved particularly suitable for use in phytochemical laboratories.

2.2.1 Agar Diffusion Method

The agar diffusion method is the most widely used in non-specialized laboratories; it is considered reliable and accurate even if it gives only semi-quantitative data. In the DST (Disc Sensitivity Test) version, a paper disc impregnated with the substance under test is applied to an agar plate (glass or, preferably, plastic Petri dish) inoculated with one strain. After incubation (overnight for rapidly growing bacteria) the zones of inhibition are measured. This method is a slight modification of that described by Bauer et al. (1966) and may be particularly useful when a large number of samples are to be tested, as, for example, in activity-guided chromatographic fractionations.

2.2.1.1 Sample Preparation

The antimicrobial agent is dissolved in a suitable solvent to a final concentration of 10–100 μg/ml, and the solution (5–10 μl) is applied to a sterile paper disc (usually 6 mm), which is dried at room temperature. Each test is at least duplicated.

2.2.1.2 Selection and Preparation of the Medium

Suitable growing media may be obtained from commercial sources (among others, BBL Microbiology Systems and Difco Laboratories). Dehydrated media should be reconstituted according to the directions given by the manufacturer, and autoclaved (usually at 121 °C for 15 min). Mueller-Hinton agar is recommended for rapidly growing bacteria (Washington 1985), but other suitable media have been used; the pH should be 7.2 to 7.4 after equilibration at room temperature. Air-sensitive compounds have been tested using Thioglycolate medium (Tringali et al. 1989). Fungi are able to grow in many media, for example Yeast Nitrogen Base (YNB) or Sabouraud agar. When bacterial contamination is expected, a commercial antibiotic lacking antifungal activity is added to the medium (for example, 100 μg/ml of gentamicin) (Holt 1975). Marine bacteria and fungi have been cultured on appropriate media (bacteria: Bacto-tryptone, 5 g; yeast extract, 3 g; glycerol 3 ml; Bacto-agar 17 g; deionized water 250 ml; filtered seawater 750 ml; fungi: yeast extract, 0.1 g; glucose, 1 g; filtered seawater 1 l (Gerwick 1981). Freshly prepared and cooled (about 50 °C) media are poured into dishes in order to obtain uniform agar plates (approximately 4 mm depth). The plates are allowed to cool to room temperature, and then stored in a refrigerator, preferably clingfilmed. Just before use, the plates should be maintained for about 15 min in an incubator (35 °C) with their lids ajar to allow excess surface moisture to evaporate.

2.2.1.3 *Preparation of Inoculated Test Plates*

Colonies of a number of microorganisms can be obtained from specialized companies like ATCC (American Type Culture Collection, Rockville, Maryland) or microbiological laboratories. Standardized inocula of pure cultures should be used. For this purpose, small portions of four or five discrete colonies are dispersed into a few ml of a suitable broth medium (for example, Trypticase soy broth). The broth cultures are then allowed to incubate at 35 °C until a turbidity appears (usually 2 to 5 h). This suspension is then adjusted with sterile water or broth until the turbidity is visually comparable to that of a standard prepared by adding 0.5 ml of 11% $BaCl_2$ (11.7 g/l $BaCl_2.2H_2O$) to 99.5 ml of 1% H_2SO_4, agitated on a Vortex mixer immediately before use. This turbidity is normally equivalent to a concentration of microorganisms (bacteria) of 10^7 CFU (Colony Forming Units) per ml. A 1/20 dilution is then prepared with broth and used within 30 min. Since these primary cultures are frequently slow-growing, it is preferable to prepare actively growing subcultures from colonies of the formerly obtained cultures. To inoculate the agar plates an applicator stick (or a cotton swab) is dipped into the standardized suspension and streaked evenly in three planes onto the entire surface of the plate to obtain a uniform culture. Plates should be allowed to stand at least 5 min before use. Inoculation of fungi follows an analogous procedure; to obtain a primary culture, spores may be suspended in a 0.01 M phosphate buffer containing 0.1% Triton ×100 as surfactant (sonication can be useful). For very filamentous fungi, the use of a suitable homogenizer or other modifications (Holt 1975) may be required to prepare the inoculum. After incubation, samples from these cultures are taken off and diluted to approximately 10^5 cells per ml.

2.2.1.4 *Test Procedure*

Within 15 min after inoculation of the plates, paper discs impregnated with the test sample are applied onto the surface of the agar with sterile forceps and gently pressed down to ensure good contact with the surface. Multiple analysis may be carried out in a single plate, provided that the zones of inhibition neither overlap nor reach the edge of the dish. Blank discs treated with the same volume of clean solvent are added. A control disc, containing a known concentration of a commercial antibiotic, should be used if comparison between different sets of assays, separately performed, is desired. Within 15 min after the application of the discs, the plates are incubated at 35 °C. For rapidly growing bacteria or fungi, a period of 16 to 22 h of incubation is used, but some fungi require much longer incubation periods (Ieven et al. 1979; Pesando et al. 1979; Gerwick 1981). During incubation, the test sample diffuses through the agar medium, and a gradient of the concentration of the substance from the edge of the disc is obtained. After incubation, zones of concentration high enough to inhibit microbial growth appear as clear round spots whose diameter can be measured. The size of the inhibition zone is affected by a multiplicity of factors (Barry and Thornsberry 1985) and consequently comparison of data obtained from different laboratories cannot be more than semi-quantitative.

2.2.2 Agar Dilution Method

When quantification is important, dilution methods can be profitably used. The results are expressed as the minimal concentration of the substance required to inhibit (MIC, minimal inhibitory concentration) or kill (MMC, minimal microbicide concentration) a microorganism.

2.2.2.1 Sample Preparation

A known amount of the substance is dissolved in the minimum volume of a suitable organic solvent (DMSO is low toxic for many microorganisms) and a serial dilution with sterile water or broth is prepared according to a fixed schedule. Twofold dilutions are normally used, so that final concentrations follow the scale 1:2, 1:4, 1:8, 1:16, and so on (100 µg/ml is a convenient starting dosage).

2.2.2.2 Preparation of Plates and Test Procedure

Preparation of agar plates is carried out essentially as described above, but test substance is added to the agar just before cooling, and thoroughly mixed. Normally, 1 volume of each concentration of the sample is added to each 9 volumes of agar; the mixture is then quickly poured into the plates. The preparation of the inoculum follows the procedure described above. A single plate can accommodate more than one microbial strain, and this is particularly useful when very small amounts of substance are available. Control plates should also be prepared.

After incubation, the MIC is determined as the lowest concentration of the sample which has completely prevented growth of the microorganism.

2.2.3 Broth Dilution Method

The standard broth serial dilution method is a modification of the agar dilution technique, the main difference being in the use of a liquid medium (broth) instead of the solid agar layer. Therefore, preparation of the sample, medium, and inoculum follows the procedures outlined above. Dilutions are made in sterilized glass tubes stoppered with sterile cotton or aluminum caps, and broth volumes of at least 1.0 ml should be used. After incubation, the turbidities in individual tubes are checked visually or turbidimetrically; the MIC is measured as the lowest concentration of the sample which shows no visible growth. If the MMC (MBC for bacteria) is desired, subcultures from the tubes with no visible growth are inoculated onto agar plates, incubated, and checked for microbial growth. The minimal concentration of the antimicrobial agent in a tube from which no agar subculture has grown is the MMC.

A more recent modification of the method, particularly suitable when the use of small amounts of sample is desired, is the so-called microdilution broth procedure (Jones et al. 1985). In this technique, small multiwell plastic trays are used instead of glass tubes. With the aid of mechanical devices, a rapid and simple determination of the MIC can be carried out on a large number of samples in volumes of 0.05 to 0.2 ml. Turbidity is more accurately observed under the microscope.

2.2.4 Bioautographic Techniques

The methods previously described are slight modifications of techniques routinely used in microbiological laboratories involved in sensitivity tests with commercial antibiotics. Some problems could arise by application of these methods to natural products with unfavorable properties, such as very low water solubility, lability in aqueous media, intense coloration. In addition, some of the above assays are time-consuming and are not particularly suitable for use in a chemical laboratory. Bioautography, a method for the detection of antimicrobial activity on a chromatography, is a rapid and reliable method which gives bioactivity data on an individual compound even for components of mixtures (for instance, chromatographic fractions). In spite of these advantages, bioautography has rarely been used to test marine metabolites, the only report on diterpenoids concerning detection of antifungal activity (Tringali et al. 1986b). This is probably due to the widespread use of the well-established agar diffusion and dilution tests, but also to the fact that only recently have attempts been made to simplify the original procedure for antibacterial bioautography [where a paper or thin layer chromatography is applied onto an agar plate similarly to the paper disc in the above cited DST (Betina 1973)] through direct observation of the chromatographic layer (Hamburger and Cordell 1987).

2.2.4.1 Test Procedure

The sample is prepared as for a usual thin layer chromatography (TLC), spotted on glass or aluminum-backed TLC (normally silica gel) plates and eluted with an appropriate solvent system. If chromatography of the test compound has not been carried out previously, general books on TLC (e.g., Stahl 1969) may be used as a guide. For quantitative evaluation of the results, a weighed amount of the substance should be used. Care should be taken to thoroughly remove the solvent(s) from the plate after developing. The plate is then sprayed with a suspension of the microorganism in a nutrient medium (avoiding excessive wetting) and incubated in a moist atmosphere (a simple way to do this is to put the plate into a Pyrex dish containing moist cheese cloth, covered with a plastic sheet during incubation). The fungus *Cladosporium cucumerinum* requires incubation for 2–3 days at 25 °C, and inhibitory zones appear as white spots on a black background; many other fungi have been used in a similar manner (Homans and Fuchs 1970). The minimum amount (μg) of compound applied to the plate required to cause a visible inhibition of growth is normally reported. When the method is used with microorganisms which do not darken the background after incubation, the zones of inhibition may be visualized with a spray reagent, e.g., a tetrazolium salt (dehydrogenase-activity-detecting reagent) (Hamburger and Cordell 1987).

2.3 Antimicrobial Diterpenoids from Seaweeds

As mentioned above, in spite of the widespread use of antimicrobial assays in the study of algal diterpenoids, no pharmacologically relevant result has been obtained so far. Data through the chemical literature are difficult to compare because only qualitative or semi-quantitative results are normally cited. DST tests are widely used, and positive responses are in many cases obtained with 100 μg/disc of a pure compound. MICs are reported only occasionally, and are frequently beyond the limit for possible pharmacological exploitation. For these reaons, only a rough comparison of activities can be made. Data on antibacterial and antifungal activity of diterpenoids from seaweeds are reported in Tables 1 and 2, respectively, where the active compounds are roughly distinguished as weak/mild (+) and good (++) with respect to a particular microorganism; only compounds active against at least one bacterial or fungal strain have been reported. Forty-one compounds (of a total of 103 herein reported, see Table 8) are reported as active against one or more bacteria, and 32 against at least one fungal strain. Data concerning marine microorganisms (namely *Serratia marinorubra, Vibrio splendida, Vibrio harveyi, Vibrio leiognathi, Vibrio anguillarum, Vibrio* sp., *Alternaria* sp., *Drechsleria haloides, Lindra thalassiae, Dendriphyella salina, Leptosphaeria* sp., *Lulworthia* sp.) are quite interesting, in view of their possible role as defense agents against both direct overwhelming by microbial epibionts (Gerwick 1981; Norris and Fenical 1982) and larval settlement, which is induced more favorably in the presence of a bacterial film (Paul and Fenical 1987). It is worth noting that the isolation of a new antimicrobial compound has, in some cases, opened the way to the isolation and structure elucidation of a group of related metabolites, some of them possessing more promising biological properties [for instance, udoteatrial (**4**) (Nakatsu et al. 1981) and linear diterpenoids from Udoteaceae (**1–17**), pachydictyol A (**31**) (Hirschfeld et al. 1973) and dictyols (**32–40**), epoxyoxodolabelladiene (**47**) (Amico et al. 1980) and dolabellanes (**46–61**)].

Table 1. Antibacterial activity of diterpenoids from seaweeds[a]

Bacteria	Compounds																				
	1	**2**	**3**	**4**	**5**	**6**	**7**	**8**	**9**	**10**	**11**	**12**	**13**	**14**	**15**	**16**	**18**	**19**	**21**	**22**	**23**
Staphylococcus aureus	+	+	+	+	+	–	–		+	+	+		+	+	+	–	+		++	+	–
Bacillus subtilis	+	+	+		+	–	–		+	+	–		+		+	+	+		++	+	+
Escherichia coli																	+	+			–
Serratia marinorubra						+	+			–	–		+		–						
Vibrio splendida						+	+			–	+		+	+	–						
Vibrio harveyi	+	+	+		+	+	+	+	–	–	+	+	+	–	+	+	+				
Vibrio leiognathi	+	–	+		+	+	+	+		+	+	–	+	+	–						
Vibrio anguillarum																	+				
Vibrio sp.		+					+			+	+		+	+	+	–	+				

Bacteria	Compounds																				
	31	**33**	**34**	**35**	**36**	**44**	**46**	**47**	**48**	**49**	**50**	**60**	**61**	**66**	**67**	**85**	**86**	**87**	**88**	**89**	**102**
Staphylococcus aureus	+	++	+	+	+	+	+	–	+	+	–	++	+	+	+	–	+	++	+	+	
Bacillus subtilis			+	+	+													++	–	+	++
Escherichia coli		–	+	+	+		+	++	–	++	–					–		++	–	++	++
Micrococcus luteus			+	+	+	+											+	++	–		
Mycobacterium smegmatis			+	+	+																
Pseudomonas aeruginosa		–	–				–	–	–	–	–					+				++	
Klebsiella pneumoniae		–	+				++	+	–	–	+					–					
Proteus mirabilis								–	–	–	–							++	–		
Enterobacter cloacae		–	–				++	–	+	–	+										
Citrobacter freundii		–	+				++	–	++	++	–										
Aeromonas hydrophyla						+											+	++	–		
Enterobacter aerogenes																				+	

[a]Measured as inhibition of bacterial growth; see Section 2.5.5 for details. ++ = good activity, + = mild or weak activity; — = no activity.

Table 2. Antifungal activity of diterpenoids from seaweeds[a]

Fungi	Compounds															
	1	**3**	**4**	**5**	**6**	**7**	**10**	**11**	**13**	**14**	**15**	**21**	**28**	**31**	**32**	**34**
Candida albicans	–	–	+	–	–	–	+	–	+	–	–	+				
Alternaria sp.	+	–		+	+	+	–	–	+	+	+		+	–	–	–
Drechsleria haloides	–			–			–	+	+	+	–					
Lindra thalassiae	–	+		–	+				+		–			–	–	+
Dendriphyella salina														++	–	–
Leptosphaeria sp.	–	–		–	–	–	+	–	–	+	–					
Lulworthia sp.	–	–		–	–	+	–	+	+	+	+			–	+	+

Fungi	Compounds															
	35	**36**	**41**	**42**	**45**	**47**	**54**	**55**	**56**	**57**	**58**	**60**	**61**	**63**	**66**	**67**
Candida albicans	+	+								–		–	–			
Cladosporium cucumerinum						++	+	+	+	+	+	–	–			
Serratia marcescens										–		++	+			
Aspergillus niger										+		–	+			
Mucor racemus										+		++	++			
Mucor mucedo															+	+
Alternaria sp.			+	+	+									–		
Lindra thalassiae			–	–	+									–		
Dendriphyella salina			–	–	+									–		
Lulworthia sp.			+	–	+									+		

[a]Measured as inhibition of fungal growth; see Sect. 2.2.5 for details. ++ = good activity; + = mild or weak activity; — = no activity.

3 Antialgal Activity

3.1 Background

In spite of commercial interest in effective anti-fouling compounds, a few reports concerning growth inhibitors of surface-fouling organisms have appeared to date (Paul 1988 and references cited therein). The observation has been made that many marine plants and animals are remarkably free of encrustation by other organisms (Hornsey and Hide 1974), and this may be due to surface diffusion of anti-fouling substances. A possible role of some antibiotic compounds produced by seaweeds as fouling retardants has been proposed (Sieburth and Conover 1965; Al-Ogily and Knight-Jones 1977). The activity of algal metabolites against marine bacteria and fungi has been mentioned above; Gerwick (1981) has investigated a number of metabolites (including diterpenoids) isolated from Dictyotaceae seaweeds for activity against microalgae. The inhibition of the growth of the pennate marine diatom *Phaeodactylum tricornutum* has been evaluated, and the method is here reported.

3.2 Description of the Method

The method is based on the determination of the inhibition of the algal growth in the presence of the test substance, with respect to a control.

3.2.1 Preparation of the Culture

An axenic culture of *P. tricornutum* is used to inoculate a culture medium of the following composition: filtered seawater 1 l; KNO_3, 500 mg; ferric ammonium citrate green 5 mg; K_2HPO_4: 100 mg. Before use, the media is tyndallized by repeated boiling on 3 subsequent days.

3.2.2 Test Procedure

The sample is dissolved in ethanol at a concentration of 40 μg/μl, and 2.5 μl are added to 1 ml of the medium in a 5 ml clear glass culture tube. In a control experiment, the concentration limit of ethanol which does not inhibit algal growth has been assessed as 0.6% (concentrations used in the assay are 0.25%). Controls are run simultaneously containing both the medium and pure 95% ethanol. One drop of inoculum is added and the tubes are covered and maintained at 22 °C for 4 days under cool white light (continuous light). After this period, the diatom growth is analyzed by means of a hemocytometer; the diatoms present in a 0.6 mm^2 area are counted both for control and test tubes. Normal growth (control) gives 30–50 individual diatom per area, i.e., 6–7 × 10^5 cells/ml). Inhibition is considered strong when less than 7.4 × 10^4 cells/ml are counted, and regarded as mild with a computation range of 8.6×10^4 to 1.7×10^5. Changes in the morphology and locomotion of the diatom cells have been observed in most assays with active compounds.

3.3 Antialgal Diterpenoids from Seaweeds

Table 3 reports the data available on the inhibition of algal growth (*P. tricornutum*) by diterpenoids from seaweeds. All data concern metabolites isolated from brown algae of the family Dictyotaceae. Only compounds which gave positive results are reported (less than 2.5×10^5 cells/ml). The final concentration of the sample in the assay medium was 100 μg/ml. The original list of compounds tested (Gerwick 1981) included 37 metabolites, 15 of which (41%) showed reasonably strong inhibition, and only 4 (10%) proved inactive. These results may have an ecological significance (inhibition of diatom growth on the surface of macroalgae), but in any case encourage the increased use of this, or similar, assay in the search for new anti-fouling agents.

Table 3. Antialgal activity of diterpenoids from seaweeds[a]

Compound	**21**	**28**	**30**	**31**	**32**	**33**	**34**	**41**	**42**	**43**	**45**	**63**	**68**	**69**	**70**
Cell count[b]	6.2	14.8	7.4	9.9	4.9	8.6	7.4	4.9	4.9	17.3	7.4	16.0	4.9	11.1	14.8

Compound	**71**	**72**	**73**	**74**	**75**	**76**	**77**	**78**	**79**	**80**	**83**	**95**	**97**	**98**	**101**
Cell count[b]	9.9	4.9	9.9	11.1	12.3	7.4	7.4	9.9	16.0	7.4	0	3.7	11.1	8.6	17.3

[a]Measured as inhibition of the growth of the marine diatom *Phaeodactylum tricornutum* at 100 μg/ml; see Sects. 3.2.2 and 3.3 for details.
[b]Times 10^4 (cells/ml).

4 Cytotoxic Activity and Other Related Activities

4.1 Background

Today, several clinically useful anticancer drugs are available, but adequate chemotherapy for adult solid tumors is still unavailable. Consequently, the search for more powerful antitumoral compounds is very active, and it is likely that a number of the drugs of the future will probably be natural products, or their synthetic analogs with enhanced activity (Munro et al. 1987; Suffness and Thompson 1988). So far, efforts to obtain new anticancer products by way of a "drug design" based on the molecular biology of the disease (the so-called rationale approach) have been less fruitful than the screening, using biological model systems, of a large number of substances (the "empirical" approach). Most of the clinically useful compounds presently available have been discovered as the result of screening programs (Suffness and Thompson 1988). Hundreds of thousands of both crude extracts and pure compounds, in part derived from marine organisms, have been tested by the National Cancer Institute (Bethesda, USA), the most important organization in the World for cancer study and therapy (Douros and Suffness 1981; Suffness and Douros 1982; Suffness and Thompson 1988). Similar screenings have been carried out by other research groups, with positive results for extracts of marine algae (Kashiwagi et al. 1980; Rinehart et al. 1981; Yamamoto et al. 1982, 1984; Hodgson 1984; Patterson et al. 1984). A number of pure bioactive compounds have been isolated, including several algal diterpenoids, whose properties will be discussed in Section 4.3.

In order to avoid confusion in the literature, NCI has recommended an appropriate use of terms describing bioactivity (Suffness and Douros 1982); *cytotoxicity* is defined as in vitro toxicity to tumor cells. The terms *antitumor* or *antineoplastic* should be strictly used in referring to in vivo experimental results and the term *anticancer* should be adopted for reporting data from clinical trials on humans. This distinction emphasizes that cytotoxic compounds include "cell poisons" without any selectivity toward tumors; there is no way in which these substances could be included in anticancer formulations, even though they could be useful as molecular probes in biochemical studies.

4.2 Description of the Methods

For a long time NCI has used in vivo tests (particularly P388 leukemia in mice) for the screening of crude plant and animal extracts and in vitro bioassays (for example using cells of the human nasopharyngeal carcinoma, KB) as a guide during the purification of the active principle(s) (Suffness and Douros 1982). Recently, NCI has developed a new approach; substances are tested in a panel of more than 60 different cell lines, representing slow-growing human solid tumors with several subtypes for each tumor type and more than one cell line for each subtype. Compounds showing significantly selective cytotoxicity are then subjected to in vivo tests using the same tumor lines as xenografts in athymic mice (Suffness and Thompson 1988). The amount of compound currently needed for routine analysis is 10 mg. The results of the in vitro tests are expressed as ED_{50} (= ID_{50}), i.e., the effective dose which reduces cell growth to 50% with respect to the control; for pure compounds an ED_{50} value less than 10 μg/ml is sufficient to invite further testing (Munro et al. 1987). The amount of sample required for in vivo testing of pure compounds is dependent on the starting dosage (mg/kg), and ranges generally from 100 to 200 mg. Results are normally given in percent as the ratio of the mean survival time of the treated group, compared to that of the control group (T/C); when P388 tests are used, T/C $\geq$ 120% is the minimum value for a compound to be considered active; compounds with T/C $\geq$ 150% become candidates for clinical trials (Munro et al. 1987).

The in vivo or in vitro tests mentioned above are rarely performed in phytochemical laboratories; the former require maintenance of animals and considerable quantities of substance for each trial; both are time-consuming and require carefully controlled conditions and should be executed by highly skilled personnel. Since generally these assays are carried out in organizations like NCI or, under contract or cooperative programs, by companies or specialized laboratories, details about these analyses will be limited here to one of the most frequently used bioassays in vitro, employing KB tumor cells. Some alternative methods have been used by chemists involved in the study of natural products as in-house tests related to antitumoral activity; the sea urchin eggs bioassay has been largely used to test compounds from marine organisms and is described in Section 4.2.2; the brine shrimp (Meyer et al. 1982) and crown-gall potato disc (Galsky et al. 1981) bioassays are useful tests, to date only occasionally employed in the marine field, and not reported here in detail.

4.2.1 KB Cytotoxicity Assay

One of the most widely used cytotoxicity assays employs the KB tumor cell culture. This assay, originally developed by Oyama and Eagle (1956), has been revised by Smith et al. (1959), and is reported in a NCI protocol (National Cancer Institute 1972). It is based on the measurement of percent cells surviving when treated with the test substance, in comparison with a control. The procedure described in the following is largely based on that by Nemanich et al. (1978), used in the course of a screening program on cytotoxic compounds in marine organisms.

4.2.1.1 Sample and Culture Preparation

A carefully weighed sample amount (approximately 1 mg) in a screw-top sterile tube is dissolved (or suspended) in growth medium (e.g., Dulbecco's Modified Eagle's Medium, Flow Laboratories), prepared with some appropriate additives (e.g., 5% v/v fetal calf serum and 1% Fungizone, a mixture of penicillin, streptomycin, and amphotericin, ISI Biological) to obtain a final concentration of 100 µg/ml. If the expected value for ED_{50} is lower than 100 µg/ml, the solution is serially diluted. KB cells, obtained from specialized companies or cell biology laboratories, are placed in tissue culture plates containing the growth medium (12 ml for 100 × 15 mm plates) and incubated at 37 °C in a 5% CO_2 high-humidity incubator. Cultures are maintained in continous logarithmic growth; a cell counter is used for quantitation.

4.2.1.2 Test Procedure

Assays are performed in multiwell tissue culture plates; culture broth (approx. 2 ml) containg $4–6 \times 10^4$ cells/ml is put into each sample well; after 24 h incubation, the medium is replaced by the same volume of the suspension with test substance. A number N of control plates without sample material are also prepared, according to the NCI recommendation, N being determined by the number of samples being assayed, n, so that $N = 2\sqrt{n}$ (National Cancer Institute 1972). A positive control, containing a known amount of a cytotoxic compound (e.g., vinblastine) should also be used. Test plates are incubated at 37 °C, 5% CO_2, controlled humidity for 3 days. At the end of the incubation period, the surviving cells are counted with the cell counter and the percentage of the surviving cells is measured as the ratio of the number of cells in a test plate to the average number of cells in the control plates, multiplied by 100. Activity of the sample tested is expressed as ED_{50} (µg/ml). Morphological changes associated with cytotoxicity may be monitored under microscopic examination; alterations in the cell membrane, cytoplasm, and nucleus are observed, depending on the mode of action of the substance under test.

4.2.2 Sea Urchin Egg Assay

The use of tests employing mammal cells in the search for antitumoral compounds has the above-mentioned disadvantages, and in addition cytotoxicity assays have been considered nonspecific in that they may lead to broad spectrum poisons, most often toxic toward mammals (Fusetani 1987). Alternative methods should combine simplicity and selectivity in their anticancer activities.

Fertilized sea urchin egg is well known to biologists as a convenient model for the study of cell proliferation, and is presently used in pharmacological and toxicological investigation (Allemand et al. 1989; Pesando et al. 1989). Inhibition of sea urchin egg cleavage has been used by Cornman (1950) in carcinogenesis studies, and Ruggieri and Nigrelli (1960) were probably the first to use this system to test the activity of a marine metabolite; more recently, Jacobs et al. (1981) have applied the sea urchin egg assay to screen a number of marine compounds for potential antitumor activity. The sea urchin embryo provides an easy-to-study system because of large cell size (100 µm) and short division cycle (1–2 h) (Jacobs et al. 1985); other advantages are that simultaneously fertilized eggs undergo synchronous cleavages, the effect of the test substance on cell division may be easily monitored, and special equipment is not required. The method is based on observation of the cell division process in recently fertilized eggs after administration of the test substance; this system permits the detection of selective antimitotic agents such as DNA or RNA synthesis inhibitors, microtubule assembly inhibitors, and protein synthesis inhibitors (Ikegami et al. 1979; Fusetani 1987); subsequent antitumoral testing of such agents may provide compounds suitable for the development of clinical trials. Among antineoplastic compounds, antimitotic agents are important because they are usually not mutagenic, unlike many alkylating agents or DNA interactive agents (Suffness and Douros 1982). On the other hand, the selectivity of this assay is a disadvantage in a search for antitumorals other than antimitotics; indeed, it is not sensitive to several classes of antineoplastic agents (Jacobs et al. 1981). A closely related assay, as yet never applied to any algal diterpenoid, is that proposed by Ikegami et al. (1979), which employs starfish oocytes.

4.2.2.1 Sample Preparation

An exact amount of the test substance is dissolved in seawater or, if highly lipophylic, in a water-miscible solvent like propylene glycol (Jacobs et al. 1981), EtOH (Gerwick 1981) or DMSO (Ikegami et al. 1979). EtOH is inert up to 25 µl/3 ml seawater, while DMSO does not affect biological assay at a concentration of 0.1%. In initial screening of pure compounds a dose of 50 µg/ml is usually a convenient one.

4.2.2.2 Preparation of Biological Material

Male and female adult sea urchins (*Lytechinus pictus* is one of the most frequently used species, but *Strongylocentrotus purpuratus* and *Paracentrotus lividus* have also been used) are maintained in separate seawater aquaria; sea lettuce may be used as food. Spawning is induced by intracoelomic injection of a small amount of 0.5 M KCl. Just before use, a few drops (0.2 ml) of sperm are diluted with seawater (10 cc). Eggs are washed three times with seawater and volumetrically dispensed into small vials or plastic Petri dishes, and treated with sperm. The occurrence of formed fertilization membranes to the 90% level is used as a criterion to proceed to the test.

4.2.2.3 Test Procedure

Five minutes after insemination, sea urchin embryo are treated with the sample solution and incubated for 1.5 h at room temperature. The cells should be frequently agitated to minimize settling, avoid contact inhibition, and improve sample dispersion. Controls both with and without the same volume of the solvent are also incubated. In the method reported by Jacobs et al. (1981; 1985), when the controls undergo the first division, the test vials are examined microscopically to determine the percentage of embryos showing complete division; a substance is considered active when it inhibits 80 to 100% of cell cleavage at a concentration of 16 µg/ml. Assays performed by Fenical and coworkers were originally described in the PhD Dissertations of Gerwick (1981) and Paul (1985). Data reported by Gerwick refers to observations made during the period when control embryos undergo division from 8 to 16/32 cells; the actual ED_{50} values are determined graphically from the data points obtained for a series of different concentrations of the sample under the 100% inhibition value. Paul reports ED_{100} values (the lowest effective concentration to cause 100% cleavage inhibition) obtained testing all metabolites at several concentrations (16, 8, 4, 2, 1 µg/ml).

4.3 Cytotoxic and Antimitotic Diterpenoids from Seaweeds

Data on some classes of bioactive compounds from marine organisms, including cytotoxic and antimitotic diterpenoids from marine algae, have been recently summarized by Fusetani (1987) and Munro et al. (1987). A more complete list of algal diterpenoids is reported in Tables 4 and 5, concerning respectively the activity toward KB or other tumor cells in culture (A549, human lung; B16-F10, murine melanoma; CBS, human colon; HCT-116, human colon; SW1271, human lung) and toward fertilized sea urchin eggs. Only positive results are here listed. High cytotoxic activity toward B16 mouse melanoma cells is shown by the xenicane and norxenicane lactons **24–27** (Ishitsuka et al. 1988); seven members of the interesting class of diterpenoids known as dolabellanes (**46–61**) have been

Table 4. Cytotoxic activity of diterpenoids from seaweeds[a]

Cell cultures[b]	Compounds													
	20	**24**	**25**	**26**	**27**	**33**	**39**	**48**	**54**	**55**	**57**	**58**	**59**	**60**
KB	1.5					20	22	40		10				
A549									23	30	>62.5	18.3	22	21.3
B16-F10		1.57	2.57	0.58	1.58				18	22	20.5	18.5	9.1	14.5
CBS									16.5	21.3	26.5	11.1	8.2	10.5
HCT-116									6.2	21.5	24	6.8	10	24
SW1271									21	24.5	33.5	16.5	17.6	21

[a]Measured as inhibition of tumoral cell growth; values are ED$_{50}$, µg/ml; see Sects. 4.2.1 and 4.3 for details.
[b]KB = human nasopharynx; A549 = human lung; B16–F10 = murine melanoma; CBS = human colon; HCT–116 = human colon; SW1271 = human lung.

Table 5. Antimitotic activity of diterpenoids from seaweeds[a]

Compound	**1**	**2**	**3**	**5**	**6**	**7**	**8**	**9**	**10**	**11**	**12**	**13**	**14**	**15**	**21**	**31**	**32**	**33**
ED$_{100}$[b,c]	8	8	8	8	1	1	16	8	16	16	8	1	16	8				
% div[b,d]															0	75	70	75
Compound	**34**	**41**	**42**	**43**	**45**	**68**	**70**	**77**	**84**	**89**	**90**	**91**	**92**	**93**	**95**	**97**	**98**	**101**
ED$_{50}$[c]						1.2[e]	16[e]			4[f]	2[g]	2[g]	2[g]	2[g]				1.1[f]
% div[b,d]	50	60	60	50	0	0	50	50	75						60	5	50	0

[a]Measured as inhibition of fertilized sea urchin cell cleavage, see Sects. 4.2.2 and 4.3 for details.
[b]Data within the same row refer to embryos of *Lytechinus pictus*.
[c]Values are reported in µg/ml.
[d]Values are reported as percentages of divided embryos at 50 µg/ml.
[e]Data refers to embryos of *Lytechinus pictus*.
[f]Data refers to embryos of *Strongylocentrotus purpuratus*.
[g]Data refers to embryos of *Paracentrotus lividus*.

tested against several tumoral cell lines, and significant activities ($ED_{50} \leq 10$ µg/ml) have been registered, the lower ED_{50} values being observed for the two isomeric acetoxy-aldehydes **54** and **58** (C. Tringali unpubl.); activities toward T242 melanoma and 224C astrocytoma neoplastic cell lines have been reported for the unusual tricyclic diepoxide spatol (**68**) (Gerwick and Fenical 1983), not included in Table 4 because of the lack of quantitative data. Since data on sea urchin egg bioassays are reported as ED_{100}, ED_{50} or percent divided cells, Table 5 is accordingly subdivided into three groups. Strong antimitotic activity is shown by the polyhaldehydes **6** (petiodial), **7** and **13** (halimedatrial) (Paul and Fenical 1987), with $ED_{100} = 1$ µg/ml, spatol (**68**), $ED_{50} = 1.2$ µg/ml) and by a diterpenoid of mixed biogenesis, stypoldione (**101**, $ED_{50} = 1.1$ µg/ml) (Gerwick and Fenical 1981; Jacobs et al. 1985); the latter compound is also a very effective inhibitor of bovine brain microtubule protein polymerization (O'Brien et al. 1983, 1984) and a potent ichthyotoxin (see Sect. 5.3).

To date, no diterpenoid isolated from marine algae appears to be a candidate for anticancer clinical trials; nevertheless, some interesting compounds have been isolated which may be useful tools in biochemical studies, and if one takes into account the limited number of compounds tested, bioassay-oriented research in the future will probably lead to potentially useful new antineoplastic compounds.

5 Ichthyotoxicity and Other Defensive Bioactivities

5.1 Background

Studies aimed at determining the role of secondary metabolites in the marine ecosystem have been growing only in recent years, and little is presently known about the functions that most of the compounds isolated from marine algae may play in the environment. The hypothesis has been put forward that sessile organisms like seaweeds or a number of benthic animals, particularly if they lack obvious mechanical devices, may have evolved defense systems based on the production of toxic or repellent metabolites which can enhance survival by discouraging predation (Baker 1976; Vadas 1979). Data supporting this hypothesis have been mounting in the last decades, due particularly to cooperative work between chemists and biologists (Paul 1988). The relationship between toxicity and fish feeding behavior has been discussed by Bakus (1981), who suggests that marine fish may learn to avoid noxious organisms by trial and error feeding. Fenical and coworkers examined chemical defense in tropical seaweeds (Sun and Fenical 1979; Gerwick 1981; Norris and Fenical 1982; Paul and Fenical 1983, 1987; Paul 1985); despite intense herbivory, certain tropical and subtropical algae grow apparently undisturbed and this has been related to distinct feeding preferences by herbivore predators observed in field and laboratory. A number of marine metabolites, among others algal diterpenoids, have been examined in laboratory trials for fish toxicity and feeding deterrence, and this section is mainly devoted to these tests; other reported activities of ecological importance, involving diterpenoids, are those toward sea urchin sperm and larvae; the relevant assay methods are briefly described in the sequel as a complement to the method of Sect. 4.2.2 (sea urchin egg assay). Toxicity against marine molluscan (abalone) larvae has also been assayed in a limited number of cases (Paul 1985; Kurata et al. 1988a,b).

5.2 Description of the Methods

Toxicity assays are based on the observation of acute toxicity symptoms in organisms maintained for a fixed period of time in a solution of the sample in seawater. In feeding deterrence assays the feeding behavior of marine fish is examined toward food pellets impregnated with the test compound.

5.2.1 Ichthyotoxicity Assay

Piscicidal or, more broadly, ichthyotoxic activity is routinely evaluated to assess toxicity of commercial products (e.g., pesticides) which may pollute fresh or seawaters; in the standard methods a cheap and easily available fish like the goldfish *Carassius auratus* is employed, and a contact method is used, examining whether fish are killed by a substance added to the water. Assays on marine products are essentially similar, but are frequently carried out on locally available marine fish. Tropical herbivorous fish (*Eupomacentrus leucostictus, Pomacentrus coeruleus*, and *Dascyllus aruanus*) have been more frequently used; *Gambusia patruelis* has been employed to test dolabellanes **52–54**; *C. auratus* can also be used, since a good correlation exists between toxicity to goldfish and other marine herbivorous fish (Bakus 1981; Gerwick 1981).

5.2.1.1 Sample Preparation

The test sample is thought to be absorbed through the gills directly into the blood of the fish; the test compound (about 2 mg) is stirred into seawater using a small amount of a carrier solvent (e.g., 25 to 100 µl EtOH in about 200 ml seawater) and this solution is diluted to various concentrations, for example with twofold serial dilution. For compounds not previously tested, starting concentrations ranging from 100 to 10 µg/ml have been used. Sample solutions are prepared in beakers of suitable size to accommodate one fish (less frequently two); each test is replicated (three or four times in most cases). The lack of aeration during the test causes a negligible effect on fish behavior or health. To economize on rare natural products, it has proved convenient to use a small goldfish for each test, which can be accommodated in a 50-ml beaker containing 20 ml of solution (Gerwick 1981).

5.2.1.2 Test Procedure

Fish are put in each beaker and observed for 1 h (in most cases); controls with and without solvent are run simultaneously. If death occurs within this period of time, the sample is defined as toxic at the concentration used; ED_{100} values (μg/ml) express the lowest concentration causing death. Other acute toxication symptoms may be observed; a compound is defined *moderately toxic* if the fish appears nearly dead by the end of the assay (e.g., lying on its side on the bottom of the beaker); *narcotic* if the fish shows definite signs of lethargy; *hyperactive agent* if the fish swims and gasps for air much more energetically than the control (Gerwick 1981). Loss of righting reflex has also been used as a criterion to assess toxicity (Schlenk and Gerwick 1987). If death does not occur within 1 h, the fish is returned to clean seawater and chronic effects are monitored; in this condition death may occur several hours later (Sun and Fenical 1979).

5.2.2 Sea Urchin Sperm and Larvae Toxicity Assay

Sea urchin grazing is one of the environmental factors affecting the distribution of marine algae (Dart 1972; Carpenter 1981); seaweed metabolites which may interfere, in natural concentrations, with the life cycle of sea urchins may be regarded as playing a defensive role for the alga producing them. In this connection, the above-mentioned cleavage inhibition of sea urchin embryo may have an ecological significance. (See Sect. 4.2.2). In addition, toxicity toward sea urchin sperm and pluteus larvae has been evaluated by Paul and Fenical (1987) for the terpenoids produced by Caulerpaceae and Udoteaceae green algae.

5.2.2.1 Sea Urchin Sperm Assay Procedure

A known amount of the sample, dissolved in seawater with the aid of a carrier solvent (EtOH), is serially diluted starting from a concentration of 16 μg/ml. Sperm of the sea urchin *Lytechinus pictus*, obtained as outlined in Sect. 4.2.2, is used at an approximate concentration of 3×10^7 sperm/ml, spectrophotometrically determined (Vacquier and Payne 1973). Sperm is left in contact with the sample solutions for 30 min; solvent and seawater controls are also assayed. Toxicity is determined microscopically as the complete loss of flagellar motility. Replicate tests are carried out to measure the lowest effective dose, ED_{100}.

5.2.2.2 Sea Urchin Larvae Assay Procedure

Sample solutions at known concentrations are prepared as above. Approximately 20–30 pluteus larvae (36 h after fertilization) of *L. pictus* for each test are treated with the sample solution and observed after 1 h (acute toxicity) and 24 h under the microscope; seawater and solvent (EtOH) controls are run simultaneously. Toxicity is defined as 100% inhibition of larval swimming and ciliary motion (ED_{100}, μg/ml). Replicate tests for each metabolite were carried out.

5.2.3 Feeding Deterrence Assay

A large number of works have been devoted to terrestrial natural products possessing antifeedant properties. In contrast, much less is known about feeding deterrence in the marine ecosystem, especially as regards algal adaptation against herbivores. The majority of algal terpenoids tested for antifeedant properties were isolated from green algae, and are included in a recent review (Paul and Fenical 1987). A complete examination of the parameters which may be evaluated in an ecological study on food avoidance properties in the marine environment is a complex work, requiring both laboratory and field experiments; anyway, simple in-home assays may give a reliable indication of the feeding deterrence activity of the test sample. Various similar assay methods have been proposed; the following is to a large extent based on the reports of Paul and Fenical.

5.2.3.1 Test Procedure

The test sample is dissolved in diethyl ether at a known concentration close to the natural one (2000–5000 ppm) and applied volumetrically to pellets of fish food (20 mg); control pellets are impregnated with the same volume of the solvent, which is evaporated at room temperature. A tank containing six to eight locally available fish is used for each test. Damselfish, *Pomacentrus coeruleus* and *Eupomacentrus leucostictus*, have been used in separate experiments [also the rabbitfish *Siganus spinus* has been used to test compound **17** (Paul et al. 1988)]; the omnivorous fish *Tilapia mossambica* has been employed to test hydroxydictyodial **4** at 1% concentration (Tanaka and Higa 1984). Ten treated and ten control food pellets are added randomly to the tank, and the number of bites taken of treated and control pellets by individual fish is counted; the results are analyzed statistically with the Mann-Whitney Test or other methods and are finally expressed as active deterrent or not active deterrent (at 5000 ppm).

5.3 Ichthyotoxic and Other Defensive Diterpenoids from Seaweeds

Tables 6 and 7 summarize the data on algal diterpenoids reported to be respectively ichthyotoxic and toxic to sea urchin sperm and larvae. Data on ichthyotoxic metabolites in the original literature are reported in either quantitative or qualitative form, the latter specified by Gerwick (1981) as outlined above, ranging from toxic to weak narcotic activity, measured at 10 ppm. Due to the different criteria used to express these activities, in Table 6 only a rough distinction between strongly toxic (+++), toxic (++) (at least for one of the fish species used) and moderately toxic or narcotic, etc. (+) is used. Mention should be made of the toxic properties of stypotriol (**99**, $ED_{50} = 0.2$ μg/ml), stypoldione (**101**, $ED_{50} = 1$ μg/ml) (Gerwick and Fenical 1983) and some aldehydes from Udoteaceae (**3, 6, 7,** and **13**, $ED_{100} = 5$ μg/ml) (Paul and Fenical 1987). Table 7 concerns data on toxicity of algal diterpenoids toward sperm and larvae of the sea urchin *Lytechinus pictus*, as reported by Paul and Fenical (1987); the original table included also sesquiterpenoid metabolites. All the compounds here reported have been isolated from green algae of the family Udoteaceae; data are reported as the 100% lowest effective dose, ED_{100}; for larval toxicity, ED_{100} after 1 h and 24 h were determined.

Values ranging from 16 to 0.2 μg/ml have been obtained, and only one of the diterpenoids tested proved inactive at 16 μg/ml. Literature data on algal diterpenoids showing fish feeding deterrence properties are limited to seven compounds, namely **5, 6, 7, 13, 15, 17,** and **22**. The concentration used in the test is 2000–5000 ppm for compounds **5–17**, and 10 000 ppm for idroxydictyodial **22**. Activity against the swimming larvae of the abalone *Haliotis discus* has been reported for compounds **15** (Paul 1985), **40** (Kurata et al. 1988a), **81**, and **82** (Kurata et al. 1988b).

Table 6. Ichthyotoxicity of diterpenoids from seaweeds[a]

Compound	**2**	**3**	**6**	**7**	**12**	**13**	**14**	**23**	**28**	**29**	**31**	**32**	**33**	**34**	**41**	**42**	**43**	**45**	**51**	**52**
Toxicity	++	++	++	++	++	++	++	+	++	++	+	++	++	++	++	++	++	++	++	++
Compound	**53**	**63**	**68**	**70**	**71**	**74**	**76**	**77**	**78**	**79**	**80**	**83**	**84**	**94**	**95**	**96**	**98**	**99**	**100**	**101**
Toxicity	++	++	++	++	++	++	+	++	++	++	+	+	+	+	+	+	+	+++	++	+++

[a]Measured in the range 0.2–20 μg/ml. See Sects. 5.2.1 and 5.3 for details.

Table 7. Activity against sea urchin sperm and larvae of diterpenoids from seaweeds[a]

Compound	**1**	**2**	**3**	**5**	**6**	**7**	**8**	**9**	**10**	**11**	**12**	**13**	**14**	**15**
Sperm toxicity[b]	-[c]	8	4	8	1	1	8	8	16	8	4	1	8	8
Larval toxicity (1 h)[d]	8	4	4	8	nt[e]	0.5	8	16	16	4	2	1	8	8
Larval toxicity (24 h)[f]	8	2	4	0.2	nt[e]	8	8	2	0.2	2	8	4	4	16

[a]See Sects. 5.2.2 and 5.3 for details; values are reported as ED_{100} (μg/ml).
[b]Measured for *Lytechinus pictus* sperm after 30 min exposure.
[c]Not active at 16 μg/ml.
[d]Measured for *L. pictus* larvae after 1 h exposure.
[e]Not tested.
[f]Measured for *L. pictus* larvae after 24 h exposure.

6 Molluscicidal Activity

6.1 Background

Schistosomiasis (bilharziasis) afflicts more than 200 million people in the world, particularly in the tropics. The etiological agents are blood flukes (schistosomes). The disease is transmitted by some snails, most notably of the genera *Biomphalaria, Bulinus*, and *Oncomelania*, which are intermediate hosts on which schistosome larvae grow and subsequently diffuse through the water as fork-tailed cercariae; these cercariae can penetrate the skin and establish infection in man and animals. At present, chemotherapy is used against this parasitic disease, but it is not a conclusive remedy against the infection (Marston and Hostettmann 1985). A useful tactic in combating the diffusion of the disease is to interrupt the life cycle of the schistosomes by killing the carrier snails, but molluscicides which are more selective against the target host and less toxic toward other living species are required (World Health Organization 1965; Marston and Hostettmann 1985). Since the discovery of molluscicidal agents in *Phytolacca dodecandra* and *Tetrapleura tetraptera*, plant molluscicides are receiving increasing attention because of their presumably lower costs and readier availability than synthetic chemicals; moreover, their quick degradation is favorable considering the widespread rigorous legislation concerning pesticides (Kloos and McCullough 1982). Molluscicidal properties of several natural products have been reported, including saponins, terpenoids, and a number of other compounds. (Marston and Hostettmann 1985; Hostettmann and Marston 1987). Marine organisms, with the exception of a single work concerning diterpenoids from the Dictyotaceae seaweed family (Tringali et al. 1986b), have not been examined for this activity, partly due to the fact that the source of the active principles should preferably be available in the endemic areas of the disease. A broader examination of the local marine flora (where present) may lead to other more unusual molluscicides and furnish candidates for application in the field. Moreover, this activity may have some ecological significance in the protection of algae by herbivorous molluscs (see Sect. 5.1). The methods reported in the literature are derived, with minor modifications, from a WHO Memorandum (World Health Organization 1965), and are based on observations of the survival of snails subjected to the test substance. The assay detailed in the sequel is simple, rapid, and requires no special equipment.

6.2 Description of the Method

6.2.1 Sample Preparation

The molluscicides may act on snails by ingestion, or, more likely, by absorption through the external membranes; thus the sample has to be brought into aqueous solution. A weighed amount of the test sample is dispersed in water by means of an ultra-sonic bath (if necessary); highly lipophilic substances may be dissolved in the minimum volume of a solvent such as DMSO (Duncan and Sturrock 1987) and the solution diluted with water. A tenfold dilution series is generally used, including 1 and 10 ppm values. In some cases, a twofold series is further used after determination of the critical range. Tests are carried out in dechlorinated tap water solutions; distilled water is not recommended because there is evidence that it significantly reduces the amount of substance taken up by snails (Duncan and Sturrock 1987).

6.2.2 Preparation of the Biological Material

Reared snails are used as test organisms; the American species *Biomphalaria glabrata* is frequently employed and kept in aquaria at 24 °C, with a continuous circulation of water through an Eheim Filter System. Young mature snails of uniform size as far as possible should be used; 9 mm is the appropriate average shell diameter for *B. glabrata.*

6.2.3 Test Procedure

Snails are placed in dechlorinated water solutions containing a known concentration of test sample. Ten snails per test should be used; each test is at least duplicated; the correct volume should be higher than 40 ml per snail. These requirements are reduced if a very limited amount of sample is available for the assay. Laboratory lighting with normal diurnal light cycles is used. Snail control in dechlorinated water and molluscicide control using a reference compound are carried out simultaneously; Bayluscide (Bayer), sodium pentachlorophenate (NaPCP, Monsanto and others), or copper sulfate may be used for this purpose (World Health Organization 1965). At regular intervals the snails are placed on a Petri dish lit from below, and examined for heart beat by a microscope. (Nakanishi and Kubo 1977). Dead snails are immobile and the body and shell may be discolored; confirmation of death may be obtained by lack of reaction to prodding the body. Dead snails should be removed as soon as possible. A 24- or 48-h recovery period is used, and the number of dead and live snails is registered. Activity for plant-derived compounds has been frequently reported as the 100% mortality dose (ppm, μg/ml) after 24 h exposure. The calculation of LC_{50} or LC_{90} values requires a congruous number of snails to be of significance, and should be obtained by means of a probit analysis (Finney 1971), now often replacing the Litchfield and Wilcoxon statistical method (1949). Crude extracts showing a 24-h 100% mortality dose ≤ 100 ppm are considered worthy of further exploitation; pure natural products with a lethal dose beyond the limit of 10 ppm are regarded as considerably active (Hostettmann and Marston 1987).

6.3 Molluscicidal Diterpenoids from Seaweeds

Seventeen diterpenoids isolated from brown algae of the family Dictyotaceae have been evaluated for molluscicidal activity (Tringali et al. 1986b), namely compounds **28, 33, 46–48, 50, 54–61,** the acetates of **60** and **61**, and the tetracyclic aldehyde fascioladienal (not reported) (Tringali et al. 1986a). Most of them are dolabellane derivatives, and the rationale for the examination of this set of structurally related compounds was the evaluation of structure-activity relationships. Two of the compounds tested, the isomeric acetoxy-aldehydes **54** and **58**, showed significant molluscicidal activity on *Biomphalaria glabrata*, respectively

7.5 and 25 ppm as the lowest 100% lethal dose after 24 h, the potency of the former being comparable to those of other natural products considered to be very active. The absence of activity in the other closely related compounds has suggested that molluscicidal properties may be associated with a definite portion of the molecule, namely that incorporating the two functional groups. Moreover, it has been observed that the double bond geometry change from *cis* in **54** to *trans* in **58** causes a marked (70%) reduction of the activity, which therefore appears strongly dependent on steric factors. The above-cited results encourage the examination of marine natural products for molluscicidal activity, also in view of the possible role of this activity in the marine ecosystem.

7 Other Bioactivity Data on Diterpenoids from Seaweeds

In addition to those discussed in the previous sections, some other bioactivity data have been reported on diterpenoids from marine algae; they concern only a few compounds, and are normally reported in qualitative form, without details of the methods used. For these reasons, the results are not described in separate sections with a description of the methods, although the compounds are included in Table 8, and a brief list of their properties is here reported.

Crinitol (**19**) possesses insect growth inhibition activity toward *Pectinophora gossypiella* (ED_{50} = 500 ppm) (Kubo et al. 1985); dictyotriols A (**37**) and B (**38**) show antiinflammatory activity in mice (Niang and Hung 1984); epoxyoxodolabelladiene (**48**) exhibits activity against virus influenza and adenovirus (Amico et al. 1982); dolabelladienes (**51–53**) are phytotoxic for *Hordenum vulgaris* seeds (De Rosa et al. 1984); at 16 μg/ml, the dolastane derivative **62** exhibits a 71% reversible histamine antagonism on guinea pig ileum preparations, the related metabolite **65** shows a 27% increase of the twitch height in a rat hemidiaphragm preparation, while the other dolastane **64** shows a 60% decrease (Crews et al. 1982). The brominated diterpene **103** is active toward *Artemia salina* (brine shrimp test; LC_{50} = 17.9 μg/ml) (De Rosa et al. 1988).

8 Concluding Remarks

The principal methods reported for the determination of bioactivity of diterpenoids from marine algae (seaweeds) have been described. These methods can also be used for the study of compounds of a different type, and may be carried out with relative ease by chemists with a minimum of training. It is hoped that this may stimulate further attention on the biological aspects of marine natural product research. Continuous accumulation of the structures of new metabolites appears to be of limited interest today, when not complemented by the knowledge of their biomedical and ecological importance.

Tables 1–8 report the essential results of the bioassays carried out on diterpenoids obtained from seaweeds about which some overall generalizations can be made. As regards their filetic distribution (Chlorophyta: **18** compounds; Phaeophyta: **83** compounds; Rhodophyta: **2** compounds) it is to be noted that many bioactive compounds isolated from red algae are halogenated sesquiterpenoids, which are not included in the present work; however, the bioactive diterpenoids appear not to be homogeneously distributed, with high incidence in the families of Udoteaceae (green algae) and Dictyotaceae (brown algae). Within this latter taxon, the majority of products have been isolated from species of the closely related genera *Dictyota* and *Dilophus*.

With reference to the chemical structures, most bioactive diterpenoids belong to a limited number of groups of related compounds, i.e., enolacetates from Chlorophyta, xenicane derivatives (**21–27**), dictyols (**31–40**), dolabellanes (**46–61**), dolastanes (**62–67**), spatanes (**68–82**), tetraprenyltoluquinols, and derivatives (**83–101**). A number of metabolites showing interesting bioactivities are mono- or polyaldehydes, frequently α, β-unsaturated; the importance of the aldehyde function (which can react variously with proteins) in displaying biological activities, has already been reported (Paul and Fenical 1987).

Acknowledgments. I am very grateful to Prof. M. Piattelli (University of Catania), who introduced me to the study of the chemistry of marine algae, both for his conviction in interdisciplinary research and for his helpful discussions and constructive comments on the manuscript. Thanks are also due to Dr. C. Geraci (CNR Institute, Catania) and Prof. K. Hostettmann (University of Lausanne) for vital information as regards, respectively, Sections 2 and 6. I also wish to thank Prof. W. Fenical, Dr. W.J. Gerwick (University of California) and Dr. V.J. Paul (University of Guam) for their gift of copies of Gerwick's and Paul's PhD Dissertations, and Prof. V. Amico (University of Catania), Dr. P. Cianci, Dr. G. Nicolosi and Mrs. C. Rocco (CNR Institute, Catania) for their kind cooperation during this work.

Table 8. Bioactivities of diterpenoids from seaweeds

Compound	Source	Bioactivity[a] ab	af	aa	ct	am	st	lt	it	fd	ot	Reference
	Chlorophyta											
1	*Tydemania expeditionis*	+	+			+		+				Paul and Fenical 1987
2	*Chlorodesmis fastigiata*	+				+	+	+	+			Paul and Fenical 1987
3	*Chlorodesmis fastigiata*	+	+			+	+	+	+			Paul and Fenical 1987
4	*Udotea flabellum*	+	+									Paul and Fenical 1987
5	*Udotea flabellum*	+	+			+	+	+		+		Paul and Fenical 1987
6	*Udotea flabellum*	+	+			+	+		+	+		Paul and Fenical 1987
7	*Udotea flabellum*	+	+			+	+	+	+	+		Paul and Fenical 1987
8	*Udotea spinulosa*	+				+	+	+				Paul and Fenical 1987
9	*Udotea argentea*	+				+	+	+				Paul and Fenical 1987
10	*Penicillus dumetosus*	+	+			+	+	+				Paul and Fenical 1987
11	*Penicillus dumetosus*	+	+			+	+	+				Paul and Fenical 1987
12	*Penicillus pyriformis*	+				+	+	+	+			Paul and Fenical 1987
13	*Halimeda* sp.	+	+			+	+	+	+	+		Paul and Fenical 1987
14	*Halimeda* sp.	+	+			+	+	+	+			Paul and Fenical 1987
15	*Halimeda* sp.	+	+			+	+	+		+		Paul and Fenical 1987
											+ [b]	Paul 1985
16	*Halimeda* sp.	+										Paul and Fenical 1987
17	*Pseudochlorodesmis furcellata*									+		Paul et al. 1988
18[c]	*Caulerpa brownii*	+										Paul and Fenical 1987
	Phaeophyta											
19	*Sargassum tortile*[d]	+									+	Kubo et al. 1985
20	*Cystoseira elegans*				+							Amico et al. 1982
21	*Dictyota crenulata*	+	+									Finer et al. 1979
	Dictyota flabellata			+		+						Gerwick 1981
22	*Dictyota spinulosa*	+								+		Tanaka and Higa 1984
23	*Dilophus guineensis*	+							+			Schlenk and Gerwick 1987
24	*Dictyota dichotoma*				+							Ishitsuka et al. 1988
25	*Dictyota dichotoma*				+							Ishitsuka et al. 1988
26	*Dictyota dichotoma*				+							Ishitsuka et al. 1988
27[c]	*Dictyota dichotoma*				+							shitsuka et al. 1988
28	*Dilophus ligulatus*		+	+					+			Gerwick 1981
29	*Dictyota masonii*								+			Gerwick 1981
30	*Dictyota acutiloba*			+								Gerwick 1981
31	*Pachydictyon coriaceum*	+										Hirschfeld et al. 1973
			+	+		+			+			Gerwick 1981
32	*Dictyota dichotoma*		+	+		+			+			Gerwick 1981
33	*Dilophus ligulatus*	+			+							Amico et al. 1982
				+		+			+			Gerwick 1981
34	*Dictyota dichotoma*	+										Amico et al. 1982
		+										Enoki et al. 1983
			+	+		+			+			Gerwick 1981
35	*Dictyota dichotoma*	+	+									Enoki et al. 1983
36	*Dictyota dichotoma*	+	+									Enoki et al. 1983
37	*Dictyota indica*										+	Niang and Hung 1984
38	*Dictyota indica*										+	Niang and Hung 1984

Table 8. *(continued)*

Compound	Source	Bioactivity[a] ab	af	aa	ct	am	st	lt	it	fd	ot	Reference
39	*Dictyota dentata*				+						+	Alvarado and Gerwick 1985
40	*Dilophus okamurai*										+[b]	Kurata et al. 1988a
41	*Dictyota* sp.		+	+		+			+			Gerwick 1981
42	*Dictyota* sp.		+	+		+			+			Gerwick 1981
43	*Dictyota crenulata*			+		+			+			Gerwick 1981
44	*Dilophus ligulatus*	+										Tringali et al. 1988
45	*Dictyota* sp.		+	+		+			+			Gerwick 1981
46	*Dictyota dichotoma*	+										Amico et al. 1982
47	*Dictyota dichotoma*	+										Amico et al. 1980
			+									Tringali et al. 1986b
48	*Dictyota dichotoma*	+										Amico et al. 1980
					+						+	Amico et al. 1982
49	*Dictyota dichotoma*	+										Amico et al. 1980
50	*Dictyota dichotoma*	+										Amico et al. 1980
51	*Dilophus fasciola*								+		+	De Rosa et al. 1984
52	*Dilophus fasciola*								+		+	De Rosa et al. 1984
53	*Dilophus fasciola*								+		+	De Rosa et al. 1984
54	*Dilophus fasciola*[e]		+								+[f]	Tringali et al. 1986b
					+							Tringali C, unpubl.
55	*Dilophus fasciola*[e]		+		+							Tringali et al. 1984a 1986b
					+							Tringali C, unpubl.
56	*Dilophus fasciola*[e]		+									Tringali et al. 1986b
57	*Dilophus fasciola*[e]		+									Tringali et al. 1984b 1986b
					+							Tringali C, unpubl.
58	*Dilophus fasciola*[e]		+								+[f]	Tringali et al. 1986b
					+							Tringali C, unpubl.
59	*Dilophus fasciola*[e]				+							Tringali C, unpubl.
60	*Dilophus fasciola*[e]	+	+									Tringali et al. 1984b 1986b
					+							Tringali C, unpubl.
61	*Dilophus fasciola*[e]	+	+									Tringali et al. 1984b 1986b
62	*Dilophus fasciola*[e]										+	Crews et al. 1982
63	*Dictyota divaricata* *Dictyota linearis*		+	+					+			Gerwick 1981
64	*Dictyota divaricata* } *Dictyota linearis*										+	Crews et al. 1982
65	*Dictyota linearis*										+	Crews et al. 1982
66	*Dictyota linearis*	+	+									Ochi et al. 1986
67	*Dictyota linearis*	+	+									Ochi et al. 1986
68	*Spatoglossum schmittii*			+	+[g]	+			+			Gerwick and Fenical 1983
69	*Spatoglossum schmittii*			+								Gerwick 1981
70	*Stoechospermum marginatum*			+		+			+			Gerwick 1981
						+						Gerwick and Fenical 1983
71	*S. marginatum*			+					+			Gerwick 1981

Table 8. *(continued)*

Compound	Source	Bioactivity[a] ab	af	aa	ct	am	st	lt	it	fd	ot	Reference
72	*S. marginatum*			+					+			Gerwick 1981
73	*S. marginatum*			+								Gerwick 1981
74	*S. marginatum*			+					+			Gerwick 1981
75	*S. marginatum*			+								Gerwick 1981
76	*S. marginatum*			+					+			Gerwick 1981
77	*S. marginatum*			+		+			+			Gerwick 1981
78	*S. marginatum*			+					+			Gerwick 1981
79	*Spatoglossum howleii*			+					+			Gerwick 1981
80[h]	*Spatoglossum howleii*			+					+			Gerwick 1981
81	*Dilophus okamurai*										+[b]	Kurata et al. 1988b
82	*Dilophus okamurai*										+[b]	Kurata et al. 1988b
83	*Stypopodium zonale*			+		+						Gerwick 1981
									+			Gerwick and Fenical 1981
84	*Stypopodium zonale*								+			Gerwick and Fenical 1981
85	*Cystoseira elegans*	+										Banaigs et al. 1983
86	*Cystoseira spinosa*	+										Amico et al. 1988
87	*Cystoseira spinosa*	+										Amico et al. 1988
88	*Cystoseira spinosa*	+										Amico et al. 1988
89	*Bifurcaria galapagensis*	+				+						Sun et al. 1980
90	*Cystoseira mediterranea*					+						Francisco et al. 1985
91	*Cystoseira mediterranea*					+						Francisco et al. 1986
92	*Cystoseira mediterranea*					+						Francisco et al. 1986
93	*Cystoseira mediterranea*					+						Francisco et al. 1986
94	*Taonia atomaria*								+			Gerwick 1981
95	*Stypopodium zonale*			+		+			+			Gerwick 1981
96	*Taonia atomaria*								+			Gerwick 1981
97[h]	*Stypopodium zonale*			+		+						Gerwick 1981
98	*Stypopodium zonale*			+		+			+			Gerwick 1981
99	*Stypopodium zonale*								+			Gerwick et al. 1979
100	*Stypopodium zonale*								+			Gerwick and Fenical 1983
101	*Stypopodium zonale*			+								Gerwick 1981
						+			+			Gerwick and Fenical 1983
											+[g]	O'Brien et al. 1983 1984
	Rhodophyta											
102	*Laurencia obtusa*	+										Caccamese et al. 1982
103	*Sphaerococcus coronopifolius*										+	De Rosa et al. 1988

[a]Abbreviations: ab = antibacterial activity; af = antifungal a.; aa = antialgal a.; ct = cytotoxicity; am = antimitotic a.; st = sperm toxicity; lt = larval toxicity; fd = fish feeding deterrence; ot = molluscicidal and other activities, see Section 7, where not differently specified.
[b]See Sect. 5.3.
[c]Norditerpene.
[d]Originally isolated from *Cystoseira crinita*.
[e]Previously erroneously referred to as a *Dictyota* sp.
[f]See Sect. 6.3.
[g]See Sect. 4.3.
[h]Semi-synthetic compound.

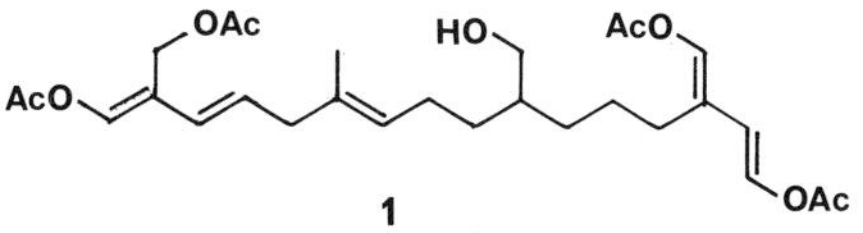

OAc HO AcO AcO OAc
1

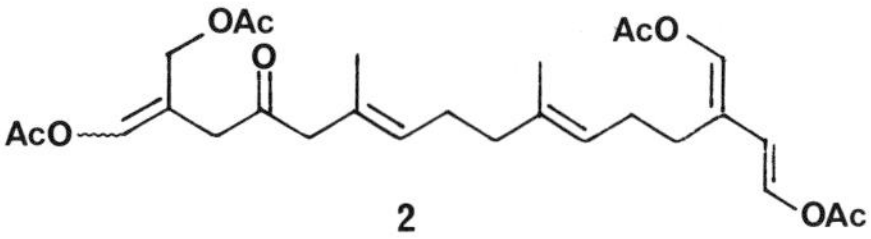

OAc O AcO AcO OAc
2

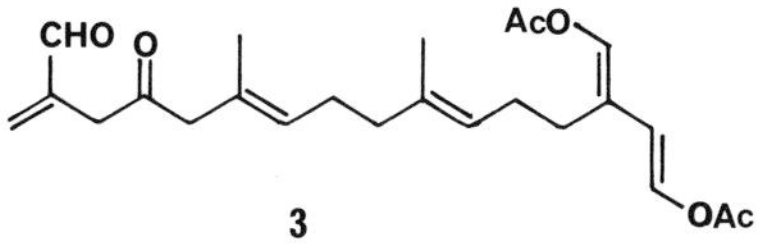

CHO O AcO OAc
3

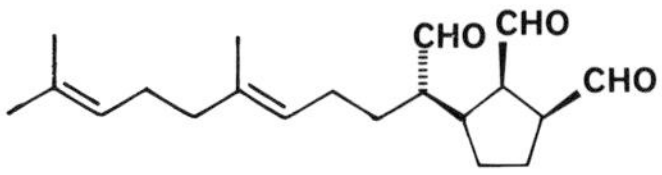

CHO CHO CHO
4

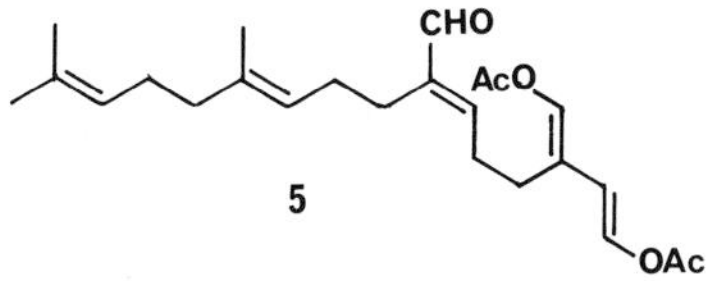

CHO AcO OAc
5

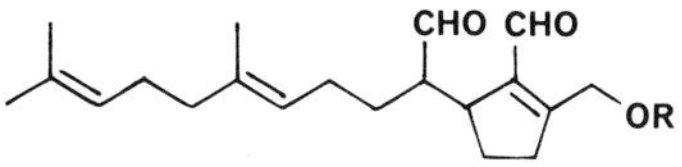

CHO CHO OR
6 R = Ac
7 R = H

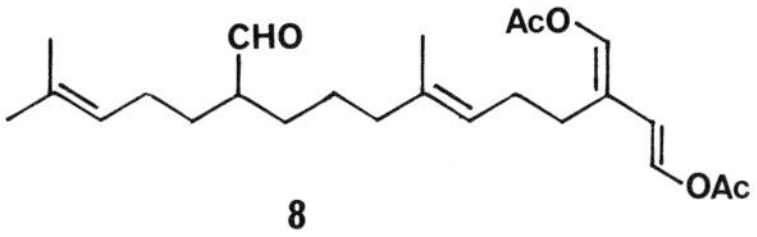

CHO AcO OAc
8

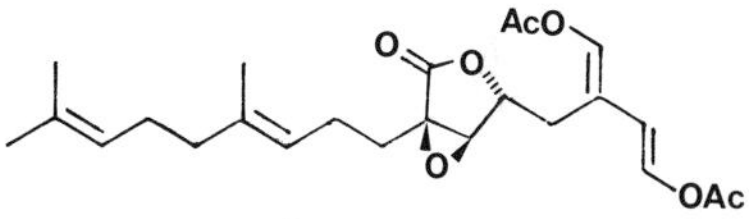

O O AcO O OAc
9

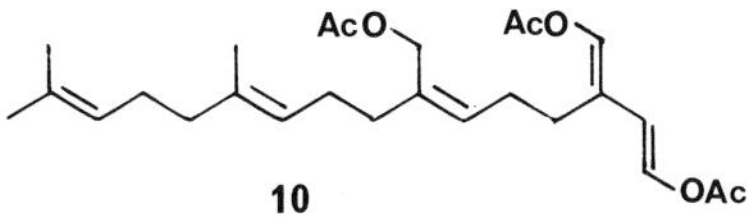

AcO AcO OAc
10

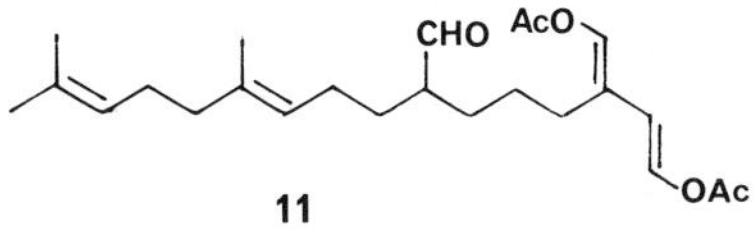

CHO AcO OAc
11

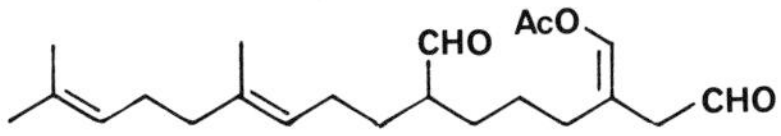

CHO AcO CHO
12

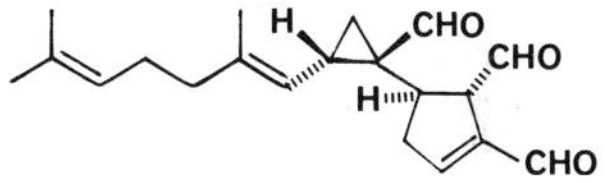

H CHO CHO H CHO
13

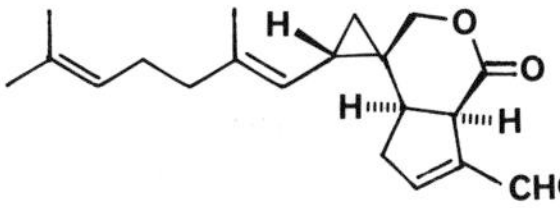

H O O H H CHO
14

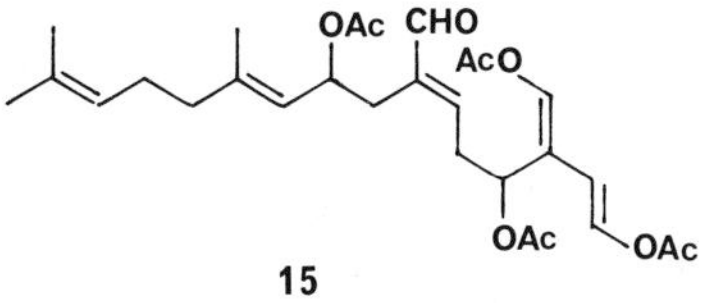

OAc CHO AcO OAc OAc
15

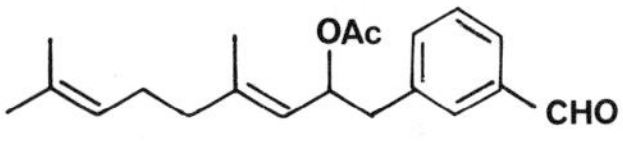

OAc CHO
16

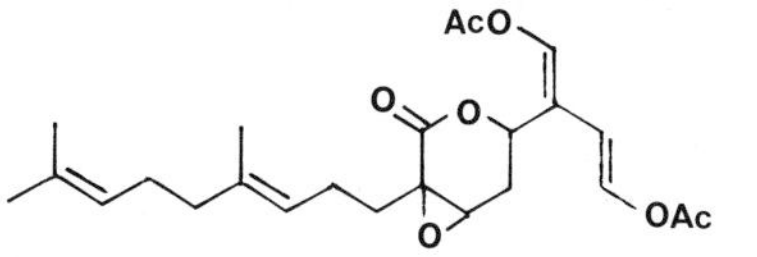

AcO O O O OAc
17

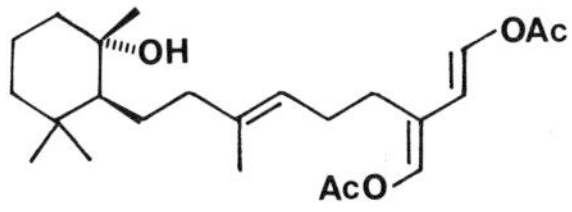

OH OAc AcO
18

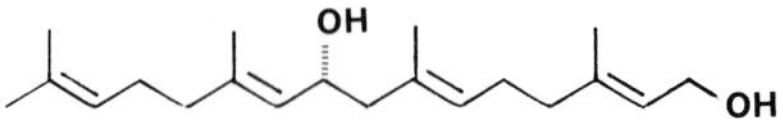

OH
OH

19

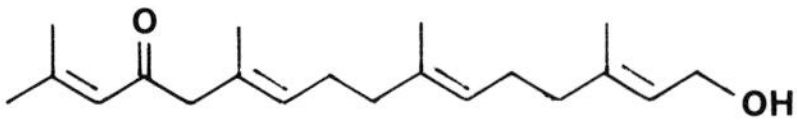

O
OH

20

R_2
H
R_1
R_3

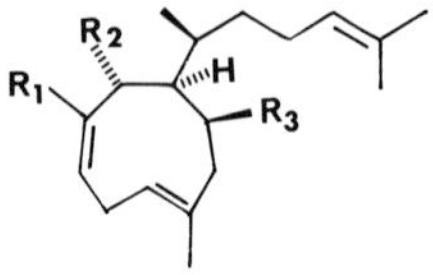

21 $R_1 = R_2 = CHO$ $R_3 = H$

22 $R_1 = R_2 = CHO$ $R_3 = OH$

23 $R_1 = Me$ $R_2 = COOH$ $R_3 = H$

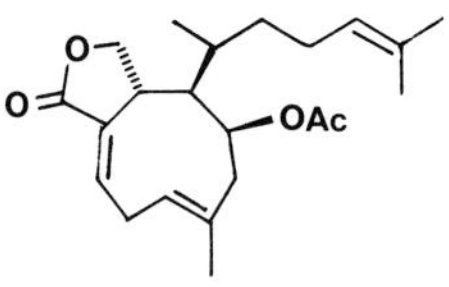

O
O
OAc

24

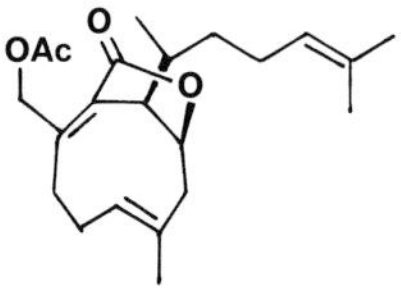

OAc
O
O

25

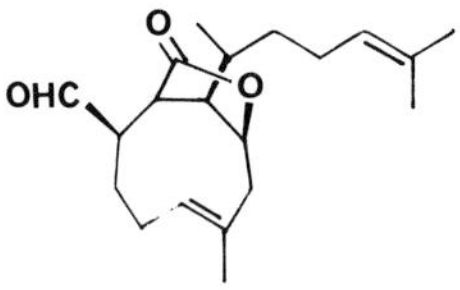

O
OHC
O

26

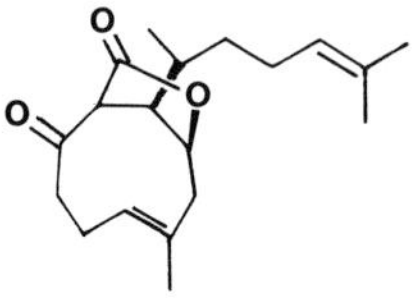

O
O
O

27

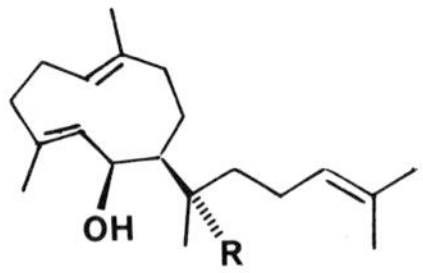

OH
R

28 R = H

29 R = OH

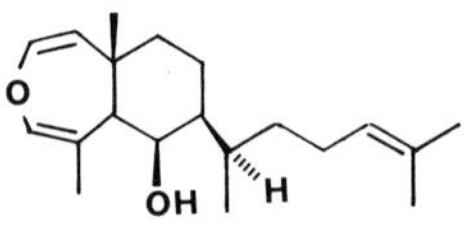

O
OH
H

30

H
R_1
H
OH
R_2

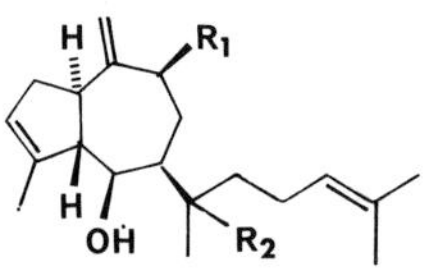

31 $R_1 = R_2 = H$

32 $R_1 = OH$ $R_2 = H$

33 $R_1 = H$ $R_2 = OH$

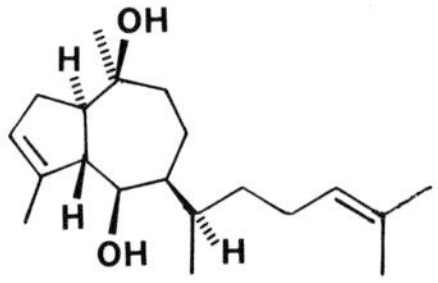

OH
H
H
OH
H

34

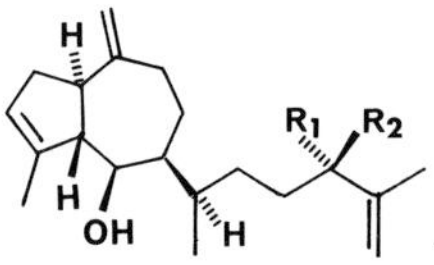

H
R_1 R_2
H
OH
H

35 $R_1 = H$ $R_2 = OH$

36 $R_2 = OH$ $R_3 = H$

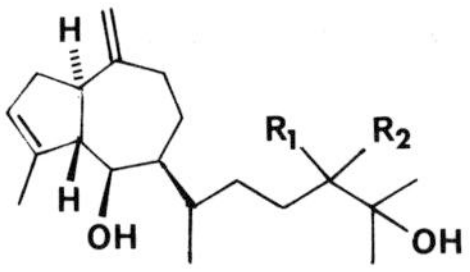

H
O
OAc
H
OH

39

H
R_1 R_2
H
OH
OH

37 $R_1 = H$ $R_2 = OH$

38 $R_1 = OH$ $R_2 = H$

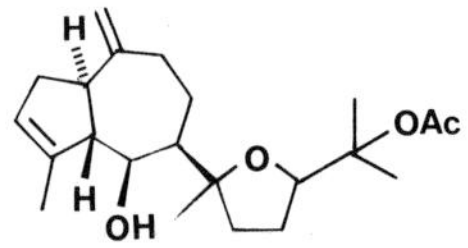

H
H
OH

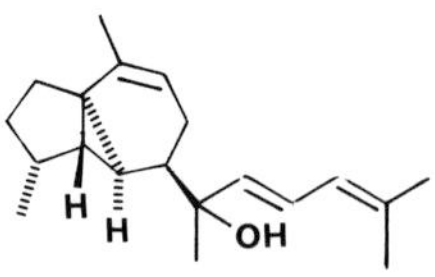

40

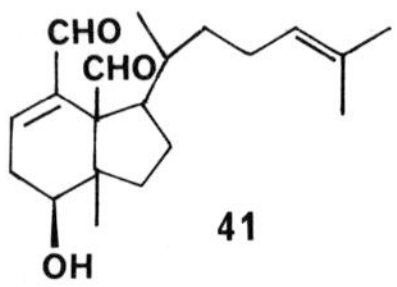

CHO
CHO
OH

41

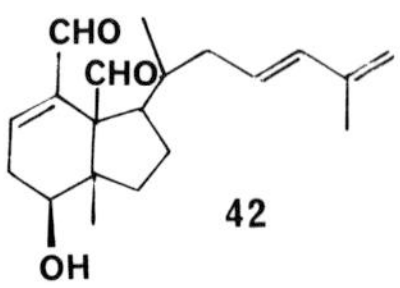

CHO
CHO
OH

42

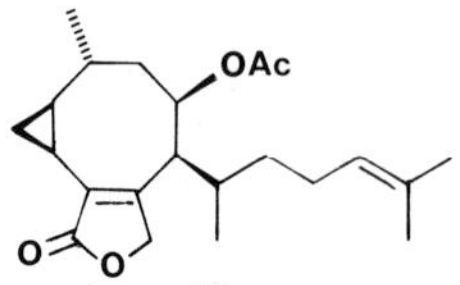

OAc
O
O

43

OAc
HO
O
OH

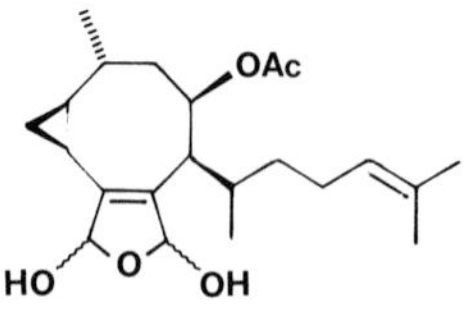

44

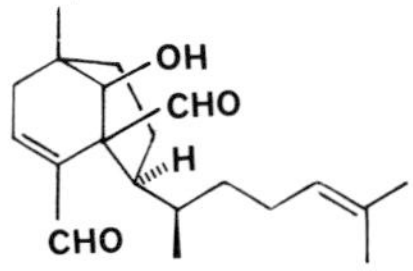

45

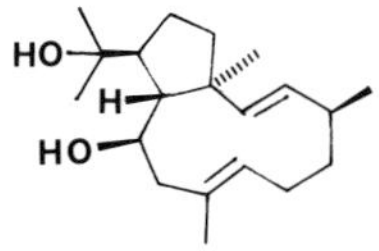

46

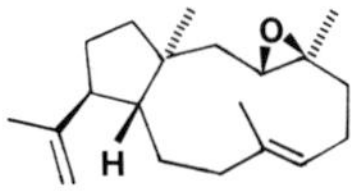

47

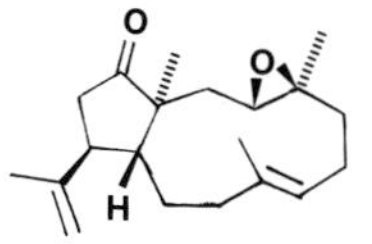

48

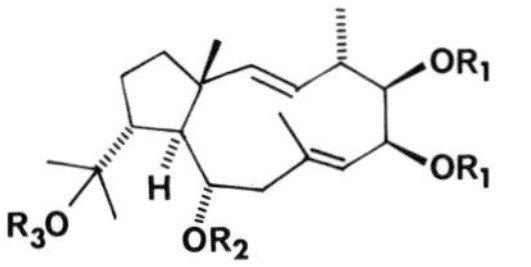

49

50

51 $R_1 = Ac$ $R_2 = R_3 = H$

52 $R_1 = R_2 = Ac$ $R_3 = H$

53 $R_1 = R_2 = R_3 = Ac$

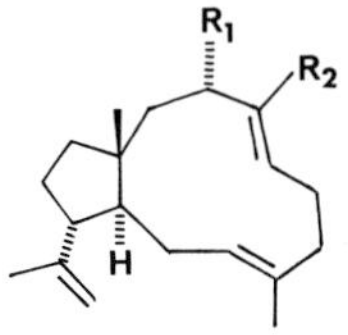

54

55 $R_1 = OAc$ $R_2 = CH_2OH$

56 $R_1 = OH$ $R_2 = CH_2OAc$

57 $R_1 = OH$ $R_2 = CH_2OH$

58 $R_1 = OAc$ $R_2 = CHO$

59 $R_1 = H$ $R_2 = CHO$

60 $R_1 = H$ $R_2 = CH_2OH$

61 $R_1 = OH$ $R_2 = Me$

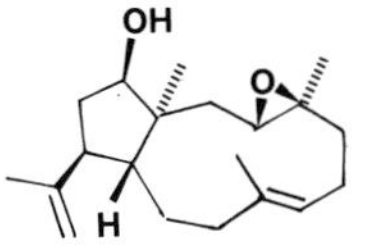

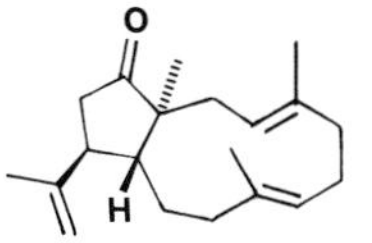

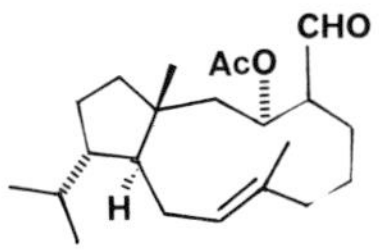

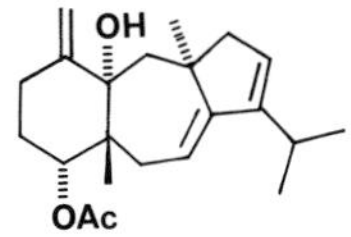

62

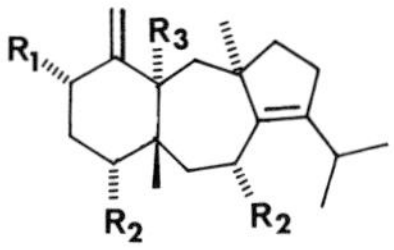

63 $R_1 = H$ $R_2 = OAc$ $R_3 = OH$

64 $R_1 = H$ $R_2 = R_3 = OH$

67 $R_1 = OH$ $R_2 = R_3 = H$

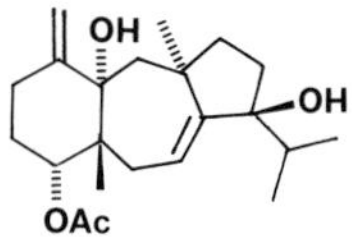

65

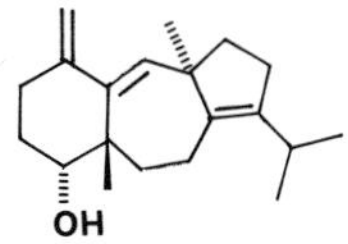

66

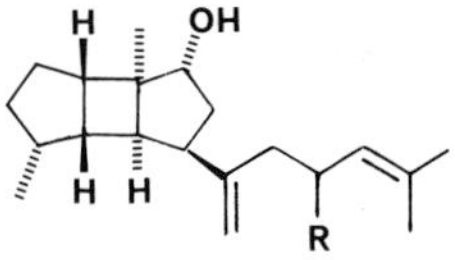

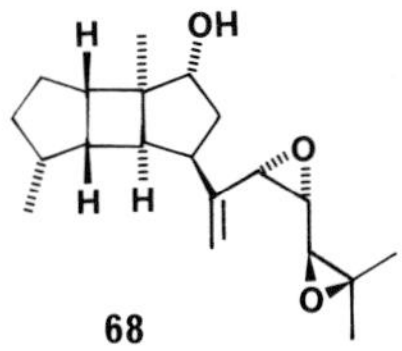

68

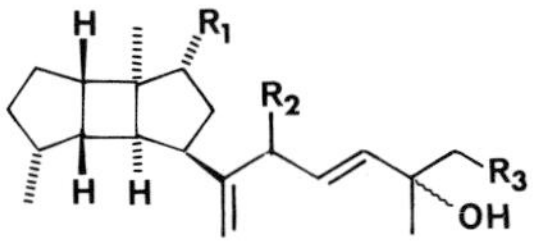

69 $R_1 = R_2 = OH$ $R_3 = H$

71 $R_1 = OH$ $R_2 = R_3 = H$

72 $R_1 = R_2 = R_3 = OH$

73 $R_1 = R_2 = OH$ $R_3 = OAc$

79 $R_1 = R_2 = R_3 = OAc$

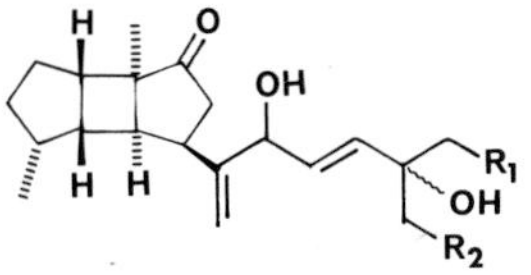

70 $R = OH$

74 $R = H$

75 $R_1 = OH$ $R_2 = H$

76 $R_1 = H$ $R_2 = OH$

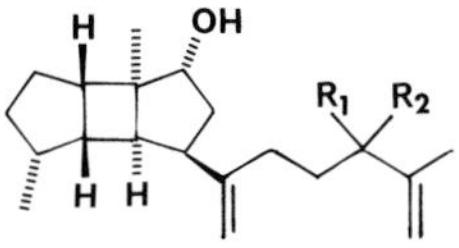

77 $R_1 = OH$ $R_2 = H$

78 $R_1 = H$ $R_2 = OH$

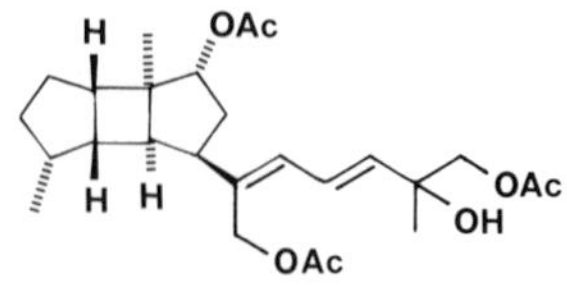

80

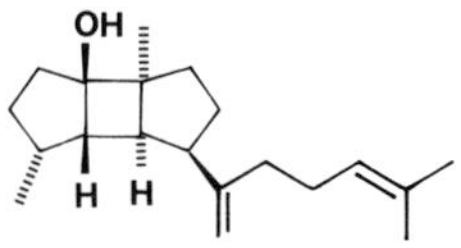

81

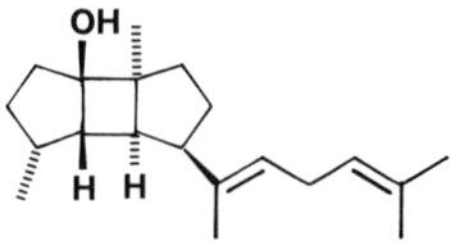

82

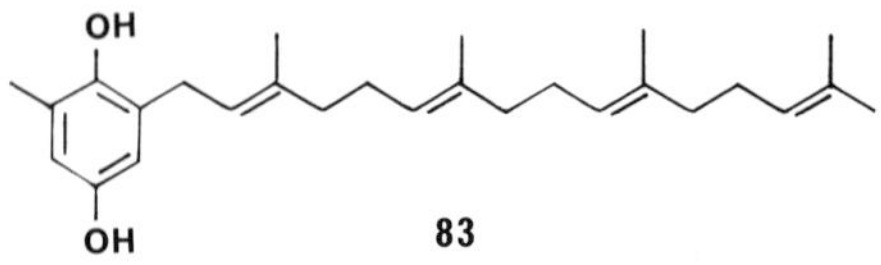

83

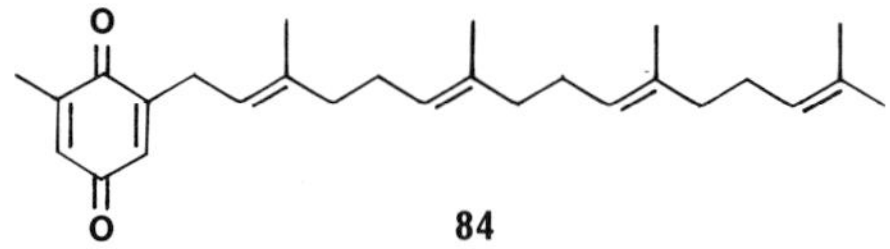

84

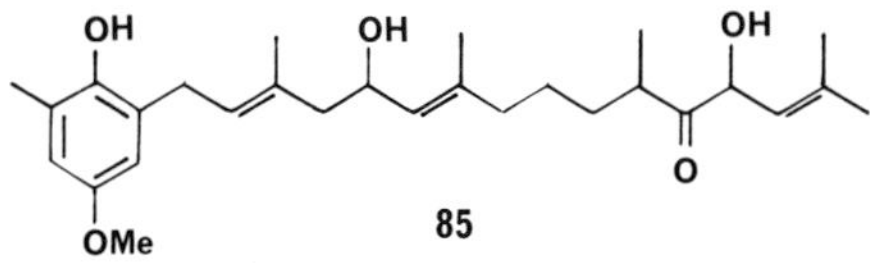

85

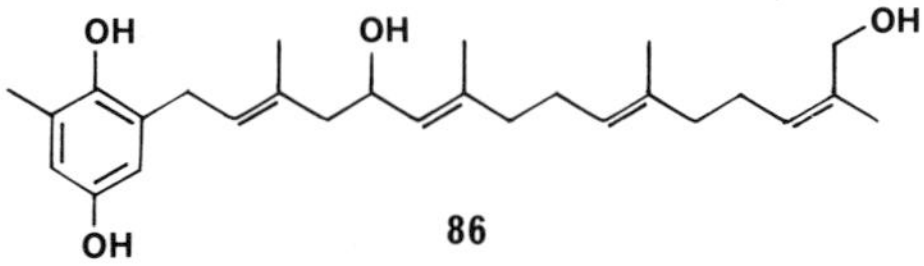

86

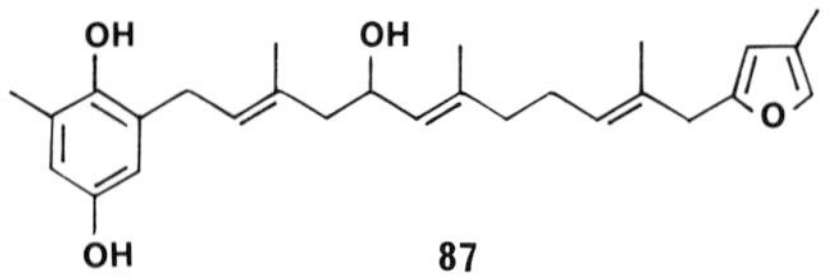

87

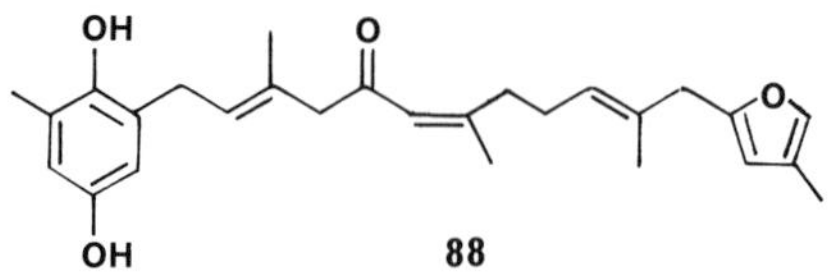

88

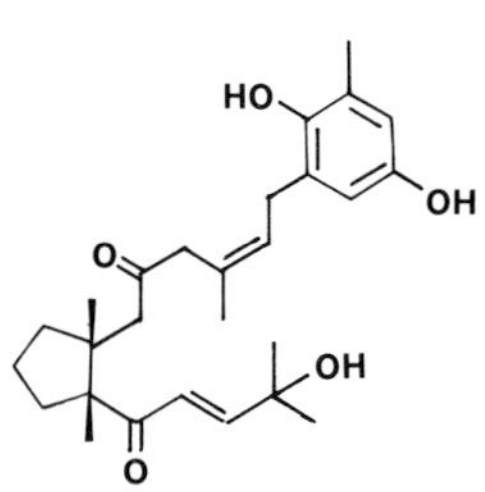

89

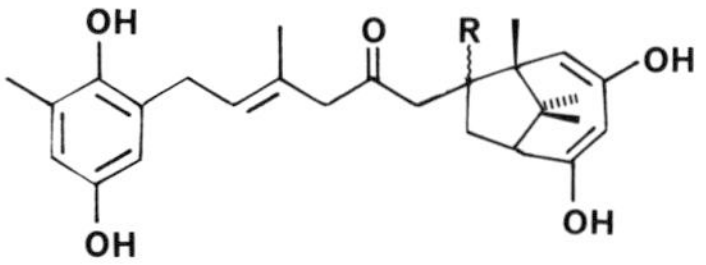

90 R = β-Me

91 R = α-Me

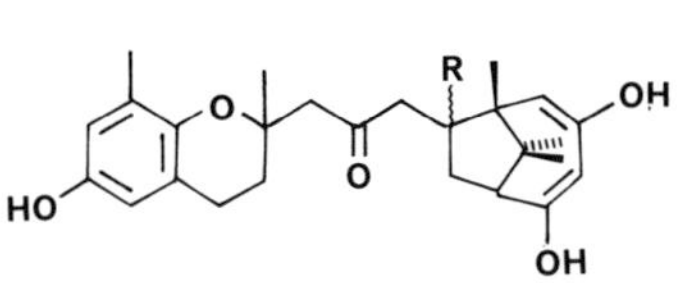

92 R = β-Me

93 R = α-Me

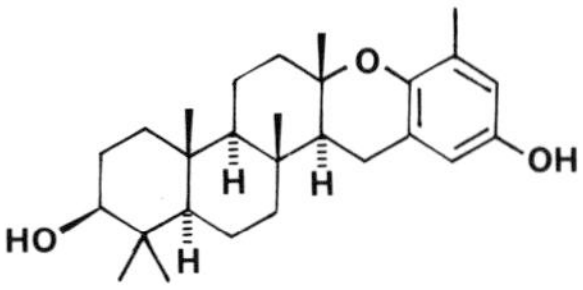

94

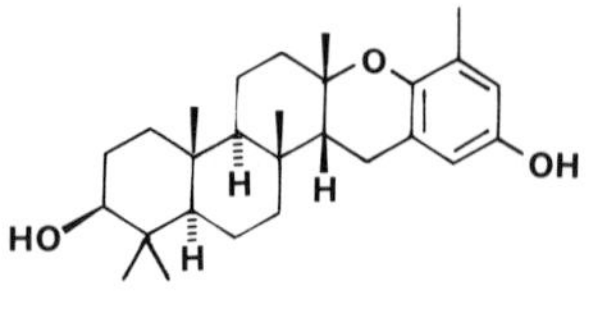

95

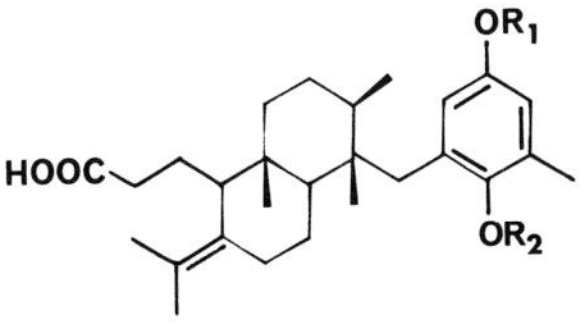

HOOC OR$_1$ OR$_2$

96 R_1 = Me R_2 = H

97 R_1 = H R_2 = Ac

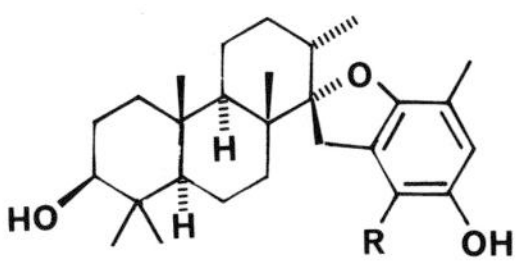

HO H H O R OH

98 R = H

99 R = OH

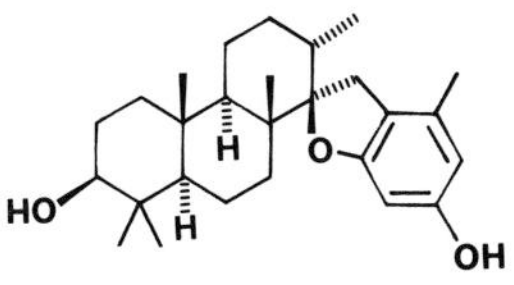

HO H H O OH

100

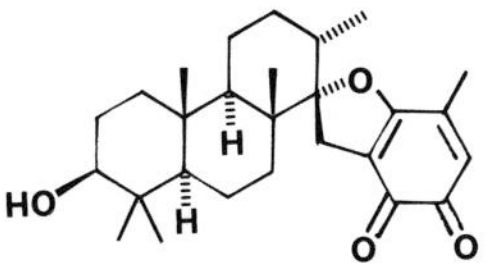

HO H H O O O

101

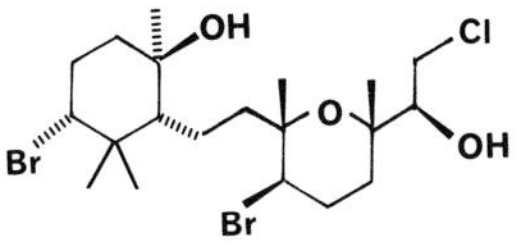

OH Cl Br O OH Br

102

HO H H Br Br

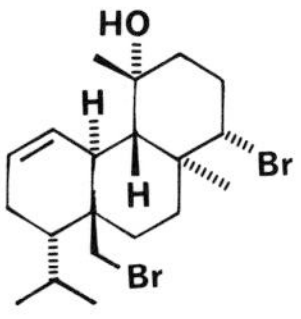

103

References

Allemand D, Pesando D, Biyiti L, De Renzis G (1989) L'oeuf d'oursin: modèle d' étude en toxicologie et pharmacologie. Vie Mar 10:216–225
Al-Ogily SM, Knight-Jones EW (1977) Anti-fouling role of antibiotics produced by marine algae and bryozoans. Nature 265:728–729
Alvarado AB, Gerwick WH (1985) Dictyol H, a new tricyclic diterpenoid from the brown seaweed *Dictyota dentata*. J Nat Prod 48:132–134
Amico V, Oriente G, Piattelli M, Tringali C (1980) Diterpenes based on the dolabellane skeleton from *Dictyota dichotoma*. Tetrahedron 36:1409–1414
Amico V, Chillemi R, Oriente G, Piattelli M, Sciuto S, Tringali C (1982) Constituenti chimici di alghe marine e loro interesse biologico. In: Atti del convegno delle Unità Operative afferenti ai sottoprogetti Risorse Biologiche e Inquinamento Marino, 10–11 Nov 1981, Roma. CNR, Roma, pp 267–279
Amico V, Cunsolo F, Neri P, Piattelli M, Ruberto G (1988) Antimicrobial tetraprenyltoluquinol derivatives from *Cystoseira spinosa* var. *squarrosa*. Phytochemistry 27:1327–1331
Baker JT (1976) Physiologically active substances from marine organisms. Aust J Pharm Sci 5:89–99
Bakus GJ (1981) Chemical defense mechanisms on the Great Barrier Reef, Australia. Science 211:497–499
Banaigs B, Francisco C, Gonzales E, Fenical W (1983) Diterpenoid metabolites from the marine alga *Cystoseira elegans*. Tetrahedron 39:629–638
Barry AL, Thornsberry C (1985) Susceptibility tests: diffusion test procedures. In: Lennette EH, Balows A, Hausler WJ, Shadomy HJ (eds) Manual of clinical microbiology, 4th edn. American Society for Microbiology, Washington, pp 978–987
Bauer AW, Kirby WMM, Sherris JC, Turck M (1966) Antibiotic susceptibility testing by a standardized single disk method. Am J Clin Pathol 45:493–496
Betina V (1973) Bioautography in paper and thin-layer chromatography and its scope in the antibiotic field. J Chromatogr 78:41–51
Betina V (1983) The chemistry and biology of antibiotics. Elsevier, Amsterdam, pp 113–114
Bhakuni DS, Silva M (1974) Biodynamic substances from marine flora. Bot Mar 17:40–51
Burkholder PR, Burkholder LM, Almodovar LR (1960) Antibiotic activity of some marine algae of Puerto Rico. Bot Mar 2:149–156
Caccamese S, Azzolina R, Furnari G, Cormaci M, Grasso S (1980) Antimicrobial and antiviral activities of extracts from Mediterranean algae. Bot Mar 23:285–288
Caccamese S, Toscano RM, Cerrini S, Gavuzzo E (1982) Laurencianol, a new halogenated diterpenoid from the marine alga *Laurencia obtusa*. Tetrahedron Lett 23:3415–3418
Carpenter RC (1981) Grazing by *Diadema antillarum* (Philippi) and its effects on the benthic algal community. J Mar Res 39:749–765
Cornman I (1950) Inhibition of sea-urchin egg cleavage by a series of substituted carbamates. J Natl Cancer Inst 10:1123–1138
Crews P, Klein TE, Hogue ER, Myers BL (1982) Tricyclic diterpenes from the brown marine algae *Dictyota divaricata* and *Dictyota linearis*. J Org Chem 47:811–815
Dart JKG (1972) Echinoids, algal lawn and coral recolonization. Nature 239:50–51
De Rosa S, De Stefano S, Macura S, Trivellone E, Zavodnik N (1984) Chemical studies of north adriatic seaweeds-I. New dolabellane diterpenes from the brown alga *Dilophus fasciola*. Tetrahedron 40:4991–4995
De Rosa S, De Stefano S, Scarpelli P, Zavodnik N (1988) Terpenes from the red alga *Sphaerococcus coronopifolius* of the north Adriatic sea. Phytochemistry 27:1875–1878
Douros J, Suffness M (1981) New antitumor substances of natural origin. Cancer Treat Rev 8:63–87
Duncan J, Sturrock RF (1987) Laboratory evaluation of potential plant molluscicides. In: Mott KE (ed) Plant molluscicides. John Wiley, New York, pp 251–265
Enoki N, Tsuzuki K, Omura S, Ishida R, Matsumoto T (1983) New antimicrobial diterpenes, dictyol F and epidictyol F, from the brown alga *Dictyota dichotoma*. Chem Lett 1627–1630
Faulkner DJ (1977) Interesting aspects of marine natural products chemistry. Tetrahedron 33:1421–1443

Faulkner DJ (1984a) Marine natural products: metabolites of marine algae and herbivorous marine molluscs. Nat Prod Rep 1:251–280

Faulkner DJ (1984b) Marine natural products: metabolites of marine invertebrates. Nat Prod Rep 1:551–598

Faulkner DJ (1986) Marine natural products. Nat Prod Rep 3:1–33

Faulkner DJ (1987) Marine natural products. Nat Prod Rep 4:539–576

Faulkner DJ (1988) Marine natural products. Nat Prod Rep 5:613–663

Fautin DG (ed) (1988) Biomedical importance of marine organisms. California Academy of Sciences, San Francisco Memoirs of the California Academy of Sciences N 13

Finer J, Clardy J, Fenical W, Minale L, Riccio R, Battaile J, Kirkup M, Moore RE (1979) Structures of dictyodial and dictyolactone, unusual marine diterpenoids. J Org Chem 44:2044–2047

Finney DJ (1971) Probit analysis, 3rd edn. Cambridge University Press, Cambridge

Francisco C, Banaigs B, Valls R, Codomier L (1985) Mediterraneol A, a novel rearranged diterpenoid-hydroquinone from the marine alga *Cystoseira mediterranea*. Tetrahedron Lett 26:2629–2632

Francisco C, Banaigs B, Teste J, Cave A (1986) Mediterraneols: a novel biologically active class of rearranged diterpenoid metabolites from *Cystoseira mediterranea* (Phaeophyta). J Org Chem 51:1115–1120

Fusetani N (1987) Marine metabolites which inhibit development of echinoderm embryos. In: Scheuer PJ (ed) Bioorganic marine chemistry vol 1. Springer, Berlin Heidelberg New York Tokyo, pp 61–92

Galsky AG, Kozimor R, Piotrowski D, Powell RG (1981) The crown-gall potato disk bioassay as a primary screen for compounds with antitumor activity. J Natl Cancer Inst 67:689–692

Gerwick WH (1981) The natural products chemistry of the Dictyotaceae. PhD Dissertation, University of California, San Diego

Gerwick WH, Fenical W (1981) Ichthyotoxic and cytotoxic metabolites of the tropical brown alga *Stypopodium zonale* (Lamouroux) Papenfuss. J Org Chem 46:22–27

Gerwick WH, Fenical W (1983) Spatane diterpenoids from the tropical marine algae *Spatoglossum schmittii* and *Spatoglossum howleii* (Dictyotaceae). J Org Chem 48:3325–3329

Gerwick WH, Fenical W, Fritsch N, Clardy J (1979) Stypotriol and stypoldione; ichthyotoxins of mixed biogenesis from the marine alga *Stypopodium zonale*. Tetrahedron Let 145–148

Hamburger MO, Cordell GA (1987) A direct bioautographic TLC assay for compounds possessing antibacterial activity. J Nat Prod 50:19–22

Henriquez P, Candia A, Norambuena R, Silva M, Zemelman R (1979) Antibiotic properties of marine algae II. Screening of Chilean marine algae for antimicrobial activity. Bot Mar 22:451–453

Hirschfeld DR, Fenical W, Lin GHY, Wing RM, Radlick P, Sims JJ (1973) Marine natural products. VIII. Pachydictyol A, an exceptional diterpene alcohol from the brown alga, *Pachydictyon coriaceum*. J Am Chem Soc 95:4049–4050

Hodgson LM (1984) Antimicrobial and antineoplastic activity in some South Florida seaweeds. Bot Mar 27:387–390

Holt RJ (1975) Laboratory tests of antifungal drugs. J Clin Pathol 28:767–774

Homans AL, Fuchs A (1970) Direct bioautography on thin-layer chromatograms as a method for detecting fungitoxic substances. J Chromatogr 51:327–329

Hoppe HA, Levring T (eds) (1982) Marine algae in pharmaceutical science vol 2. Walter de Gruyter, Berlin

Hoppe HA, Levring T, Tanaka Y (eds) (1979) Marine algae in pharmaceutical science, vol 1. Walter de Gruyter, Berlin

Hornsey IS, Hide D (1974) The production of antimicrobial compounds by British marine algae. I. Antibiotic-producing marine algae. Br Phycol J 9:353–361

Hostettmann K, Marston A (1987) Plant molluscicide research—an update. In: Mott KE (ed) Plant molluscicides. John Wiley, New York, pp 299–320

Ieven M, Vanden Berghe DA, Mertens F, Vlietinck A, Lammens E (1979) Screening of higher plants for biological activities I. Antimicrobial activity. Planta Med 36:311–321

Ikegami S, Kawada K, Kimura Y, Suzuki A (1979) A rapid and convenient procedure for the detection of inhibitors of DNA synthesis using starfish oocytes and sea urchin embryos. Agric Biol Chem 43:161–166

Ireland CM, Roll DM, Molinski TF, McKee TC, Zabriskie TM, Swersey JC (1988) Uniqueness of the marine chemical environment: categories of marine natural products from invertebrates. In: Fautin DG (ed) Biomedical importance of marine organisms, California Academy of Sciences, San Francisco, pp 41–57 Memoirs of the California Academy of Sciences N 13

Ishitsuka MO, Kusumi T, Kakisawa H (1988) Antitumor xenicane and norxenicane lactones from the brown alga *Dictyota dichotoma*. J Org Chem 53:5010–50139

Jacobs RS, White S, Wilson L (1981) Selective compounds derived from marine organisms: effects on cell division in fertilized sea urchin eggs. Fed Proc Am Soc Exp Biol 40:26–29

Jacobs RS, Culver P, Langdon R, O'Brien T, White S (1985) Some pharmacological observations on marine natural products. Tetrahedron 41:981–984

Jones RN, Barry AL, Gavan TL, Washington JA II (1985) Susceptibility tests: microdilution and macrodilution broth procedures. In: Lennette EH, Balows A, Hausler WJ, Shadomy HJ (eds) Manual of clinical microbiology, 4th edn. American Society for Microbiology, Washington, pp 972–977

Kashiwagi M, Mynderse JS, Moore RE, Norton TR (1980) Antineoplastic evaluation of pacific basin marine algae. J Pharm Sci 69:735–738

Kaul PN, Sindermann CJ (eds) (1978) Drugs and food from the sea. University of Oklahoma, Norman

Kloos H, McCullough FS (1982) Plant molluscicides. Planta Med 46:195–209

Kubo I, Matsumoto T, Ichikawa N (1985) Absolute configuration of crinitol. An acyclic diterpene insect growth inhibitor from the brown alga *Sargassum tortile*. Chem Lett 249–252

Kurata K, Shiraishi K, Takato T, Taniguchi K, Suzuki M (1988a) A new feeding-deterrent diterpenoid from the brown alga *Dilophus okamurai*. Chem Lett 1629–1632

Kurata K, Suzuki M, Shiraishi K, Taniguchi K (1988b) Spatane-type diterpenes with biological activity from the brown alga *Dilophus okamurai*. Phytochemistry 27:1321–1324

Lennette EH, Balows A, Hausler WJJr, Shadomy HJ (eds) (1985) Manual of clinical microbiology, 4th edn. American Society for Microbiology, Washington

Litchfield JT, Wilcoxon F (1949) A simplified method of evaluating dose-effect experiments. J Pharm Exp Ther 96:99–113

Mackie AM, Grant PT (1974) Interspecies and Intraspecies chemoreception by marine invertebrates. In: Grant PT, Mackie AM (eds) Chemoreception in marine organisms. Academic Press, London, pp 105–141

Marston A, Hostettmann K (1985) Plant molluscicides. Phytochemistry 24:639–652

Meyer BN, Ferrigni NR, Putnam JE, Jacobsen LB, Nichols DE, McLaughlin JL (1982) Brine shrimp: a convenient general bioassay for active plant constituents. Planta Med 45:31–34

Munro MHG, Luibrand RT, Blunt JW (1987) The search for antiviral and anticancer compounds from marine organisms. In: Scheuer PJ (ed) Bioorganic marine chemistry vol 1, Springer, Berlin Heidelberg New York Tokyo, pp 93–176

Nakanishi K, Kubo I (1977) Studies on waburganal, muzigadial and related compounds. Isr J Chem 16:28–31

Nakatsu T, Ravi BN, Faulkner DJ (1981) Antimicrobial constituents of *Udotea flabellum*. J Org Chem 46:2435–2438

Naqvi SWA, Solimabi, Kamat SY, Fernandes L, Reddy CVG, Bhakuni DS, Dhawan BN (1980) Screening of some marine plants from the Indian coast for biological activity. Bot Mar 24:51–55

National Cancer Institute, Protocol (1972) Cell culture screen, KB. Cancer Chemother Rep 3:17

Nemanich JW, Theiler RF, Hager LP (1978) The occurence of cytotoxic compounds in marine organisms. In: Kaul PN, Sindermann CJ (eds) Drugs and food from the sea. The University of Oklahoma, Norman, pp 123–136

Niang LL, Hung X (1984) Studies on the biologically active compounds of the algae from the Yellow Sea. Hydrobiologia 116/117:168–170

Norris JN, Fenical W (1982) Chemical defense in tropical marine algae. In: Rutzler K, Macintyre IG (eds) The Atlantic Barrier Reef ecosystem at Carrie Bow Cay, Belize 1: structure and communities. Smithsonian Contr Mar Sci 12:417–431

O'Brien ET, Jacobs RS, Wilson L (1983) Inhibition of bovine braine microtubule assembly in vitro by stypoldione. Mol Pharm 24:493–499

O'Brien ET, White S, Jacobs RS, Boder GB, Wilson L (1984) Pharmacological properties of a marine natural product, stypoldione, obtained from the brown alga *Stypopodium zonale*. Hydrobiologia 116/117:141–145

Ochi M, Asao K, Kotsuki H, Miura I, Shibata K (1986) Amijitrienol and 14-deoxyisoamijiol, two new diterpenoids from the brown seaweed *Dictyota linearis*. Bull Chem Soc Jpn 59:661–662

Olesen PE, Maretzki A, Almodovar LA (1964) An investigation of antimicrobial substances from marine algae. Bot Mar 6:224–232

Oyama VI, Eagle H (1956) Measurement of cell growth in tissue culture with a phenol reagent (Folin-Ciocalteau). Proc Soc Exp Biol Med 91:305–307

Patterson GML, Norton TR, Furusawa E, Furusawa S, Kashiwagi M, Moore RE (1984) Antineoplastic evaluation of marine algal extracts. Bot Mar 27:485–488

Paul VJ (1985) The natural products chemistry and chemical ecology of tropical green algae of the order Caulerpales. PhD Dissertation, University of California, San Diego

Paul VJ (1988) Marine chemical ecology and natural products research. In: Fautin DG (ed) Biomedical importance of marine organisms. California Academy of Sciences, S. Francisco, pp 23–27 Memoirs of the California Academy of Sciences N 13

Paul VJ, Fenical W (1983) Isolation of halimedatrial: chemical defense adaptation in the calcareous reef-building alga *Halimeda*. Science 221:747–749

Paul VJ, Fenical W (1987) Natural products chemistry and chemical defense in tropical marine algae of the phylum Chlorophyta. In: Scheuer PJ (ed) Biorganic marine chemistry vol 1, Springer, Berlin Heidelberg New York Tokyo, pp 1–29

Paul VJ, Ciminiello P, Fenical W (1988) Diterpenoid feeding deterrents from the pacific green alga *Pseudochlorodesmis furcellata*. Phytochemistry 27:1011–1014

Pesando D, Caram B (1984) Screening of marine algae from the French Mediterranean coast for antibacterial and antifungal activity. Bot Mar 27:381–386

Pesando D, Gnassia-Barelli M, Gueho E (1979) Antifungal properties of some marine planktonic algae. In: Hoppe HA, Levring T, Tanaka Y (eds) Marine algae in pharmaceutical science vol 1. Walter de Gruyter, Berlin, pp 461–472

Pesando D, Amade P, Rogeau D, Durand-Clement M, Puiseux-Dao S, Guyot M, Kondracki ML, Litaudon M, Berreur J, Payan P, Girard JP (1989) Sea urchin egg, a model for the search of new antimitotic marine drugs. In: Proceedings of the Sixth International Symposium on Marine Natural Products, 3–7 July 1989, Dakar, Senegal. C3

Pratt R, Mautner H, Gardner GM, Sha YH, Dufrenoy J (1951) Report on antibiotic activity of seaweed extracts. J Am Pharm Assoc 40:575–579

Rinehart KL, Shaw PD, Shield LS, Gloer JB, Harbour GC, Koker MES, Samain D, Schwartz RE, Tymiak AA, Weller DL, Carter GT, Munro MHG, Hughes RG, Renis HE, Swynenberg EB, Stringfellow DA, Vavra JJ, Coats JH, Zurenko GE, Kuentzel SL, Li LH, Bakus GJ, Brusca RC, Craft LL, Young DN, Connor JL (1981) Marine natural products as sources of antiviral, antimicrobial, and antineoplastic agents. Pure Appl Chem 53:795–817

Ruggieri GD, Nigrelli RF (1960) The effects of Holothurin, a steroid saponin from the sea cucumber, on the development of the sea urchin. Zoologica 45:1–20

Schlenk D, Gerwick WH (1987) Dilophic acid, a diterpenoid from the tropical brown seaweed *Dilophus guineensis*. Phytochemistry 26:1081–1084

Sieburth JMcN, Conover JT (1965) *Sargassum* tannin, an antibiotic which retards fouling. Nature 208:52–53

Smith CG, Lummis WL, Grady JE (1959) An improved tissue culture assay I. Methodology and cytotoxicity of anti-tumor agents. Cancer Res 19:843–846

Stahl E (1969) Thin-layer chromatography. A laboratory handbook. Springer, Berlin Heidelberg New York

Suffness M, Douros J (1982) Current status of the NCI plant and animal product program. J Nat Prod 45:1–14

Suffness M, Thompson JE (1988) National Cancer Institute's role in the discovery of new antineoplastic agents. In: Fautin DG (ed) Biomedical importance of marine organisms. California Academy of Science, San Francisco, pp 151–157 Memoirs of the California Academy of Sciences N 13

Sun HH, Fenical W (1979) Rhipocephalin and rhipocephenal; toxic feeding deterrents from the tropical marine alga *Rhipocephalus phoenix*. Tetrahedron Lett 685–688

Sun HH, Ferrara NM, McConnell OJ, Fenical W (1980) Bifurcarenone, an inhibitor of mitotic cell division from the brown alga *Bifurcaria galapagensis*. Tetrahedron Lett 21:3123–3126

Tanaka J, Higa T (1984) Hydroxydictyodial, a new antifeedant diterpene from the brown alga *Dictyota spinulosa*. Chem Lett 231–232
Tringali C, Piattelli M, Nicolosi G (1984a) Structure and conformation of new diterpenes based on the dolabellane skeleton from a *Dictyota* species. Tetrahedron 40:799–803
Tringali C, Piattelli M, Rocco C, Nicolosi G (1984b) Three further dolabellane diterpenoids from *Dictyota* sp. Phytochemistry 23:1681–1684
Tringali C, Piattelli M, Nicolosi G (1986a) Fasciola-7,18-dien-17-al, a diterpenoid with a new tetracyclic ring system from the brown alga *Dilophus fasciola*. J Nat Prod 49:236–243
Tringali C, Piattelli M, Nicolosi G, Hostettmann K (1986b) Molluscicidal and antifungal activity of diterpenoids from brown algae of the family Dictyotaceae. Planta Med 404–406
Tringali C, Oriente G, Piattelli M, Geraci C, Nicolosi G, Breitmaier E (1988) Crenuladial, an antimicrobial diterpenoid from the brown alga *Dilophus ligulatus*. Can J Chem 66:2799–2802
Tringali C, Piattelli M, Geraci C, Nicolosi G (1989) Antimicrobial tetraprenylphenols from *Suillus granulatus*. J Nat Prod 52:941–947
Vacquier VD, Payne JE (1973) Methods for quantitating sea urchin sperm-egg binding. Exp Cell Res 82:227–235
Vadas RL (1979) Seaweeds: an overview; ecological and economic importance. Experientia 35:429–433
Washington JA II (1985) Susceptibility tests: agar dilution. In: Lennette EH, Balows A, Hausler WJ, Shadomy HJ (eds) Manual of clinical microbiology, 4th edn. American Society for Microbiology, Washington, pp 967–971
World Health Organization (1965) Molluscicide screening and evaluation. Bull WHO 33:567–581
Yamamoto I, Takahashi M, Tamura E, Maruyama H (1982) Antitumor activity of crude extracts from edible marine algae against L-1210 leukemia. Bot Mar 25:455–457
Yamamoto I, Takahashi M, Tamura E, Maruyama H, Mori H (1984) Antitumor activity of edible marine algae: effect of crude fucoidan fractions prepared from edible brown seaweeds against L-1210 leukemia. Hydrobiologia 116/117:145–148

Determination of Waxes Causing Water Repellency in Sandy Soils

J.F. JACKSON and H.F. LINSKENS

1 Introduction

Water repellency in soils was first recognised and described by Schreiner and Shorey (1910) in California. The phenomenon was investigated by Jamison (1946, 1947) in Californian citrus groves, and found to be a surface property caused by organic matter coating the soil particles. The condition appeared to be built up over time as a result of the presence of the citrus trees in this case. Water repellency in soils occurs in both the USA and Australia over large areas (Bond 1969), and is known in other countries as well.

There have been suggestions that the water repellency of these soils is due to micro-organisms, especially spore-forming fungi or products secreted by them (Savage et al. 1969; Wilkinson and Miller 1978). However, McGhie and Posner (1980) showed that spore-bearing fungi often actually reduced water repellency in soils. These authors proposed that comminuted surface plant litter coated the soil particles, leading to hydrophobicity. The effectiveness of this finely divided plant material and break-down products derived from it in producing water repellency depended on the plant species and the particle size of the litter. However, it is generally recognised that whatever the origin of the material, organic compounds actually produced the hydrophobicity.

Attempts have been made over the years to identify the organic compounds involved, but this has proven to be a difficult matter, so much so, that a wide range of organic substances have been put forward as the causative agents, including substituted phenols, humic acids (Savage et al. 1969), lipids (Wander 1949), and others, with no real proof that any of these actually brought about the problem. This was the situation until recently, when Ma'Shum et al. (1988) were able to apply superior extraction procedures to isolate the organic compounds and analyse them by chromatographic and spectroscopic techniques. The causative agents (the so-called waxes) were shown to be compounds with extensive polymethylene chains including both long chain fatty acids and esters. This chapter deals with the methods involved in assessment of water repellency, and extraction of causative agents, and an explanation as to why the extraction procedures of Ma'Shum et al. (1988) have been successful when others before them had met with only limited success.

2 Assessment of Water Repellency of Soils

The degree of water repellency of a soil can be determined by a number of methods; we will describe two which have been well tried and tested and give reliable results.

2.1 Molarity of the Ethanol Droplet (MED) Method

The starting point of this method is that it is known that pure ethanol is able to wet all surfaces, and by adding more and more water to the ethanol a point is reached with water-repellent soils when the aqueous ethanol is no longer able to wet the surface of the soil. The surface tension of an aqueous ethanol drop which just readily wetted the soil surface provides a simple measure for ranking hydrophobic soils. Because the molarity of ethanol in the droplet also determines the surface tension of the water-ethanol mixture, then the hydrophobicity of the soil can be expressed as the molarity of the ethanol droplet (MED) which will just penetrate or wet the soil surface.

In practice then, the molarity of the ethanol droplet which will just penetrate the soil surface in 10 s is referred to as the MED value. It has been used by King (1981) to measure the degree of water repellency of various South Australian soils. Droplets of aqueous ethanol solutions (0 to 7 M) are placed on the soil surface and the time recorded for the droplet to be completely absorbed. The degree of water repellency is found to be low with MED values in the range 0–1.0 M, moderate in the range 1.0–2.4 M, severe between 2.4–3.0 M and very severe at MED greater than 3.0 M. The method is rapid and simple, and as little as 1 g of soil is required. MED values of 4 M or more are rarely met with in the field. Sands with MED values of more than 2 M are regarded as sufficiently water-repellent to cause poor crop establishment and to be a wind-erosion hazard.

2.2 Capillary Rise Technique

Difficulty in wettability of the soil implies that the advancing contact angle of water on the soil is appreciably greater than zero. As described by Emerson and Bond (1962), this angle can be determined by comparing the heights of capillary rise of water into columns of the dry sandy soil to be tested before and after ignition at 500 °C. Ignition removes the water-repellent coating and renders the soil completely wettable. The advancing contact angle of water on sand, O, is given by the equation

$$\text{Cos O} = h/h_0, \quad (1)$$

where h_0 is the height of capillary rise in the sand column when the contact angle is zero degrees (i.e., when the sand is wettable after ignition), and where h is the height of capillary rise in the sand column before ignition of the sand.

To carry out the measurement, a plastic tube of 6 mm internal diameter is closed at one end with brass gauze and a transparent scale subdivided into mm intervals attached to the tube. With the tube vertical, a weighed quantity of sand is poured in, small quantities at a time to prevent grading, and the tube tapped at the same time to give uniform packing. The tube is then quickly immersed to a known depth in a large reservoir and the average position of the wetting front measured on the attached scale at convenient time intervals. This procedure was repeated for ignited sand. The capillary rise height, when measured with time, was found to give a linear relationship when the rate of rise was plotted against the reciprocal of the height of rise, for a period up to about 15 min from the start of measurements. Extrapolation of this linear portion to zero rate of rise gave the final heights of capillary rise. The hydraulic head is then subtracted from this figure to give h and h_0 values (Emerson and Bond 1962). A comparison of MED values and capillary rise heights is given by Ma'Shum et al. (1988), for a variety of Australian soils.

2.3 Other Test Methods

For the simple water droplet test (WD) 40 µl droplets entry time in less than 4 min is measured. By the SRI test the average infiltration rate of water into soil from a small ring infiltrometer is determined (King 1981).

The soil-water repellency based upon contact angle-surface tension relationship can be characterized by certain indices (Watson and Letey 1970).

3 Extraction of Water-Repellent Waxes

The best method for extraction of water-repellent materials from soils involves the use of a Soxhlet apparatus with isopropanol/15.7 M ammonia (7:3, v:v) as solvent (Ma'Shum et al. 1988). A soil from the Tintinara district of South Australia, known to exhibit a high degree of water repellency (MED = 3.5 M), gave an MED value of 0 after 16 h extraction as above. An amount of 1880 mg/kg of soil was extracted during this process. Infrared spectra of the extracts showed the presence of extended paraffinic chains somewhat akin to the cuticular waxes of eucalypt or products formed from them by the action of micro-organisms. Spectroscopic examination by a variety of techniques suggested the presence of extensive polymethylene chains including long-chain fatty acids and esters, with 16 to 32 carbon atoms (Ma'Shum, et al. 1988). Addition of the extracted material back to acid-washed sand resulted in a significant rise in water repellency of the sand. When put on a quantitative basis, addition of extracted material restored water repellency to the original figure (MED = 3.5 M).

4 Significance and Conclusions

The breakthrough made by Ma'Shum et al. (1988) was in the use of an amphiphilic mixture of isopropanol-15.7 M ammonia as solvent for the extraction of organic compounds ("waxes") from the nonwettable Australian sands. This extraction rendered the sands wettable and yielded compounds which could be extensively analysed. The extraction technique used does not detectably cleave long-chain ester bonds, and therefore their work has shown that the bulk of the hydrophobicity in these soils is *not* covalently linked to the surface of the sand.

Furthermore, it now seems that the inability of less polar solvents (ether, chloroform, etc.) to remove long-chain fatty acids, esters, and other lipids from water-repellent soils, a fact well documented by many earlier workers, is not due to covalent linkage of the lipids with other compounds. This was an explanation often put forward by these earlier investigators to explain the lack of solubility of soil-bound lipids in the nonpolar solvents. Rather, it now seems from the work of Ma'Shum et al. (1988) that the lack of solubility of soil bound lipids in nonpolar solvents is due to the strengthening of hydrogen bonds which hold the lipids to the soil particles by the nonpolar solvents. For effective extraction the need is to rupture these hydrogen bands so as to remove the lipid from the particle surface and this can be done with isopropanol/ammonia. It is surprising that this approach has not been used before on the nonwettable soils, as Kates and Eberhardt (1957) reported the total extraction of lipids from leaves with hot isopropanol many years earlier.

The way is now open for further analysis of the organic compounds responsible for the water repellency of these and other nonwetting soils and sands.

Acknowledgement. In preparing this chapter, the authors acknowledge the cooperation and advice of the scientists leading this research, Prof. M. Oades and Dr. M. Tate. We believe the analyses described herein represent an interesting aspect of plant product analysis and one which will now expand and open up the whole question of water repellency in sands and soils.

References

Bond RD (1969) The occurrence of water repellent soils in Australia. In: DeBang LF, Letey J (eds) Water-repellent soils. Proc Sympos on water repellent soils. University of California, Riverside, California, pp 1–6

Emerson WW, Bond RD (1962) The rate of water entry into dry sand and calculation of the advancing contact angle. Aust J Soil Sci Res 1:9–16

Jamison VC (1946) The penetration of irrigation and rain water into sandy soils of Central Florida. Soil Sci Soc Am Proc 10:25–29

Jamison VC (1947) Resistance of wetting in the surface and sandy soils under citrus trees in Central Florida and its effect on penetration and the efficiency of irrigation. Soil Sci Soc Am Proc 11:103–109

Kates M, Eberhardt FM (1957) Isolation and fractionation of leaf phosphatides. Can J Bot 35:895–905

King PM (1981) Comparison of methods for measuring severity of water-repellency of sandy soils and assessment of some factors that affect its measurement. Aust J Soil Sci Res 19:275–285

Ma'Shum M, Tate ME, Jones GP, Oades JM (1988) Extraction and characterization of water-repellent materials from Australian soils. J Soil Sci 39:99–110

McGhie DA, Posner AM (1980) Water repellence of a heavy textured Western Australian surface soil. Aust J Soil Sci Res 18:309–323

Savage SM, Martin JP, Letey J (1969) Contribution of some soil fungi to natural and heat induced water-repellency in sand. Soil Sci Soc Am Proc 33:405–409

Schreiner O, Shorey EC (1910) Chemical nature of soil organic matter. USDA Bureau of Soils Bull No 74

Wander IW (1949) An interpretation of the cause of water-repellent sandy soils found in the citrus groves of Central Florida. Science 110:299–300

Watson CL, Letey J (1970) Indices for characterizing soil-water repellency based upon contact angle-surface tension relationship. Soil Sci Soc Am Proc 34:841–844

Wilkinson JF, Miller RH (1978) Investigation of localized dry spots on sand golf greens. Agro J 70:299–304

Analysis of Monoterpene Hydrocarbons in the Atmosphere

Y. YOKOUCHI

1 Introduction

Monoterpenes as well as isoprene are important biogenic organics in the atmosphere. Monoterpenes are mainly emitted from coniferous trees, and are often responsible for the aroma of forest air, while isoprene is dominantly emitted from deciduous trees. Since they are chemically reactive, the concern about their role in atmospheric chemistry has prompted several investigations to determine their ambient air concentrations (Rasmussen and Went 1965; Holdren et al. 1979; Yokouchi et al. 1981b, 1983; Roberts et al. 1983b, 1985; Riba et al. 1987; Yokouchi and Ambe 1988).

The detection of monoterpenes in the atmosphere is one type of trace analysis of organic gases. There have been a large number of analytical techniques involving trace analysis, and most of them rely on gas chromatography (GC) and gas chromatography/mass spectrometry (GS/MS) methods coupled with some sample concentration procedure.

The difficulty in analyzing monoterpenes in the atmosphere is that (1) many monoterpenes are chemically and physically similar to each other so that complete chromatographic separation is difficult and (2) they are often present at very low concentrations even in forest air because of their high reactivity. Therefore, methods of high sensitivity and high selectivity are required.

In this chapter, I describe some suitable methods for the determination of atmospheric monoterpenes by GC or GC/MS. In addition to detailing the analytical procedures, some features of monoterpenes in the atmosphere are described.

2 Sampling and Concentration

There have been a large number of sampling and concentration techniques utilized for ambient trace organics (Stern 1976). Among them, the adsorptive trapping method has been successfully utilized for the collection and concentration of atmospheric monoterpenes. Grab sampling using stainless steel or glass bulbs has also been used for the collection in the field followed by cryogenic concentration in the laboratory.

2.1 Adsorption Method

Sampling tubes filled with adsorbents have been widely used for concentrating organics in the atmosphere (Russell 1975; Pellizzari et al. 1975, 1976; Wathne 1983). A known volume of air is drawn through a sampling tube maintained at ambient or subambient temperatures, and the organics are trapped on the adsorbent. After sampling, the tube is connected to a GC or GC/MS and the trapped organics are thermally desorbed directly into an analytical column for analysis. Among the many adsorbents available, Tenax GC (2,6-diphenyl-p-phenylene oxide polymer) is excellent for monoterpenes as well as for many other volatile organics, because of its inertness and its excellent retentive property.

Figure 1 shows the two types of sampling tubes used for the collection of atmospheric monoterpenes. Type A is a Pyrex tube (4.5 mm ID × 15 cm long) packed with 0.3 g of Tenax GC (80/100 mesh) to form a bed approximately 10 cm long and secured at both ends with quartz-wool plugs. This tube is connected to the analytical GC by inserting the needle end into the injection port of the GC. Type B is a stainless steel tubing (1/4" OD × 12 cm long) with Swagelok fittings filled with Tenax GC adsorbent approximately 10 cm in length. This tube can be tightly sealed and is good for several days of storage. The basic features of the interfacing of these tubes and GC are schematically shown in Fig. 2. They involve valves to switch over the carrier gas and some type of heating system for desorption of the sample from the sampling tube.

Prior to sampling, the sampling tubes are conditioned overnight at 220–280 °C (higher than the desorption temperature) with passage of inert gas (nitrogen or helium) and are reconditioned after each use.

Collection is usually accomplished by passing a known volume (usually 100–3000 ml) of ambient air through the Tenax GC tube by a portable pump. The sampling flow rate (10–500 ml/min) is controlled by a needle valve or flow controlling unit. The sampling volume is determined by the expected concentration and the sensitivity of the analytical system is use. However, it should not be more

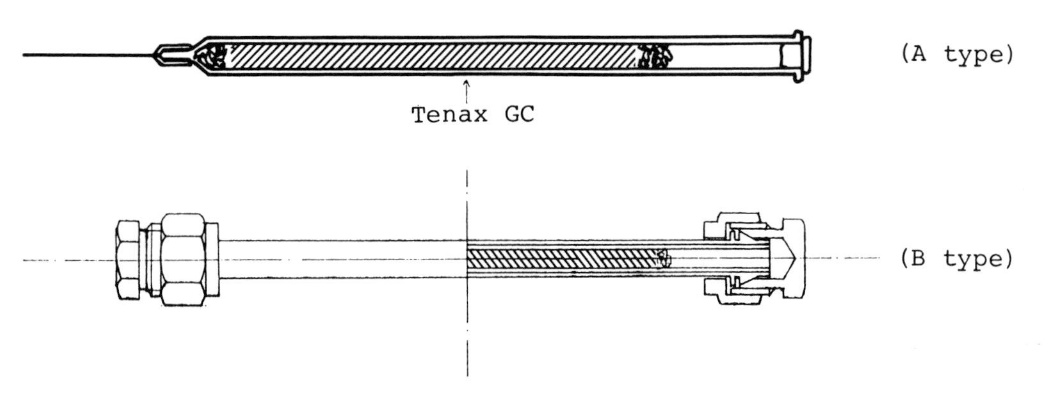

Fig. 1. Adsorption tubes for the sampling of trace organic gases

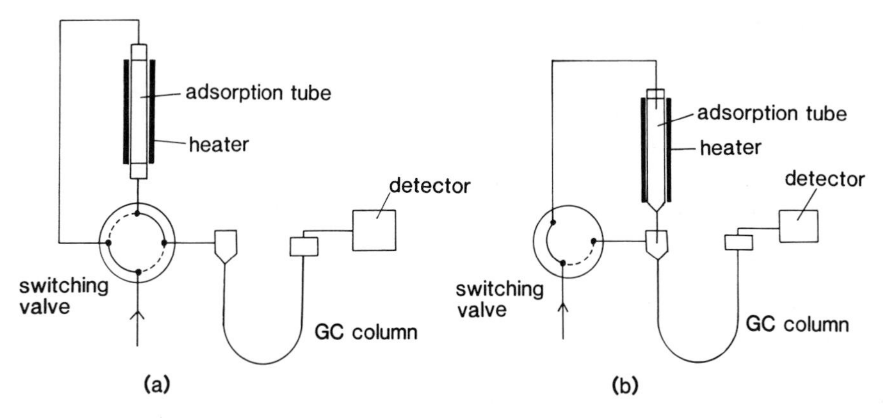

Fig. 2. Schematic diagrams of the analytical systems interfaced with adsorption tubes

than the breakthrough volume, which is the maximum volume of air that may be sampled before losses of the compound of interest can occur. The breakthrough volume of monoterpenes on Tenax GC has been found to be more than 12 liters per 1 g Tenax GC at ambient temperature. It should be noted that the breakthrough volume decreases in accordance with a higher temperature, higher humidity, and higher sampling flow rate.

If a large sampling volume is required because of very low terpene concentrations or if there is a necessity to trap more volatile organics such as isoprene, the adsorptive collection at subambient temperature is recommended. Ice/water, dry ice/acetone, and liquid nitrogen are popular coolants. However, cryogenic trapping has some disadvantages for field study. Furthermore, all sub-0 °C coolants tend to condense large quantities of water that can plug the trapping system, or interfere with the analysis. This problem can be minimized by using a desiccant filter such as calcium chloride at the entrance to the trap. The amount of the desiccant should be as small as possible, since most deciccants may possibly retain some terpenes.

2.2 Grab Sampling

Containers made of stainless steel or glass have also been used for the collection of monoterpenes in the field (Holdren et al. 1979). Samples are usually gathered with an evacuated container by opening a valve or, filled to a positive pressure with a pump. In the laboratory, a known volume of sample is drawn from the container with a syringe or a pump, and is concentrated using the adsorption method described above or using the cryogenic method.

3 GC, GC/MS Method

As described in the Section 1, complete chromatographic separation of many monoterpenes is difficult. In addition, there are many nonterpene hydrocarbons in the atmosphere. Using long and efficient capillary columns is a powerful analytical technique, and capillary GC/MS is usually the preferred method for identification of monoterpenes. However, capillary GC needs to be considered when combined with the adsorption technique as described below. In contrast to capillary GC, packed GC is more easily interfaced to an adsorption tube. Thus, packed GC combined with MS has also been utilized for the analysis of monoterpenes.

Here, I will describe three methods based on packed GC/MS (selected ion monitoring, SIM), capillary GC (flame ionization detector, FID), and capillary GC/MS in combination with the adsorption method at ambient or subambient temperatures.

3.1 Packed GC/MS (SIM)

Interfacing of a sampling tube and packed GC is performed as shown in Fig. 2 so that carrier gas flow rate in the order of 20–50 ml/min can be introduced through the tube. This method was utilized for the analysis of forest air (Yokouchi et al. 1981b).

One liter of air was collected on the sampling tube containing Tenax GC (Fig. 1). The tube was connected to the analytical column as shown in Fig. 2. The GC column was 5% Silicone DS-200 + 5% Bentone 34 on Chromosorb W AW DMCS (60/80 mesh) packed in a glass column (8 ft. × 2 mm ID). Desorption of the sample from the adsorbent was carried out at 200 °C with the carrier gas (He) flowing at a rate of 16 ml/min for 2 min. The monoterpenes transferred onto the analytical column were temporarily trapped at the top of it for 2 min at a temperature of 0 °C and the carrier flow was then switched from the bypass to the GC. The selected ion monitoring (SIM) chromatograms of the ions having m/z 93, 136, 68, and 41 were obtained by temperature programming at 16 °C/min to 100 °C. Ion of m/z 136 is the molecular ion and the other three ions are major fragment ions of many monoterpenes. Positive identifications of monoterpenes were made on the basis of a combination of retention times and the ratio of peak heights of the four monitored ions.

Figure 3 shows the SIM chromatogram at m/z 93 ion of a 1 liter sample collected in a pine forest. α-Pinene, camphene, β-pinene, myrcene, and β-phellandrene were detected. The detection limit of monoterpenes by this method was found to be 0.1 ng at a signal-to-noise (S/N) ratio of >3.

3.2 Capillary GC (FID)

The problem associated with the interface of the adsorption method and capillary GC is that usually more than 10 ml of carrier gas is required to transfer 100% of

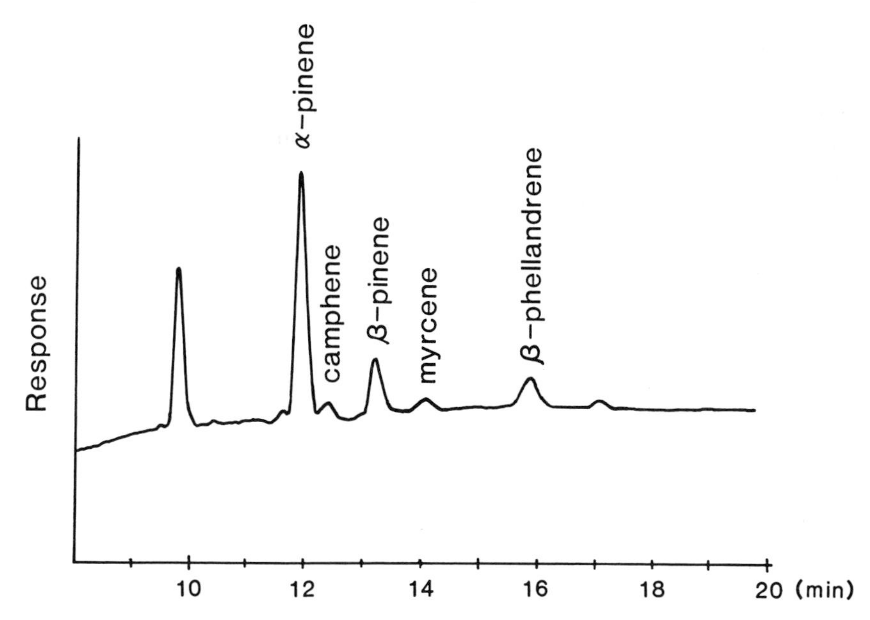

Fig. 3. Selected ion chromatogram (m/z 93) of an air sample collected in a pine forest (Yokouchi et al. 1982b)

the desorbed volatiles into a capillary GC column with a flow rate as low as a few ml/min. Therefore, a focusing step on the capillary column is needed in order to make the best use of high resolution chromatography. Refocusing on the column is usually done by lowering its temperature (cryofocus), and for this technique, there are two different ways: (1) immersion of the tip of a capillary column in a suitable coolant held in a dewar followed by temperature programming of the column; (2) cooling the oven temperature using a cryogenic oven. We used the latter method, since it is more convenient and monoterpenes are effectively focused at the lowest possible temperature of a cryogenic oven (–50 °C). Minimizing the size of the adsorption tube and the dead volume is necessary for rapid desorption and rapid transfer from the trap to the capillary column.

Figures 4 and 5 show systems A and B which were used for the capillary GC analysis of monoterpenes and isoprene in the atmosphere. System A was originally developed for the continuous monitoring of trace organic volatiles at a field station (Yokouchi et al. 1986; Yokouchi and Ambe 1988). System B was used for the analysis of the samples collected on the adsorption tube as shown in Fig. 1B.

System A. The diagram in Fig. 4 is shown in the sampling mode. Two hundred ml of sample air was flown through the $CaCl_2$ trap at a flow rate of 40 ml/min, where most of the water vapor is removed. The volatile organics are then collected in a small trap (T) made of stainless steel (1/8" OD × 4 cm long) packed with Tenax GC(80/100), which is kept at –50 °C by flushing with liquid CO_2. Before purging the trapped sample into the column, the GC oven is cooled to –50 °C. On purging, the valve (VL1) is switched over, and the trap (T) is heated to a preset temperature of 230 °C. Helium

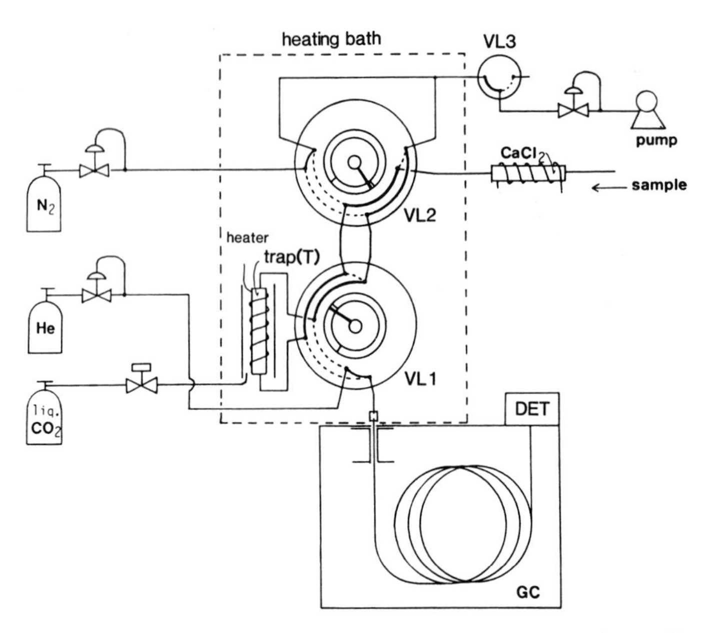

Fig. 4. Schematic diagram of system A for adsorptive concentration/capillary gas chromatography

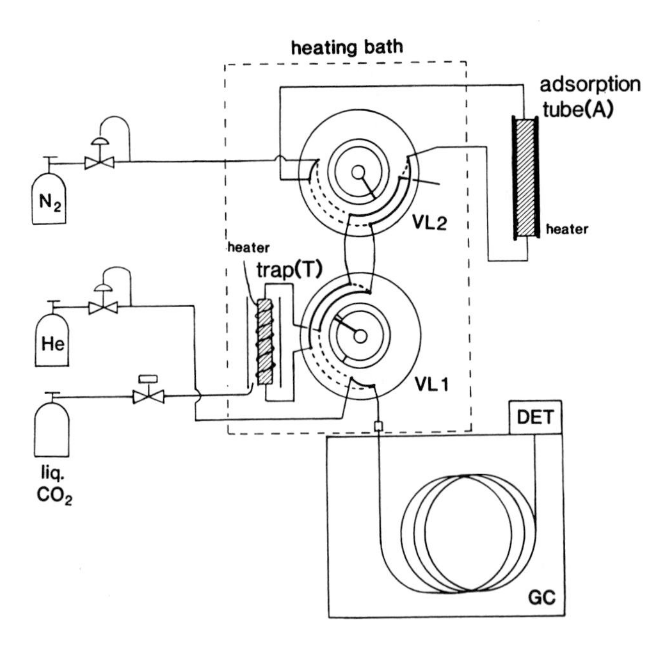

Fig. 5. Schematic diagram of system B for adsorptive concentration/capillary gas chromatography

carrier gas with a flow rate of 2 ml/min sweeps the trapped compounds into the analytical capillary column (methyl silicone, 0.3 mm ID × 25 m long, 0.52 μm thickness) for 5 min. At the end of the purging time, the valve (VL1) is switched back. At this moment, the temperature program (–50 to 280 °C at 8°/min) of the GC is started. During the subsequent analysis, the valves VL2 and VL3 are switched over to allow pure N_2 to flow through them and the temperature of the Tenax GC trap (T) and the water filter remains at 230 °C for 10 and 30 min, respectively. This enables the Tenax trap to be cleaned and the $CaCl_2$ in the water filter to be activated.

Collection and recovery efficiencies for isoprene and monoterpene hydrocarbons were 100% with a precision of ±5%.

Figure 6 shows a chromatogram of the air sample, where the four monoterpenes (α-pinine, camphene, β-pinene, and β-phellandrene) and isoprene were found.

All processes can be performed automatically, and the total automation is effective for precise and continuous analysis.

System B. An adsorption tube containing the air sample is connected to the valve (VL2) as shown in Fig. 5. Organics in the tube are thermally (at 230 °C) desorbed under the flow of N_2 gas and are transferred into the small trap (T) cooled at –50 °C. The

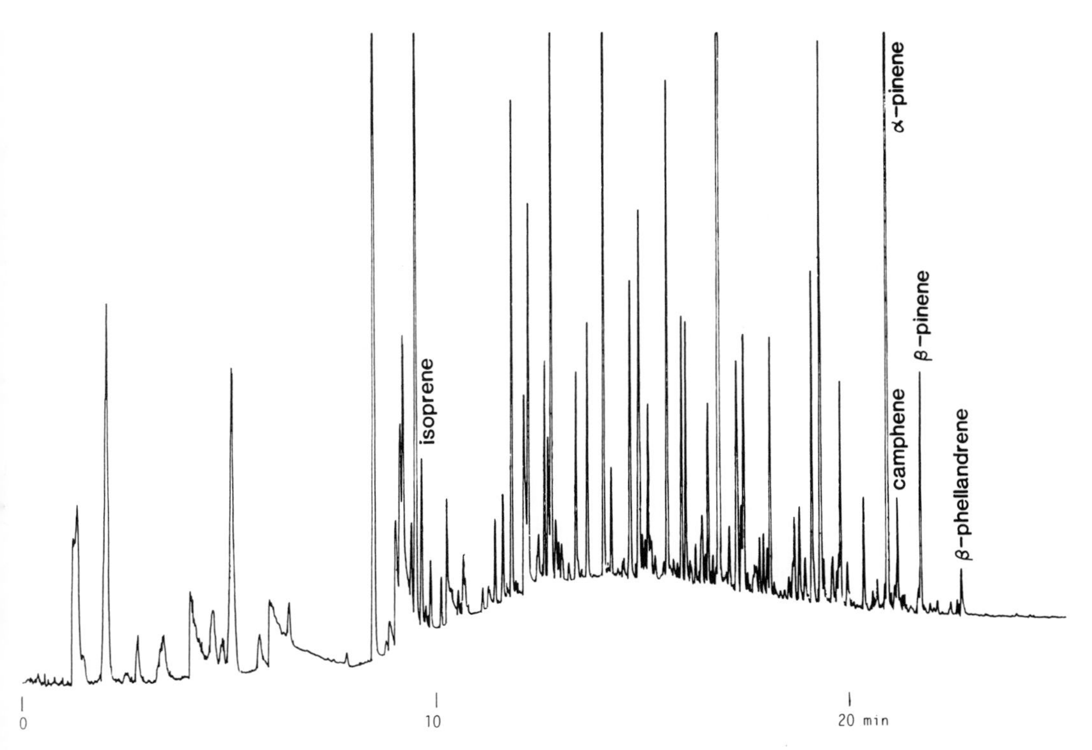

Fig. 6. Gas chromatograms obtained for the forest atmosphere (Yokouchi and Ambe 1988)

following procedure is the same as for System A. This technique is more practical than that of system A, if the sampling sites are not fixed or when a large sampling volume is necessary.

3.3 Capillary GC/MS

The combination of capillary gas chromatography and mass spectrometry is the most powerful method for the identification of volatile organics. The systems described above for capillary GC can be applied to capillary GC/MS without any modification.

Figure 7 shows the reconstructed ion chromatogram (RIC) and the trace of m/z 93 ion of 1 liter of air which was sampled in a green house with an adsorption tube (Fig. 1B) and was analyzed with capillary GC/MS using system B.

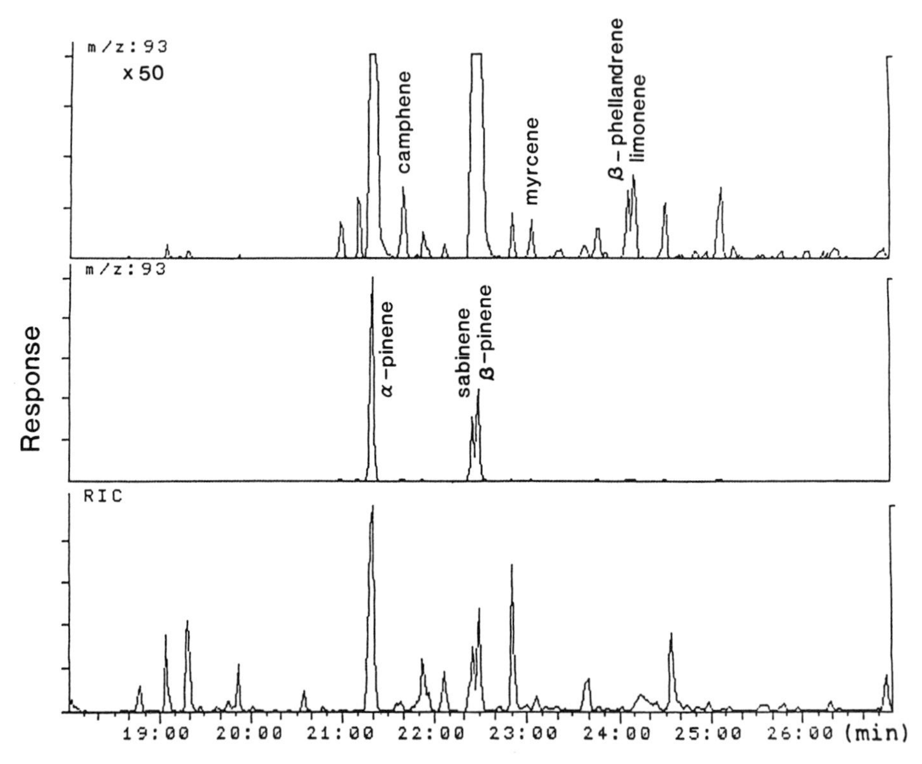

Fig. 7. Reconstructed ion chromatogram (lowest) and m/z 93 ion chromatogram of air sample collected in a greenhouse

4 Calibration

It is desirable that calibration is performed using monoterpene standards at a similar concentration as that of the samples. Monoterpene concentration in the atmosphere is usually in the low-ppb range. Standard gases in the ppb range can be prepared dynamically using the permeation method (O'Keeffe and Ortman 1966; Scaringelli et al. 1970) or the diffusion cell method (Altshuller and Cohen 1960).

More conventionally, standard gases in the ppm range can be used for the purpose of calibration, or for collection and recovery tests by using the following method. The appropriate amount of pure monoterpene is injected into a fixed volume of air in a glass bulb. A sample of gas from this bulb is subsequently diluted until the ppm level is reached. Usually, several milliliters is analyzed in the same manner as that of the air samples.

Standard mixtures in liquid solvents have also been used (Yokouchi et al. 1981b; Roberts et al. 1983a). Several microliters of standard monoterpenes (approximately 1 ng/μl) are injected onto the adsorption tube, and analyzed in the same manner as that of the atmospheric samples after eluting the solvent by flushing with pure air. If the sampling and recovery efficiencies of the method used are ascertained to be quantitative, a calibration curve prepared from the direct injection of standard solutions to the GC can be used.

5 Features of Atmospheric Monoterpenes

The most noticeable feature of monoterpene concentration in the atmosphere is that it varies greatly from day to day, and from season to season (Yokouchi et al. 1983; Roberts et al. 1985; Riba et al. 1987; Yokouchi and Ambe 1988), reflecting the variation of their emission rate and their disappearing rate through dilution and reaction. Thus, numerous measurements should be performed to accurately assess atmospheric levels.

Monoterpene compositions in the atmosphere can be characterized by the relative ratios of individual constituents and, moreover, characteristic patterns for particular classes of forests have been found. For example, for a pine forest, α-pinene is dominant, while β-pinene, β-phellandrene, and myrcene are also present.

Figure 8 shows the monoterpene composition of (a) the leaf oil; (b) the foliar emission gas; (c) the atmosphere in the pine forest (Yokouchi et al. 1981a). Comparing (a) and (b), monoterpenes with higher volatility such as α-pinene are more dominant in the emission gas than in the leaf oil. This suggests that the emission of monoterpenes is controlled by vapor pressure. Comparing (b) and (c), myrcene, one of the most reactive monoterpenes, is much reduced in the atmosphere. This is most likely because reactivity is an important factor for the fate of monoterpenes in forest air. Therefore, the relative ratios of monoterpenes in the forest might be roughly estimated from that in the leaf oil when the vapor pressure and the reaction rate of each monoterpene are known, and vice versa.

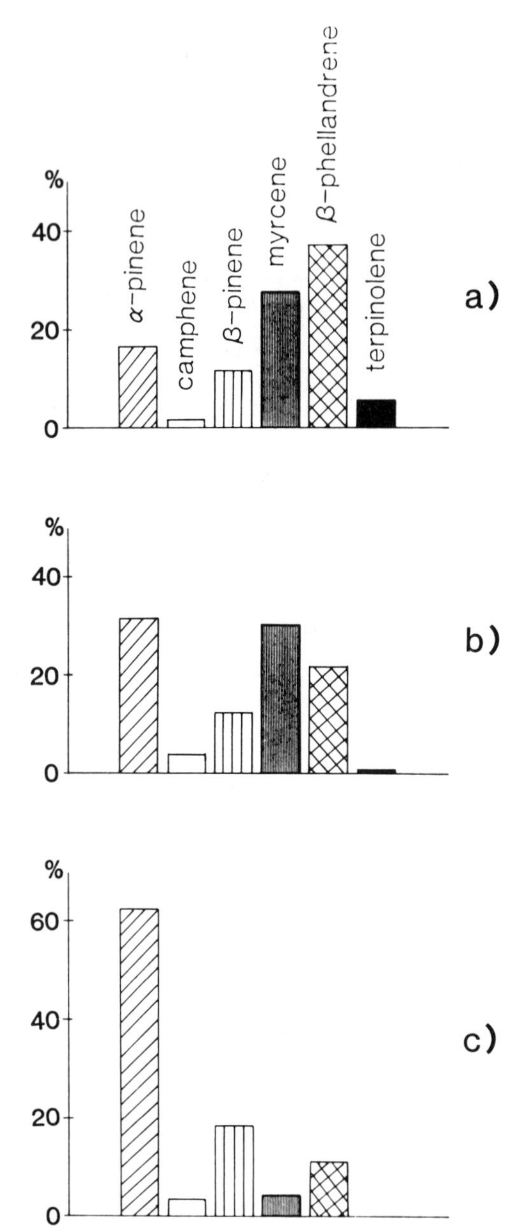

Fig. 8. Monoterpene composition of **a** the pine leaf (needle) oil; **b** the foliar emission gas; and **c** the atmosphere in the pine forest (Yokouchi et al. 1981a)

References

Altshuller AP, Cohen IR (1960) Application of diffusion cells to the production of known concentrations of gaseous hydrocarbons. Anal Chem 32:802–810

Holdren MW, Westberg HH, Zimmerman PR (1979) Analysis of monoterpene hydrocarbons in rural atmosphere. J Geophys Res 84:5083–5087

O'Keeffe AE, Ortman GC (1966) Primary standard for trace gas analysis. Anal Chem 38:760–763

Pellizzari ED, Bunch JE, Carpenter H (1975) Collection and analysis of trace organic vapor pollutants in ambient atmosphere. Environ Sci Technol 9:552–555

Pellizzari ED, Bunch JE, Berkley RE, McRae J (1976) Determination of trace hazardous organic vapor pollutants in ambient atmospheres by gas chromatography/mass spectrometry/computer. Anal Chem 48:803–807

Rasmussen RA, Went FW (1965) Volatile organic material of plant origin in the atmosphere. Proc Natl Acad Sci USA 53:215–220

Riba ML, Tathy JP, Tsiropoulos N, Monsarrat B, Torres L (1987) Diurnal variation in the concentration of α- and β-pinene in the Landes Forest (France). Atmos Environ 21:191–193

Roberts JM, Fehsenfeld DL, Albritton DL, Sievers RE (1983a) Sampling and analysis of monoterpene hydrocarbons in the atmosphere with Tenax gas chromatographic porous polymer. In: Keith LH (ed) Identification and analysis of organic pollutants in air. Butterworth, Boston, MA, pp 371–387

Roberts JM, Fehsenfeld DL, Albritton DL, Sievers RE (1983b) Measurement of monoterpene hydrocarbons at Niwot Ridge, Colorado. J Geophys Res 88:10667–10678

Roberts JM, Hahn CJ, Fehsenfeld FC, Warnock JM, Albritton DL, Sievers RE (1985) Monoterpene hydrocarbons in the nighttime troposphere. Environ Sci Technol 19:364–369

Russell JW (1975) Analysis of air pollutants using sampling tubes and gas chromatography. Environ Sci Technol 9:1175–1178

Scaringelli FP, O'Keeffe AE, Rosenberg E, Bell JP (1970) Preparation of known concentrations of gases and vapors with permeation devices caliberated gravimetrically. Anal Chem 42:871–876

Stern AC (ed) (1976) Air pollution. Measuring, monitoring and surveillance of air pollution. 3rd edn, vol 3. Academic Press, New York

Wathne BM (1983) Measurements of benzene, toluene and xylenes in urban air. Atmos Environ 17:1713–1722

Yokouchi Y, Ambe Y (1988) Diurnal variations of atmospheric isoprene and monoterpene hydrocarbons in an agricultural area in summertime. J Geophys Res 93:3751–3759

Yokouchi Y, Ambe Y, Fuwa K (1981a) The relationship of the monoterpene composition in the atmosphere, the foliar emission gas and the leaf oil of *Pinus densiflora*. Chemosphere 10:209–213

Yokouchi Y, Ambe Y, Maeda T (1986) Automated analysis of C_3-C_{13} hydrocarbons in the atmosphere by capillary gas chromatography with a cryogenic preconcentration. Anal Sci 2:571–575

Yokouchi Y, Fujii T, Ambe Y, Fuwa K (1981b) Determination of monoterpene hydrocarbons in the atmosphere. J Chromatogr 209:293–298

Yokouchi Y, Okaniwa M, Ambe Y, Fuwa K (1983) Seasonal variation of monoterpenes in the atmosphere of a pine forest. Atmos Environ 17:743–750

Evaluation of Antimicrobial Activity of Essential (Volatile) Oils

S.G. DEANS

1 Introduction

Higher plants have been exploited as a source of biologically active compounds since antiquity. In particular, the ability to inhibit the growth of spoilage and food poisoning bacteria, human and animal pathogens and a number of filamentous fungi has been of immense importance to man over the centuries (Zaika 1989; Deans and Svoboda 1990a; Deans et al. 1990). It is worth noting that even with today's battery of synthetic and semi-synthetic antibiotics, over 25% of pharmaceutical preparations in the West contain at least one component originating from plant sources: in the East this percentage is far higher. That some of these plant antimicrobials also possess antioxidant properties is a welcome bonus in the quest to preserve the food reserves of the world.

Natural antimicrobial compounds have been detected in a number of plant genera, with the Labiatae being particularly well represented. Attempts have been made by several research groups to isolate and identify the individual constituents present in the plant and to relate the chemical composition to biological activity. In many plants exhibiting biological activities, the greatest antimicrobial power lies with the *essential* or *volatile oil* fraction. The volatile oil profile is subject to variability due to a number of intrinsic and extrinsic factors. The geographical location of aromatic plants is of considerable importance as the constituents of the volatile oils vary with seasonal, climatic and ontogenic changes (Rhyu 1979; Deans and Svoboda 1988).

2 Extraction of Plant Volatile Oil

There is a variety of different methods of obtaining volatile oils ranging from cold or hot expression, aqueous infusion, organic solvent extraction, steam distillation and supercritical carbon dioxide (CO_2). Of these, steam distillation and organic solvent extraction are the more commonly used. The chemical composition, both quantitative and qualitative, will differ according to the technique adopted: some procedures will not remove compounds that are insoluble in water. The volatile oil fractions obtained by steam distillation is chemically complex: in the case of *Artemisia dracunculus* (French tarragon) there were over 50 compounds present (Lawrence 1979; Balza et al. 1985). The basic principle involved in steam distillation is to boil the fresh or dried plant material in distilled water, a process that results in the disruption of the gland cells (Figs. 1, 2) and release of the volatile

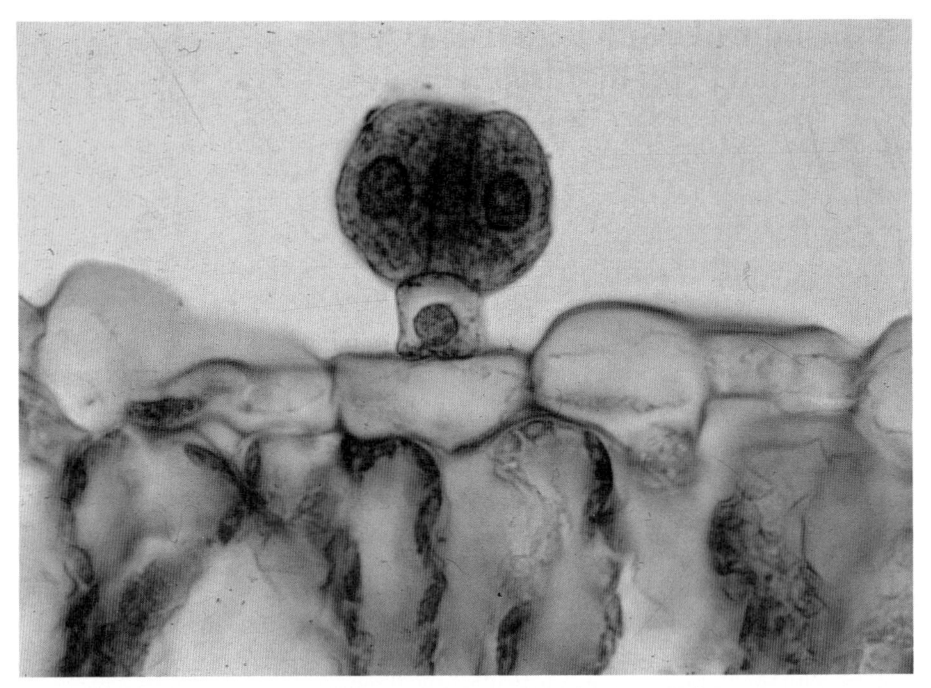

Fig. 1. Stalked glandular hair of *Rosmarinus officinalis* (magnification ×1000)

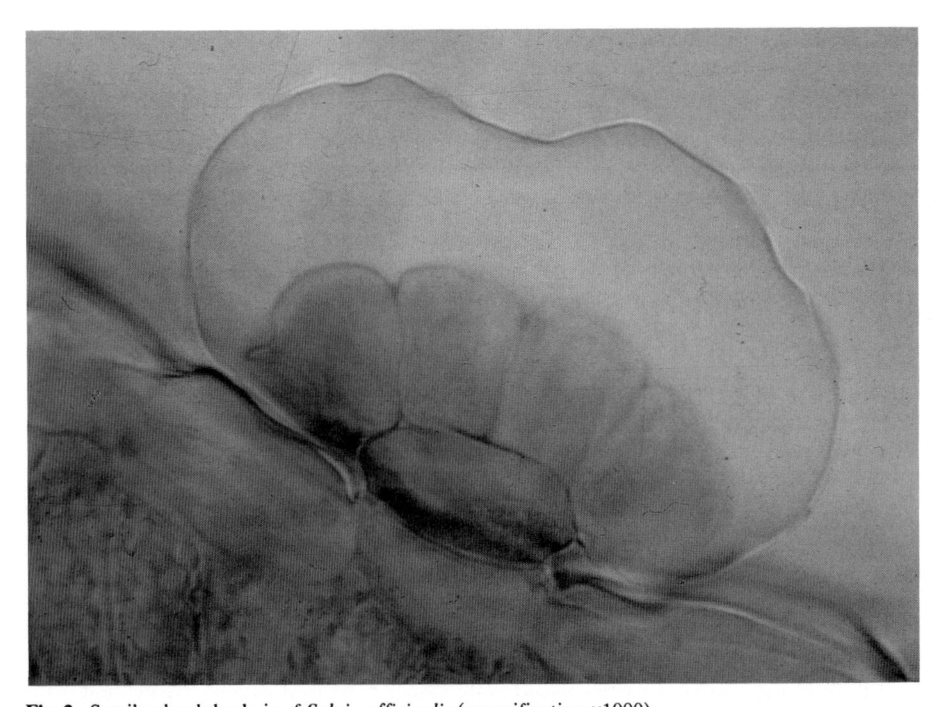

Fig. 2. Sessile glandular hair of *Salvia officinalis* (magnification ×1000)

substances contained therein. The steam plus volatile oil passes through a cold-water condenser allowing the volatile oil fraction to float on top of the water. Collection can be made by drawing out the water followed by the volatile oil. Typically, steam distillation for 3 to 5 h is required, while the oil yield is around 1–2% (v/w) with a few plants having higher values (*Rosmarinus officinalis* 3–5%, *Origanum officinalis* 3–7%).

The choice of organic solvent for extracting volatile oils is usually restricted to alcohols (MeOH and ETOH) but other solvents are also possible, although for food or pharmaceutical applications, the problem of residual solvent usually precludes their choice (Evans et al. 1986; Hethelyi et al. 1988).

The use of supercritical CO_2 to extract volatile oils and oleoresins from aromatic plants is a recent innovation, but one that has many advantages: it behaves like a solvent and can be manipulated to obtain differential or sequential fractions; all traces of the gas are removed from the volatile oil; it is particularly useful for heat-labile compounds. Studies on *Origanum majorana* and *Artemisia absintheum* have shown that by using this technique, a higher yield of specific oil constituents can be realized. In addition, it is possible using this technique to selectively deterpinate citrus volatile oils with a high (90%) monoterpene hydrocarbon content. This technique allows for the retention of colour and results in a high value product (Stahl and Quirin 1984).

3 Evaluation of Antimicrobial Properties of Volatile Oils

3.1 Antibacterial Testing

There exist a number of tests aimed at evaluating the potency of volatile oils and their constituents in inhibiting the growth of bacterial cells. Prior to the test procedure being undertaken, it is important to ensure that the bacterial culture is in a highly active growth phase. To achieve this, a culture should be started from a stock slope 3 days before the test proper, and placed in a suitable broth matching the medium constituents of the agar to be used. IsoSensitest Broth and Agar (Oxoid) are very suited to this purpose having been designed for antibiotic screening. IsoSensitest medium is extremely nutritious and supplies the total requirements for all but the most fastidious organisms. An alternative medium is Tryptic Soya Broth and Agar (Difco), but is slightly less well defined in terms of chemical constituents. The broth culture is subcultured into fresh broth 24 h before the test and incubated at the temperature at which the test will be carried out. This should give a very vigorous culture ready to test, and there will be very little lag time when the seeded agar is prepared.

The *paper disc method* is a well-established test wherein filter paper discs impregnated with the test solution are applied to the surface of a seeded agar plate. Following incubation at an appropriate temperature to suit the test bacterium, a zone of inhibition may form around the disc where antibacterial activity is present. This technique has the drawback that should the plates be even slightly moist, the test substance can spread over an area greater than the disc giving the impression that the compound is very effective at inhibition of growth. Accuracy of application of the disc is of paramount importance since the disc has only to touch the surface of the seeded plate to exert an inhibitory effect. Equally important is the accuracy of determination of the volume of the test substance: it should be applied using a micropipette. Time must be allowed to permit the test solution to diffuse into the agar before the plate is inverted for incubation.

A modification of this technique is the *well test* whereby a plug of agar is removed from the seeded plate and the test substance placed in the well so formed. This has the advantage that all the sample is confined and must undergo radial diffusion into the agar. The volume in the well can be very accurately controlled if a fixed volume of agar is used each time: for 25 ml agar and using a 4-mm well punch, between 12 and 15 µl can be added to this well. It is important to allow time for the test solution to diffuse into the agar, and between 30 and 60 min should be allowed prior to plate inversion. Plates should be incubated in the dark at a temperature between 25 and 30 °C: any higher than this may encourage volatilization of the test solutions. Inspection of the plates following incubation should show zones of inhibition around the wells and their diameter can be measured with vernier calipers (Deans and Ritchie 1987).

Using a technique borrowed from disinfectant testing, the *suspension test* aims to give a mathematical expression of the killing power of the test solution. This is possible by removing samples, at specified times, from a reaction mixture of a suspension of bacterial cells and test solution in order to estimate the number of survivors at that given time. By plotting survivors against contact time, a linear relationship reveals the nature of the logarithmic death (Fig. 3). From this graph, the *decimal reduction time* can be calculated, being the time taken for the bacterial population to be reduced by 90% (or one logarithmic cycle). This is a direct measure of the effectiveness of the test solution, and can be carried out with a number of different variables (concentration, test organism, temperature and pH). The accuracy of this particular test resides with a number of factors, the important ones being the ability to dilute out the effect of the test compound immediately following a timed sample. With disinfectants, this simply involves making the first serial dilution in an inactivator buffer, but this may not be possible with volatile oils and their constituents. Effective mixing must be maintained in order to prevent the formation of areas in the flask where poor interaction between organism and test solution occurs.

A simplified version of the suspension test involves inoculating a broth with an actively growing culture to which is added the test compound. Following incubation for a defined period, *turbidity* in the broth is assessed. There can be problems with this test if the volatile oil does not mix easily into the broth: this problem may also be present in the suspension test, but the constant agitation alleviates this.

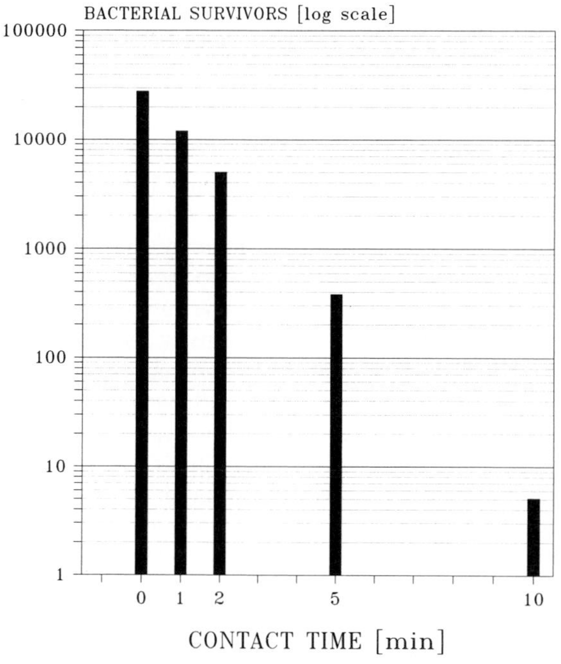

Fig. 3. Graphical representation of reduction in numbers of bacteria on exposure to plant volatile oil. Decimal reduction time is calculated from gradient of line passing through points on histogram

3.2 Antifungal Testing

Attempts have been made by research workers to assess antifungal activity by the *paper disc* technique described above (Connor and Beuchat 1984). The major problems with this approach centre around the slow-growing habit of filamentous fungi and some yeast in comparison to bacteria. Typically, it can require incubation for 7 to 10 days to obtain sufficient mycelial growth for assessment: during this prolonged incubation period, a proportion of the test substance will have volatilized making the assessment difficult to quantify. Other workers have deliberately utilized this volatility by placing impregnated discs on the lid of a Petri dish and inverting the inoculated agar plate. Following incubation, the extent of fungal inhibition by vapours from the disc was determined by inspection (Maruzzella and Sicurella 1960; Benjilali et al. 1984).

To alleviate some of these problems, an alternative approach is to use reductions in dry weight of mycelium following incubation of the fungi in contact with volatile oils. Continual shaking is required in order that good contact be maintained between the fungus and the volatile oil. Flasks without the addition of volatile oils serve as controls (Deans and Svoboda 1990b). Following incubation for up to 10 days, the contents of the flasks are passed through a preweighed dried glass fibre filter (Whatman GF/C) and washed several times with sterile distilled water. The filters are dried at 105 °C for 5 h, or until constant weight, and the index of inhibition calculated from the following formula:

$$\text{Percentage Inhibition} = \frac{C - T}{C} \times 100,$$

where C = mean weight of mycelium from *Control* flasks; T = mean weight of mycelium from *Test* flasks. Studies on spoilage and mycotoxigenic fungi (Buchanan and Shepherd 1981; Azzouz and Bullerman 1982; Bahk and Marth 1983; Karapinar 1985; Graham and Graham 1987; Garg and Dengre 1988; Farag et al. 1989; Knobloch et al. 1989; Deans et al. 1990) have revealed a pronounced effect upon toxin synthesis by the presence of volatile oils. Aflatoxin production by *Aspergillus* species was greatly reduced when volatile oils were present even at low concentrations. In a study of the volatile oil from *Pimpinella anisum* seed, significant antifungal activity against members of the genera *Alternaria, Aspergillus, Cladosporium, Fusarium* and *Penicillium* was recorded at concentrations of 500 ppm, the active constituent having been identified as anethole (Shukla and Tripathi 1987).

It is important to note that mycotoxins are not formed during exposure to volatile oils since, as secondary metabolites, their synthesis may have been encouraged in view of the restrictions to primary metabolism. However, there is a requirement for a biochemical committment to their synthesis which is not realized with such restricted mycelial growth.

Exploitation of volatile oils as fungicides against field crop pathogens has yet to receive widespread acceptance, despite the fact that in laboratory tests, several fungi responsible for significant crop damage can be severely inhibited in their growth by low concentrations of these oils. There are a number of problems to be addressed prior to their general use as crop protection agents. Resistance to washing off must be fully achievable, but in view of their restricted or non-solubility in water, this should not be insurmountable. Prolonged action over several weeks may present difficulties with a volatile substance, and little study has been undertaken into development of resistance in the target pathogen. The site of action of volatile oils and their constituents at the cellular and molecular levels represents an important area requiring fundamental scientific investigation. This will best be achieved by molecular biology and biotechnology studies into enzymic pathways and radiolabelled intermediates. Electron microscopy may also have a part to play since morphological changes often accompany inhibition of growth in fungi (Deans et al. 1980). In a chemotaxonomic study of *Ocimum basilicum*, Reuveni et al. (1984) highlighted the importance of choosing the most appropriate chemotype since the major constituent may be variable. They recorded that the European types had concentrations of linalool > methyl chavicol > cineol > eugenol, while Reunion types had methyl chavicol > linalool > cineol > eugenol with the most active antifungal oil coming from Israeli plants with oil composition linalool > methyl chavicol > eugenol > cineol when tested against *Fusarium oxysporum* and *Rhizopus nigricans*.

4 Antimicrobial Activity of Volatile Oils

From the various studies into the bioactivity of plant volatile oils, there are several findings which merit note. Despite the differences in methods of assessment, it is clear that many higher plants contain compounds of interest and value to man. Plant secondary metabolites have been shown to be active against food poisoning bacteria such as *Clostridium botulinum, Salmonella enteritidis, Staphylococcus aureus* and *Yersinia enterocolitica* (Dabbah et al. 1970; Beuchat 1976; Huhtanen 1980; Tharib et al. 1983; Aktug and Karapinar 1986; Deans and Ritchie 1987; Deans and Svoboda 1988, 1989), in addition to food spoilage organisms such as lactic acid bacteria, *Saccharomyces* sp., *Bacillus cereus* and *Pseudomonas fluorescens* (MacNeill et al. 1973; Nadal et al. 1973; Shelef et al. 1980; Zaika et al. 1983; Connor and Beuchat 1984; Janssen et al. 1985). Culinary compounds such as garlic, vanillin and eugenol from clove have been demonstrated to inhibit the growth of the pathogenic and dimorphic yeasts *Candida albicans, Cryptococcus neoformans* and *Histoplasma capsulatum* (Fliermans 1973; Fromtling and Bulmer 1978; Boonchird and Flegel 1982; Ghannoum 1988). Animal pathogenic fungi, *Microsporum gypseum, Trichophyton equinum* and *Trichophyton rubrum*, were shown to be inhibited in their growth by a number of Indian plants including *Artemisia maritima, Cymbopogon flexuosus, Santalum album, Trachyspermum ammi* (the most active) and *Vetiveria zizanoides* (Dikshit and Husain 1984). Viruses have also been shown to be susceptible to volatile oils, although the mechanism of action is not fully understood. Plant, animal and human viruses have all been tested as to their susceptibility to these oils, and there is even some evidence that plant extracts can inhibit the HIV virus in vitro (A. Bell, pers. commun.; van den Berghe et al. 1978; Pompei et al. 1979; Ieven et al. 1982; Kaul et al. 1985; Romero et al. 1989).

The volatile oil obtained from *Origanum marjorana* gives a good illustration of the extent of biological activity located in a single species. It was found to be inhibitory to the growth of 25 test bacteria, and these findings are summarized in Table 1 (Deans and Svoboda 1990b). *Beneckea natriegens, Erwinia carotovora* and *Moraxella* sp. were very susceptible to marjoram oil, having inhibition zones with diameters in excess of 20 mm. Several organisms of public health significance including *Pseudomonas aeruginosa, Salmonella pullorum* and *Yersinia enterocolitica* were inhibited in their growth to some extent, while *Escherichia coli* and *Staphylococcus aureus* were less affected. Table 2 shows the *antifungal activity* of this oil against spoilage and mycotoxigenic organisms. There were differing degrees of inhibition, the most susceptible of these organisms, at all dilutions of the oil, being *Aspergillus niger*. At concentrations of 10 $\mu l\ ml^{-1}$, the five fungi were inhibited to a similar extent, while at a concentration of 1 $\mu l\ ml^{-1}$, both the mycotoxin-producing strains *A. flavus* and *A. ochraceus* were least inhibited by the oil. The major constituents of *Origanum marjorana* volatile oil were determined to be alpha-terpinene, gamma-terpinene, sabinene, *cis*-sabinene hydrate, linalool, carvacrol and eugenol (Janssen et al. 1988). Several of these components have been previously tested for biological activity and found to be active against

Table 1. Antibacterial properties of marjoram volatile oil at four dilutions (inhibition zone diameter in mm; diameter of well, 4 mm, included)

Organism	Dilution of marjoram volatile oil			
	1:1	1:2	1:5	1:10
Acinetobacter calcoacetica	14.3	14.8	11.8	8.6
Aeromonas hydrophila	11.3	11.5	8.5	7.6
Alcaligenes faecalis	10.4	11.1	9.6	8.1
Bacillus subtilis	13.6	12.5	11.9	11.7
Beneckea natriegens	20.1	10.9	10.6	8.2
Brevibacterium linens	13.9	13.0	9.9	5.3
Brocothrix thermosphacta	8.0	8.3	8.7	6.9
Citrobacter freundii	10.0	9.9	9.0	7.2
Clostridium sporogenes	8.7	5.7	4.0	4.0
Enterobacter aerogenes	11.6	11.4	7.4	6.2
Enterococcus faecalis	8.7	4.8	4.2	4.2
Erwinia carotovora	22.4	10.9	7.8	6.6
Escherichia coli	9.4	8.8	8.2	6.3
Flavobacterium suaveolens	18.9	14.5	16.6	11.9
Klebsiella pneumoniae	9.2	8.2	7.4	6.8
Lactobacillus plantarum	17.7	9.1	7.9	6.7
Leuconostoc cremoris	8.1	7.6	5.9	5.2
Micrococcus luteus	10.6	8.3	7.2	5.6
Moraxella sp.	21.7	13.6	11.0	10.0
Proteus vulgaris	12.9	11.9	9.6	8.2
Pseudomonas aeruginosa	19.2	12.8	8.9	6.8
Salmonella pullorum	14.3	12.9	11.2	9.1
Serratia marcescens	11.9	12.1	11.6	7.5
Staphylococcus aureus	5.5	5.3	4.6	4.2
Yersinia enterocolitica	15.7	9.2	11.3	8.1

Table 2. Antifungal properties of marjoram volatile oil (inhibition index as % relative to control flasks)

Organism	Marjoram volatile oil ($\mu l\ ml^{-1}$)			
	1	2	5	10
Aspergillus flavus	9	72	74	83
Aspergillus niger	82	83	88	89
Aspergillus ochraceus	20	83	83	84
Aspergillus parasiticus	75	72	75	81
Trichoderma viride	74	77	79	79

a number of bacteria (Deans and Svoboda 1988, 1989) and fungi (McDowell et al. 1988). The potency of the oil against filamentous fungi is significant, in particular against the mycotoxigenic strains and there may well be a future for their greater exploitation as natural fungicides.

5 Future Developments in Volatile Oils

For utilization of plant volatile oils on an industrial scale, it may be necessary to obtain such secondary metabolites from tissue culture-derived material. Several laboratories have used cell suspension cultures as a source of fine chemicals, but suffered from the inherent genetic instability in both the growth and secondary metabolites profiles of these cell lines (Deus-Neumann and Zenk 1984; D'Amato 1985). The genetic transformation of dicotyledenous plants with the plant pathogenic bacterium *Agrobacterium rhizogenes* can result in the formation of stable differentiated root cultures (Riker 1930; Ooms et al. 1985). Upon infection at the wound site, two strands of bacterial plasmid DNA are inserted into the plant genome where they randomly integrate (Huffmann et al. 1984; Ambros et al. 1986). This has the effect of altering the auxin metabolism with the resultant production of 'hairy' roots (Fig. 4). The root cultures are fast growing and are

Fig. 4. Transformed root culture of *Osimum basilicum* in flask culture

biochemically as well as genetically stable, and in general possess the same level and spectrum of products as would the normal root system in vivo (Tepfer 1984; Hamill et al. 1986; Parr and Hamill 1987; Aird et al. 1988; Kennedy et al. 1990). This culture system offers the exciting prospect of production of biologically active secondary metabolites from plant cells which can be handled in a manner similar to microorganisms in fermenters. There are numerous advantages to this method of production, including immediate response to an increase in demand irrespective of season, freedom from climatic vagaries, pests and disease, and product formation in a clean, sterile environment.

Acknowledgements. The author wishes to express his gratitude to colleagues involved in aromatic plant research for assistance and guidance, especially Professor P.G. Waterman and Dr. A.I. Gray of Strathclyde University Phytochemistry Research Laboratory; Dr. K.P. Svoboda, Mr. A.I. Kennedy and Mrs. E. Eaglesham of Scottish Agricultural College, Auchincruive. Support from DAFS Increased Flexibility Scheme is gratefully acknowledged.

References

Aird ELH, Hamill JD, Rhodes MJC (1988) Cytogenetic analysis of hairy root cultures from a number of plant species transformed by *Agrobacterium rhizogenes*. Plant Cell Tissue Organ Cult 15:47–57

Aktug SE, Karapinar M (1986) Sensitivity of some common food-poisoning bacteria to thyme, mint and bay leaves. Int J Food Microbiol 3:349–354

D'Amato F (1985) Cytogenetics of plant cell and tissue cultures and their regenerates. CRC Crit Rev Plant Sci 3:73–112

Ambros PF, Matzke AJM, Matzke MA (1986) Localisation of *Agrobacterium rhizogenes* T-DNA in plant chromosomes by in situ hybridisation. EMBO J 5:2037–2077

Azzouz MA, Bullerman LB (1982) Comparative antimycotic effects of selected herbs, spices, plant components and commercial antifungal agents. J Food Prot 45:1298–1301

Bahk J, Marth EH (1983) Growth and synthesis of aflatoxin by *Aspergillus parasiticus* in the presence of ginseng products. J Food Prot 46:210–215

Balza F, Jamieson L, Towers GHN (1985) Chemical constituents of the arial parts of *Artemisia dracunculus*. J Natural Prod 48:339–340

Benjilali B, Tantaoui-Elaraki A, Ayadi A, Ihlal M (1984) Method to study antimicrobial effects of essential oils: application to the antifungal activity of six Moroccan essences. J Food Prot 47:748–752

Beuchat LR (1976) Sensitivity of *Vibrio parahaemolyticus* to spices and organic acids. J Food Sci 41:899–902

Boonchird C, Flegel T (1982) In vitro antifungal activity of eugenol and vanillin against *Candida albicans* and *Cryptococcus neoformans*. Can J Microbiol 28:1235–1241

Buchanan RL, Shepherd AJ (1981) Inhibition of *Aspergillus parasiticus* by thymol. J Food Sci 46:976–977

Connor DE, Beuchat LR (1984) Effects of essential oils from plants on growth of food spoilage yeasts. J Food Sci 49:429–434

Dabbah R, Edwards VM, Moats WA (1970) Antimicrobial action of some citrus fruit oils on selected food-borne bacteria. Appl Microbiol 19:27–31

Deans SG, Ritchie GA (1987) Antibacterial activity of plant essential oils. Int J Food Microbiol 5:165–180

Deans SG, Svoboda KP (1988) Antibacterial activity of French tarragon (*Artemisia dracunculus* L.) essential oil and its constituents during ontogeny. J Hortic Sci 63:135–140

Deans SG, Svoboda KP (1989) Antibacterial activity of summer savory (*Satureja hortensis*) essential oil and its constituents. J Hortic Sci 64:205–211

Deans SG, Svoboda KP (1990a) Biotechnology and bioactivity of culinary and medicinal plants. AgBiotech News Info 2:211–216
Deans SG, Svoboda KP (1990b) The antimicrobial properties of marjoram (*Origanum majorana* L.) volatile oil. Flavour Fragnance J 5:187–190
Deans SG, Gull K, Smith JE (1980) Ultrastructural changes during microcycle conidiation of *Aspergillus niger*. Trans Br Mycol Soc 74:493–502
Deans SG, Svoboda KP, Gundidza M, Brechany EY (1990) Essential oil profiles of several temperate and tropical aromatic plants: their antimicrobial and antioxidant activities. Acta Horticulturae (in press)
Deus-Neumann B, Zenk MH (1984) Instability of indole alkaloid production in *Catharanthus roseus* cell suspension cultures. Planta Med 50:427–431
Dikshit A, Husain A (1984) Antifungal action of some essential oils against animal pathogens. Fitoterapia 55:171–176
Evans JS, Pattison E, Morris P (1986) Antimicrobial agents from plant cell cultures. In: Morris P, Scragg AH, Stafford A, Fowler MW (eds) Secondary metabolism in plant cell cultures. Cambridge, Cambridge Univ Press, pp 47–53
Farag RS, Daw ZY, Abo-Raya SH (1989) Influence of some spice essential oils on *Aspergillus parasiticus* growth and production of aflatoxin in a synthetic medium. J Food Sci 54:74–77
Fliermans CB (1973) Inhibition of *Histoplasma capsulatum* by garlic. Appl Mycopathol Mycol 50:227–231
Fromtling RA, Bulmer GS (1978) In vitro effect of aqueous extract of garlic (*Allium sativum*) on the growth and viability of *Cryptococcus neoformans*. Mycologia 70:397–405
Garg SC, Dengre SL (1988) Antifungal activity of the essential oil of *Mytus communis* var. *microphylla*. Herba Hung 27:123–124
Ghannoum MA (1988) Studies on the anticandidal mode of action of *Allium sativum* (garlic). J Gen Microbiol 134:2917–2924
Graham HD, Graham EJF (1987) Inhibition of *Aspergillus parasiticus* growth and toxin production by garlic. J Food Safety 8:101–108
Hamill JD, Parr AJ, Robins RJ, Rhodes MJC (1986) Secondary product formation by cultures of *Beta vulgaris* and *Nicotiana rustica* transformed with *Agrobacterium rhizogenes*. Plant Cell Rep 5:111–114
Hethelyi E, Tetenyi P, Kaposi P, Danos B, Kernoczi Z, Buki G, Koczka I (1988) GC-MS investigation of antimicrobial and repellant compounds. Herba Hung 27:89–94
Huffmann GA, White FF, Gordon MP, Nester EW (1984) Hairy root inducing plasmid: physical map and homology to tumor inducing plasmids. J Bacteriol 157:269–276
Huhtanen CN (1980) Inhibition of *Clostridium botulinum* by spice extracts and aliphatic alcohols. J Food Prot 43:195–196
Ieven M, Vlietinck AJ, Van den Berghe DA, Totte J (1982) Plant antiviral agents. III Isolation of alkaloids from *Clivia miniata* Regel (Amaryllidaceae). J Nat Prod 45:564–573
Janssen AM, Scheffer JJC, Baerheim Svendsen A, Aynehchi Y (1985) Composition and antimicrobial activity of the essential oil of *Ducrosia anethifolia*. In: Baerheim Svendsen A, Scheffer JJC (eds) Essential oils and aromatic plants. Martinus Nijhoff, Dordrecht, pp 213–216
Janssen AM, Scheffer JJJ, Parhan-Van Atten AW, Baerheim Svendsen A (1988) Screening of some essential oils for their activities on dermatophytes. Pharm Weekbl (Sci) 10:277–280
Karapinar M (1985) The effects of citrus oils and some spices on growth and aflatoxin production by *Aspergillus parasiticus* NRRL 2999. Int J Food Microbiol 2:239–245
Kaul TN, Middleton E, Ogra PL (1985) Antiviral effects of flavonoids on human viruses. J Med Virol 15:71–79
Kennedy AI, Deans SG, Svoboda KP, Gray AI, Waterman PG (1990) Comparison of the volatile oil from fermenter-grown transformed ('hairy') roots and field-grown roots of *Artemisia absinthium* (wormwood). Phytochemistry (submitted)
Knobloch K, Pauli A, Iberl B, Weigand H, Weis N (1989) Antibacterial and antifungal properties of essential oil components. J Essential Oil Res 1:119–128
Lawrence BM (1979) Progress in essential oils. Perfumer Flavorist 4:53–54
MacNeill JH, Dimick PS, Mast MG (1973) Use of chemical compounds and a rosemary spice extract in quality maintainance of deboned poultry meat. J Food Sci 38:1080–1081

McDowell PG, Lwande W, Deans SG, Waterman PG (1988) The volatile resin exudate from the stem bark of *Commiphora rostrata*: potential role in plant defence. Phytochemistry 27:2519–2521

Maruzzella JC, Sicurella NA (1960) Antibacterial activity of essential oil vapours. J Am Pharm Assoc 49:692–694

Nadal NGM, Montalvo AE, Seda M (1973) Antimicrobial properties of bay and other phenolic essential oils. Cosmetics Perfumery 88:37–39

Ooms G, Karp A, Burrel MM, Twell D, Roberts J (1985) Genetic modification of potato development using Ri T-DNA. Theor Appl Genet 70:440–446

Parr AJ, Hamill JD (1987) Relationship between *Agrobacterium rhizogenes* transformed hairy roots and intact, uninfected *Nicotiana* plants. Phytochemistry 26:3241–3245

Pompei R, Flore O, Marcialis MA, Pania A, Loddo B (1979) Glycyrrhizic acid inhibits virus growth and inactivates virus particles. Nature 281:689–690

Reuveni R, Fleischer A, Putievsky E (1984) Fungistatic activity of essential oils from *Ocimum basilicum* chemotypes. Phytopathol Z 10:20–22

Riker AJ (1930) Studies on infectious hairy root of nursery apple trees. J Agric Res 41:507–540

Rhyu HY (1979) Gas chromatographic characterisation of sages of various geographic origins. J Food Sci 44:758–762

Romero E, Tateo F, Debiaggi M (1989) Antiviral activity of a *Rosmarinus officinalis* L. extract. Mitt Geb Lebensmittelunters 80:113–119

Shelef LA, Naglik OA, Bogen DW (1980) Sensitivity of some common food-borne bacteria to the spices sage, rosemary and allspice. J Food Sci 45:1042–1044

Shukla HS, Tripathi SC (1987) Antifungal substance in the essential oil of anise (*Pimpinella anisum* L.). Agric Biol Chem 51:1991–1993

Stahl E, Quirin KW (1984) Extraction of natural substances with dense gases. Pharm Res 1984:189–194

Tepfer D (1984) Transformation of several species of higher plants by *Agrobacterium rhizogenes*: sexual transmission of the transformed genotype and phenotype. Cell 37:959–967

Tharib SM, Gnan SO, Veitch, GBA (1983) Antimicrobial activity of compounds from *Artemisia campestris*. J Food Prot 46:185–187

Van den Berghe DA, Ieven M, Mertens F, Vlietinck AJ, Lammens E (1978) Screening of higher plants for biological activities. Antiviral activity. Lloydia 41:463–471

Zaika LL (1989) Spices and herbs: their antimicrobial activity and its determination. J Food Sci 9:97–118

Zaika LL, Kissinger JC, Wasserman AE (1983) Inhibition of lactic acid bacteria by herbs. J Food Sci 48:1455–1459

Organization of Rapid Analysis of Lipids in Many Individual Plants

E.G. HAMMOND

1 Introduction

There is considerable interest in changing the composition of oilseed plants through plant breeding and genetic engineering (Greiner 1990). This may involve increasing or decreasing the total lipid in the seed or the composition of the lipid. Modern plant breeding is largely empirical and produces large numbers of plants whose oilseed composition must be evaluated. For example, two plants may be crossed and give rise to many offspring that are a genetic mixture of the two parents (Graef et al. 1988). Each of the offspring may differ genetically, and those with interesting oilseed compositions must be identified. Or, a batch of seed may be treated with a chemical or physical agent to induce mutations (Hammond and Fehr 1985). The surviving seed will give rise to plants that may have random mutations and must be evaluated to identify those with interesting compositions. In these situations plant breeders turn to those skilled in oilseed analysis, and close cooperation between breeder and analyst is needed for success.

For the typical chemist, cooperation with plant breeders may require some adjustment of attitude. It is often easy for the breeder to produce an overwhelming number of plants to be analyzed. Most of the plants that will be examined are of no interest; generally, those with rare and unusual compositions are sought. It is impractical to run replicate analyses and determine the composition of any one plant with great accuracy. Each plant must be analyzed once with sufficient accuracy so that there is a reasonable probability that no plant of interest will escape notice. Where possible, a screening test with minimal accuracy may be used to eliminate a portion of the test population that is unlikely to contain interesting specimens (Bubeck et al. 1990). The remaining plant population that is selected by the screening test may then be analyzed by a more accurate and time-consuming procedure. However, in all these tests, throughput is emphasized. Sample preparation and analysis time must be minimized, and any shortcut that does not involve too much sacrifice of accuracy must be adopted.

At times it will be necessary to analyze a seed or a portion of a seed; at other times it will be desirable to analyze a number of seeds from the same plant. The seeds of an individual plant often can show considerable variation in composition, and the analysis of a pooled sample of several seeds from that plant can provide a much better estimate of its probable composition that the analysis of a single seed. The number of seeds to be analyzed is often a compromise between accuracy and the availability of sufficient seeds for planting. For a particular oilseed plant, the minimum number of seeds that will give a satisfactory estimate must be determined (Hawkins et al. 1983). Individual seeds or a portion of seed may be analyzed when, for example, a cross has been made, and possibly a quarter of the population may have desirable characteristics. To reduce the number of plants and field space, the breeder may wish to do a preliminary analysis to determine which of the seeds to plant. If a suitable, nondestructive method is available, individual seed can be analyzed (Rutar et al. 1989). If only destructive analyses are available, a portion of the seed can be analyzed (Fehr et al. 1991), and the remainder can be planted if it is found to have an interesting composition.

One must be aware that not only genetics but a number of environmental factors (Hammond and Fehr 1985) can influence the composition of oilseed. Temperature can have a marked effect on oilseed composition (Rennie and Tanner 1989), and the temperature at which a plant matures can be influenced, e.g., by planting date and rainfall. A plant such as soybeans does not flower at one time but as the plant grows; flowers blossom first at the base and progress to the top. In such a plant, the seeds from different portions of the plant can mature under quite different conditions, and this adds to the biological variation in seed composition. Thus, in any comparison of the performance of plant lines, it is necessary to grow the plants under similar circumstances. Since the line that will perform best may depend on the environment in which it is grown, it is often desirable to compare lines from several environments (Hawkins et al. 1983).

2 Analyses for Total Lipid Content

A host of methods have been proposed for measuring the lipid content of grains and other agricultural products. Wet chemical methods based on the solubility of lipids in solvents such as ether and hexane are usually the primary methods used to standardize other procedures, but they are much too slow and laborious to use for the numerous samples generated in plant breeding.

Near-infrared reflectance instruments are available that are capable of rapid analysis of grain with respect to moisture, protein and lipid content (Hurburgh et al. 1987; Williams and Norris 1987; Panford et al. 1988; Hartwig and Hurburgh 1990; Panford and deMan 1990). For many of these instruments, the samples must be finely ground, thus proper sample preparation may be the most time-consuming step in the analysis. Other instruments are capable of analyzing whole seed. Moreover, they are more rapid and nondestructive. Some instruments require more seed than is available from a single plant. Such instruments are suitable for measuring bulk grain quality but not for plant breeding.

Wide-line nuclear magnetic resonance (NMR) has been used successfully to measure the oil content (Conway and Earle 1963). This method has been applied to corn (Alexander et al. 1967), soybeans (Collins et al. 1967), oats (Frey and Hammond 1975), and sunflower (Robertson and Morrison 1979). This method is nondestructive and requires about 30 s/analysis. It can be applied to single seeds or multiple seed samples. It is necessary to reduce the moisture of the sample to less than 4% to avoid interference with the oil determination.

3 Analyses for Fatty Acid Composition

Gas chromatography has been the method commonly used for the analysis of fatty acid composition in seed oils. Methods have been described by Hammond and Fehr (1985), Dahmer et al. (1989), and Bubeck et al. (1991). For such an analysis, a representative oil sample must be extracted from the seed. For hard seeds such as soybeans it is time-consuming to reduce the seed to a mesh size sufficiently fine to extract most of the lipid. Generally, the seed is crushed in a press, and the lipid in the crushed seed is partly extracted with hexane. Comparisons of the oils obtained from finely ground seed and crushed seed have shown that the fatty acid composition of such partly extracted seeds is representative of the whole oil (E.G. Hammond and W.R. Fehr unpubl. data). Similar results have been obtained by Dahmer et al. (1989). The time of extraction may be long for such crushed seed; therefore, overnight extractions are often used. However, such steps in the process require no attention from the analyst and generally do not reduce to overall rate at which analyses can be produced. The only requirement is that each day enough samples must be crushed to keep personnel and equipment occupied the following day. If the crushed beans are left to extract for longer periods, e.g. over the weekend, this causes no problems in the analysis.

The soybean oil may be trans-esterified to methyl esters by a number of methods. Sodium methoxide solutions are efficient and inexpensive catalysts but are quite sensitive to any moisture or free fatty acids in the samples. Methanolic solutions of boron trifluoride, hydrogen chloride, or sulfuric acid can esterify free fatty acids and are less sensitive to moisture. Manipulations such as evaporation of solvents and extractions should be minimized.

The gas chromatographic separation itself may be achieved on packed or capillary columns. Capillary columns are generally capable of slightly more rapid analyses and give superior separations, but require more sophisticated injection systems. Currently, the best stationary phases for such analyses are the 50% cyanopropyl-50% cyanomethylpolysiloxane phases. These have temperature stability along with sufficient polarity to give good separations of the unsaturated fatty acids. For many of these analyses isothermal operation is possible, but if there are great differences in fatty acid chain lengths in the sample, temperature programming may be desirable.

For reasonable throughput the gas chromatograph should have automatic injection and integration of the peaks. Unfortunately, many of the computerized systems available on the market are not well designed for a very high number of similar analyses under isothermal conditions. Often automatic injectors require too much time to pick up the sample and inject it and cannot be persuaded to start their task until the previous sample has been completed. Often the integration programs are designed for flexibility rather than routine throughput and take too long to carry out the integration and produce an analytical report, which is often too detailed and requires too much paper. Usually, the program for these systems is not accessible to the user. Thus, it cannot be modified to meet particular needs.

The system currently used in our laboratory illustrates many of the principles that have been discussed and is described as follows. The soybean seeds to be analyzed are crushed between two aluminum plates. The plates are rectangular, 22 × 23 cm and are 1.3 cm thick. One of the plates has 40 groves, 8 mm wide and 1.6 mm deep, each of which holds a five-seed sample. The groves are filled with seed samples, the second aluminum plate is place on top of the seeds, and the seed samples are crushed in a hydraulic press at 1500 to 3000 kg/cm. Each sample of crushed seed is transferred to a test tube and extracted for about 18 h with sufficient hexane to just cover the seed. The hexane is mixed and 0.1 ml is transferred to a 1.5-ml autosampler vial and reacted with 0.5 ml of 1 M sodium methoxide solution in methanol for 30 min at 40 °C with gentle mixing every 10 min. Next, 0.8 ml of water is added and after 3–5 min the floating oil phase is diluted in 1.5 ml additional hexane. The samples are analyzed in a Hewlett Packard (Avondale, PA) 5890 gas chromatograph fitted with flame detectors and 15-M Durabond-23 capillary columns (J&W Scientific, Deerfield, IL) that have a 0.25 mm i.d. and a film thickness of 0.25 μ. The column temperature is 200 °C. The needles of the automatic sample injector are cut off so that they penetrate the upper hexane phase of the sample vial but do not extend into the lower aqueous layer. Percentages are calculated from electronically integrated peak areas corrected by factors based on the number of C-H bonds. Ackman and Sipos (1964), Bannon et al. (1986), and Craske and Bannon (1987) have shown that such theoretical correction factors are more reliable than empirically derived ones in well-operated flame detectors. With this apparatus, operating with two capillary columns and two people to prepare samples, one can analyze seeds from 250 plants/day.

Some alternatives to gas chromatography for the fatty acid analysis are available but at present are seldom used. It is possible to separate fatty acids by high performance liquid chromatography columns (Ottenstein et al. 1984), but ultraviolet detectors are not very sensitive to fatty acids or their methyl esters. Furthermore, refractive index detectors limit the solvents that can be used in the separation. The availability of detectors based on light scattering may make such separations more appealing (Christie 1986). Nuclear magnetic resonance (NMR) is capable of providing helpful analyses on single or multiple seed samples (Rutar et al. 1989). Natural abundance C^{13} NMR can give a complete and reliable analysis of the oleic, linoleic, linolenic, and total saturated acids in a sample, but the analysis takes significantly longer than gas chromatography. It is possible to make a similar analysis in quite short times using proton NMR, but because of the overlapping of bands, the accuracy of the analysis is less satisfactory (V. Rutar and E.G. Hammond unpubl. data). The inability of NMR to distinguish saturated fatty acids of different chain lengths can be a distinct disadvantage in some situations. Also, NMR equipment is considerably more expensive than gas chromatographs and means for the routine introduction of many samples are not readily available.

The separation step in the analysis of seed to determine the fatty acid composition is the chief barrier to higher analytical throughput, and the equipment for the analysis is so expensive that its duplication is not an acceptable option. Theoretically, methods such as paper or thin-layer chromatography, which allow concurrent separation of several samples on one plate, should accelerate analysis.

However, methods to reveal the fatty acid on such plates that are reliably quantitative are not available. Also, the separation of the "critical pairs", i.e., oleic and palmitic acids, linoleic and myristic acids, is a serious problem (Kaufmann and Schnurbusch 1958). D.N. Duvick and E.G. Hammond (unpubl. data) separated the red N,N-dimethyl-p-aminobenzene-azophenacyl derivatives (Churacek 1970; Churacek et al. 1965) of soybean oil fatty acids on C18 silanized silica gel plates developed in N,N-dimethylformamide. They could separate the derivatives into four components, but could not separate the palmitic and oleic derivatives. Such methods might be useful for screening samples under some circumstances. Even if successful separations could be obtained, for quantification each sample would have to be scanned with a densitometer, and it remains to be seen whether the separation time plus the scanning time for the densitometer will be less than current analysis time on a gas chromatograph.

Methods based on the 2-thiobarbituric acid (TBA) reaction have been used to screen seed oil samples that are high in linolenic acid. Linolenic acid is the chief source of malonaldehyde and other carbonyls that give a red color with TBA. McGregor (1974) developed a test and used it to screen rapeseed varieties for linolenic acid. Bubeck et al. (1990) modified this test and applied it to soybean samples.

To use the TBA test, seeds are crushed as described for the gas chromatography analysis except that a sheet of Whatman No. 1 filter paper is placed between the two aluminum plates so that some of the oil in the beans is absorbed by the paper. The paper with the oil samples is irradiated in a hood for 50 min with a Westinghouse 630T8 ultraviolet source at a distance of 60 cm to accelerate the oxidation of the unsaturated fatty acids. The paper is then sprayed with a freshly prepared solution of TBA (0.9 g TBA and 6 g trichloroacetic acid in 90 ml ethanol-water 1:2, v:v) and placed in a steam chamber for 15 min. The red-orange spots are rated visually for color intensity on a one to five scale. Correlations between score and gas chromatographic analyses were 0.71 and 0.6 in two tests on soybeans, and this was sufficiently accurate to select the portion of the population that was low in linolenic acid for further analysis by gas chromatography (Bubeck et al. 1990). It was possible to run 150 samples/h which was about seven times faster than gas chromatography.

4 Analyses for Glyceride Structure

There have been few systematic studies on the glyceride structure variation in plant lipids, and the studies that have been made have not revealed evidence on a great deal of genetic variation (Ohlson et al. 1975; Fatemi and Hammond 1977a,b; Pan and Hammond 1983). A number of methods are available to analyze glyceride structure. Among the more rapid methods are gas chromatography (Litchfield 1972), which separates triglycerides according to their chain length, and high performance liquid chromatography (El-Hamdy and Perkins 1981a,b), which partly separates triglycerides according to their chain length and unsaturation. However, stereospecific analyses that determine the acyl composition of the sn-1, -2, and -3 positions of glycerol give more insight into the factors that seem to control triglyceride composition. Traditional stereospecific analysis (Litchfield 1972) was simplified by Pan and Hammond (1983) with minimum loss of accuracy. Ng (1985) and Wollenberg (1990) were able to measure the concentration of oleyl, linoleyl, and saturated acyl groups on the sn-2 position of glycerol by using C^{13} NMR. Their methods did not distinguish the sn-1 and sn-3 positions. Frost et al. (1975) were able to distinguish linoleyl and linolenyl groups on the sn-2 from those on sn-1 or -3 by using proton NMR with a shift reagent. Natale (1977) noted that in mass spectroscopy the acyl groups fragmented more readily from the sn-1 and -3 positions than from sn-2 and suggested that mass spectroscopy coupled with gas chromatographic separation of the triglycerides would give valuable information on the acyl group distribution.

In general, stereospecific analysis of plant seed triglycerides has revealed that saturated fatty acids are concentrated on the sn-1 and -3 positions of the glycerol, while linoleic and oleic acids are concentrated on sn-2. Linolenic acid is usually distributed fairly equally on the three glycerol positions, and erucic acid is distributed as if it were a saturated fatty acid on sn-1 and -3. Often, the acyl group composition of the sn-1 and -3 positions is quite similar. If a number of individuals of one species are examined, plots of the percentage of an acyl group on any one of the glycerol positions versus the percentage on all three glycerol positions fall on straight lines. Fatemi and Hammond (1977a) suggested that genetic variation in the triglyceride structure would cause an individual sample to deviate from this linear relation and that this would be a good strategy to determine triglyceride variation. The analysis of a number of individual soybean and oat plants and their wild relatives, *Glycine soja* and *Avena sterilis*, revealed individuals with modest deviations from these linear relations (Pan and Hammond 1983), but it has not been demonstrated that such deviations are reproducible from generation to generation. Similar results have been reported for Cruciferae species (Ohlson et al. 1975).

5 Analyses for Other Lipid Constituents

Breeding programs for other lipid constituents of oilseeds, such as phosphatides, sterols, and tocopherols, have seldom been attempted. Phosphatides must be removed from vegetable oils during refining, and much more is produced than can readily be marketed. Suggestions that the amount of phosphatides in oilseeds might be reduced by plant breeders (Greiner 1990) are probably not realistic because these materials are constituents of cell membranes, and their amounts are probably fixed by the membrane area. Sterols and tocopherols are partly removed from vegetable oils by deodorization process that removes undesirable flavors from vegetable oils. The sterols and tocopherols recovered from deodorizer distillates find a ready market in pharmaceutical preparations. There has been concern on the part of those who deal with these products that their concentrations might be lowered inadvertently by plant breeders as they seek higher yields or altered lipid composition. Such changes seem unlikely because, like phosphatides, the minimum amounts of these constituents are probably fixed by their physiological role in plants. However, there is probably some genetic variation in all of these minor lipid constituents, and it might be possible to vary their concentrations in oil seeds to some extent if there was sufficient economic incentive. Particularly, it might be possible to increase the concentration of some of these constituents above the minimum required for their physiological roles without damage to the plants.

If breeding programs were undertaken to alter the proportion of phosphatides, high performance liquid chromatography would most likely be the best analytical method (Aitzmueller 1982, 1984a,b; Ritchie and Jee 1985). For sterols and tocopherol both gas chromatography and high performance liquid chromatography may be used (Maerker and Unruh 1986; Weber 1987). Because the proportions of sterol and tocopherol in vegetable oils are so small, the oil is usually saponified and the unsaponifiables are isolated for analysis. Such a long preparation would have to be avoided to obtain a reasonable sample throughput.

6 Analyses for Lipoxygenase

Although not a lipid component, many of the investigators who are interested in changing the lipid composition of oilseed are also interested in the enzyme lipoxygenase which brings about the rapid oxidation of linoleic and linolenic acids to hydroperoxides when plant tissue is damaged. Many plants have more than one lipoxygenase isozyme; soybeans, for example, have been reported to have at least three (Hildebrand and Hymowitz 1982). Lipoxygenase is believed to be responsible for the development of beany off-flavors in soybean meal, and some believe that lipoxygenase oxidation products produced during the extraction of soybean meal cause the refined oil to be less stable than it might be otherwise (Frankel et al. 1988). Soybean lines lacking one or more of the lipoxygenase lines have been discovered (Hildebrand and Hymowitz 1982; Kitamura et al. 1983; Davies and Nielsen 1986). Seemingly, lipoxygenase-1 is less important in generating oxidized flavors than lipoxygenase-2 and -3 (Frankel et al. 1988; Davies et al. 1987).

An assay to determine the presence or absence of lipoxygenase has been done by electrophoresis and immunological diffusion assay (Hildebrand and Hymowitz 1982; Kitamura et al. 1983). More recently, rapid screening tests for lipoxygenase-2 and -3 have been worked out by D.N. Duvick, E.G. Hammond, and W.R. Fehr (unpubl. data) based on the detection of hydroperoxides by a color reaction with ferrous thiocyanate. In these tests, soybean seeds are crushed in the same type of apparatus described for the fatty acid analysis in Section 4. The crushed seeds are left on the plate and three to five drops of a soybean oil emulsion (10 g refined soybean oil blended in 100 ml water containing 1.25 g gum arabic) are added to the seeds. When all of the seeds have been so treated, a sheet of Whatman No. 1 filter paper is pressed onto the wetted seeds so that moist spots are transferred to the paper corresponding to each of the wet seeds. Any seed pieces adhering to the paper are brushed off, and the paper is allowed to dry until most of the water has evaporated. However, the spots are still slightly moist. The paper is then sprayed lightly with a freshly prepared solution of ferrous thiocyanate (4 g ferrous sulfate and 4 g ammonium thiocyanate in 100 ml water) and dried with hot air from a hair dryer. The lipoxygenase will oxidize the polyunsaturated fatty acids in the soybean oil to form hydroperoxides, and the hydroperoxides will oxidize the ferrous ion to ferric ion which will form an intense red color with thiocyanate. The test is so sensitive that generally there will be considerable red color in the ferrous thiocyanate solution. The spots from seed with lipoxygenase-2 will turn dark red, while those lacking the isozyme will remain cream colored. The colors should be recorded as soon as they are apparent, as all the spots will turn dark red in 0.5 to 1 h. It is helpful to run control seeds with and without lipoxygenase-2 on each test paper. The test is quite reliable in differentiating strains with and without lipoxygenase-2.

A similar color test may be used to test for lipoxygenase-3 if lipoxygenase-2 is absent. There is generally much more lipoxygenase-2 than -3 in soybeans, so the former must be absent to test for the latter and ore time must be allowed for the oxidation to occur. The beans are crushed in the same way as before but a few drops of water are added to the crushed beans and they are allowed to soak in the water for 15 min. Sufficient water is added to the beans so that free water may be observed on the bean surface. Additional drops of water may be added during the 15-min wait in order to maintain free water on the surface. Paper is applied to the moist beans, and water from the beans is smeared onto the paper. One drop of the soybean oil emulsion is added to the moist spot, representing each bean. The paper is covered with a plastic film to prevent it from drying out for 25–30 min. The paper is then sprayed with the ferrous thiocyanate solution and dried slowly with a hair drier at low heat. About 20–30 min after spraying, color can be observed. Beans with lipoxygenase-2 will be dark red, those without lipoxygenase-2 and with lipoxygenase-3 will be light pink, those without lipoxygenase-2 or -3 will be light yellow.

References

Ackman RG, Sipos JC (1964) Application of specific response factors in gas chromatographic analysis of methyl esters of fatty acids with flame ionization detectors. J Am Oil Chem Soc 41:377–378

Aitzmueller K (1982) Recent progress in the high performance liquid chromatography of lipids. Progr Lipid Res 23:171–193

Aitzmueller K (1984a) HPLC of phospholipids. Part I. General considerations. Fette Seifen Anstrichmittel 86:318–322

Aitzmueller K (1984b) HPLC and phospholipids. Part II. Determination of phosphatidylcholine (PC) and lysophosphatidylcholine (LPC) in defatted soybean lecithin. Fette Seifen Anstrichmittel 86:322–325

Alexander DE, Silvela L, Collins FI, Rodgers RC (1967) Analysis of oil content of maize by wide-line NMR. J Am Oil Chem Soc 44:555–558

Bannon CD, Craske JD, Hilliker AE (1986) Gas liquid chromatography analysis of the fatty acid composition of fats and oils: a total system for high accuracy. J Am Oil Chem Soc 63:105–110
Bubeck DM, Duvick DN, Fehr WR, Hammond EG (1990) Linolenic acid content of soybean seed estimated with 2-thiobarbituric acids test. Crop Sci 30:950–952
Bubeck DM, Fehr WR, Hammond EG (1991) Inheritance of palmitic and stearic acid mutants of soybean. Crop Sci (in press)
Christie WW (1986) Separation of lipid classes by high-performance liquid chromatography with the "mass detector." J Chromatogr 361:396–399
Churacek J (1970) Einige neue Reagenzien zur chromatographischen Identifizierung von Säuren, Alkoholen und Aminen. J Chromatogr 48:241–249
Churacek J, Kopecny F, Kulhay M, Jurecek M (1965) Papierchromatographie der Carbonsäuren. V. Identifizierung der Fettsäuren C_1 bis C_{16} als N,N-Dimethyl-p-aminobenzolazophenacylester. Z Anal Chem 208:102–116
Collins FI, Alexander DE, Rodgers KC, Silvela L (1967) Analysis of oil content of soybeans by wide-line NMR. J Am Oil Chem Soc 44:708–710
Conway TF, Earle FR (1963) Nuclear magnetic resonance for determining oil content of seeds. J Am Oil Chem Soc 40:265–268
Craske JD, Bannon CD (1987) Analysis of fatty acid methyl esters with high accuracy and reliability. V. Validation of theoretical relative response factors of unsaturated esters in the flame ionization detector. J Am Oil Chem Soc 64:1413–1417
Dahmer ML, Fleming PD, Collins GB, Hildebrand DF (1989) A rapid screening technique for determining the lipid composition of soybean seed. J Am Oil Chem Soc 66:543–548
Davies CS, Nielsen NC (1986) Genetic analysis of a null-allele for lipoxygenase-2 in soybean. Crop Sci 26:460–463
Davies CS, Nielsen SS, Nielsen NC (1987) Flavor improvement of soybean preparations by genetic removal of lipoxygenase-2. J Am Oil Chem Soc 64:1428–1433
El-Hamdy AH, Perkins EG (1981a) High performance reverse phase chromatography of natural triglyceride mixtures: carbon number separation. J Am Oil Chem Soc 58:49–53
El-Hamdy AH, Perkins EG (1981b) High performance reverse phase chromatography of natural triglyceride mixtures: critical pair separation. J Am Oil Chem Soc 58:867–873
Fatemi SH, Hammond EG (1977a) Glyceride structure variation in soybean varieties. I. Stereospecific analysis. Lipids 12:1032–1036
Fatemi SH, Hammond EG (1977b) Glyceride structure variation in soybean varieties. II. Silver ion chromatographic analysis. Lipids 12:1037–1042
Fehr WR, Welke GA, Hammond EG, Duvick DN, Cianzio SR (1991) Inheritance of reduced palmitic acid content in the seed oil of soybean. Crop Sci (in press)
Frankel EN, Warner K, Klein BP (1988) Flavor and oxidative stability of oil processed from null lipoxygenase-1 soybeans. J Am Oil Chem Soc 65:147–150
Frey KJ, Hammond EG (1975) Genetics, characteristics and utilization of oil in caryopses of oat species. J Am Oil Chem Soc 52:358–362
Frost DJ, Bus J, Keunig R, Sies I (1975) PMR analysis of unsaturated triglycerides using shift reagents. Chem Phys Lipids 14:189–192
Graef GL, Fehr WR, Miller LA, Hammond EG, Cianzio SR (1988) Inheritance of fatty acid composition in a soybean mutant with low linolenic acid. Crop Sci 28:55–58
Greiner CA (ed) (1990) Economic implications of modified soybean traits. Iowa State University, Ames, Iowa
Hammond EG, Fehr WR (1985) Progress in breeding for low-linolenic acid soybean oil. In: Rattray JBM, Ratledge C (eds) Biotechnology for the oils and fats industry. American Oil Chemists' Society, Champaign, IL, pp 89–96
Hartwig RA, Hurburgh CR Jr (1990) Near-infrared reflectance measurement of moisture, protein, and oil content of ground crambe seed. J Am Oil Chem Soc 67:435–437
Hawkins SE, Fehr WR, Hammond EG (1983) Resource allocation in breeding for fatty acid composition in soybean oil. Crop Sci 23:900–904
Hildebrand DF, Hymowitz T (1982) Inheritance of lipoxygenase-1 activity in soybean seeds. Crop Sci 22:851–853

Hurburgh CR Jr, Paynter LN, Schmitt SG (1987) Quality characteristics of midwestern soybeans. Appl Eng Agric 3:159–165

Kaufmann HP, Schnurbusch H (1958) Die Papier-Chromatographie auf dem Fettgebiet. XXIX. Die pc-Analyse von Fettsäure-Gemischen mit Hilfe des Kupfer-Quecksilber-Verfahrens. Fette Seifen Anstrichmittel 60:1046–1050

Kitamura K, Davies CS, Kaizuma N, Nielsen NC (1983) Genetic analysis of a null-allele for lipoxygenase-3 in soybean seeds. Crop Sci 23:924–927

Litchfield C (1972) Analysis of triglycerides, Academic Press, New York

Maerker G, Unruh J (1986) Cholesterol oxides. I. Isolation and determination of some cholesterol oxidation products. J Am Oil Chem Soc 63:767–771

McGregor DI (1974) A rapid and sensitive spot test for linolenic acid levels in rape seed. Can J Plant Sci 54:211–213

Natale N (1977) A mass spectrometric survey of some biologically important lipids. Lipids 12:847–856

Ng S (1985) Analysis of positional distribution of fatty acids in palm oil by C^{13} NMR spectroscopy. Lipids 20:778–782

Ohlson R, Podlaha O, Toregard B (1975) Stereospecific analysis of some *Cruciferae* species. Lipids 10:732–735

Ottenstein DM, Witting LA, Silvis PH, Hometchko DJ, Pelick N (1984) Column types for the chromatographic analysis of oleochemicals. J Am Oil Chem Soc 61:390–394

Pan WP, Hammond EG (1983) Stereospecific analysis of triglycerides of *Glycine max, Glycine soja, Avena sativa, Avena sterilis* strains. Lipids 18:882–888

Panford JA, deMan JM (1990) Determination of oil content of seeds by NIR: influence of fatty acid composition on wavelength selection. J Am Oil Chem Soc 67:473–486

Panford JA, Williams PC, deMan JM (1988) Analysis of oilseeds for protein, oil, fiber and moisture by near-infrared reflectance spectroscopy. J Am Oil Chem Soc 65:1627–1634

Rennie BD, Tanner JW (1989) Fatty acid composition of oil from soybean seeds grown at extreme temperatures. J Am Oil Chem Soc 66:1622–1624

Ritchie AS, Jee MH (1985) High performance chromatographic technique for the separation of lipid classes. J Chromatogr 329:273–280

Robertson A, Morrison WH (1979) Analysis of oil content of sunflower seed by wide line NMR. J Am Oil Chem Soc 56:961–964

Rutar V, Kovac M, Lahajnar G (1989) Nondestructive study of lipids in single fir seeds using nuclear magnetic resonance and magic angle sample spinning. J Am Oil Chem Soc 66:961–965

Weber EJ (1987) Carotenoids and tocols of corn germ determined by HPLC. J Am Oil Chem Soc 64:1129–1134

Williams P, Norris K (eds) (1987) Near-infrared technology in the agricultural and food industries. American Association of Cereal Chemists, St Paul, Minnesota

Wollenberg KF (1990) Quantitative high resolutin C^{13} nuclear magnetic resonance of the olefinic and carbonyl carbons of edible vegetable oils. J Am Oil Chem Soc 67:487:494

Subject Index

Printing: Druckerei Zechner, Speyer
Binding: Buchbinderei Schäffer, Grünstadt

Essential oils and waxes.